CARBON NANOTUBES FOR A GREEN ENVIRONMENT

Balancing the Risks and Rewards

AAP Research Notes on Nanoscience and Nanotechnology

CARBON NANOTUBES FOR A GREEN ENVIRONMENT

Balancing the Risks and Rewards

Edited by
Shrikaant Kulkarni, PhD
Iuliana Stoica, PhD
A. K. Haghi, PhD

AAP APPLE
ACADEMIC
PRESS

First edition published 2022

Apple Academic Press Inc.
1265 Goldenrod Circle, NE,
Palm Bay, FL 32905 USA
4164 Lakeshore Road, Burlington,
ON, L7L 1A4 Canada

CRC Press
6000 Broken Sound Parkway NW,
Suite 300, Boca Raton, FL 33487-2742 USA
4 Park Square, Milton Park,
Abingdon, Oxon, OX14 4RN UK

© 2022 by Apple Academic Press, Inc.

Apple Academic Press exclusively co-publishes with CRC Press, an imprint of Taylor & Francis Group, LLC

Library and Archives Canada Cataloguing in Publication

Title: Carbon nanotubes for a green environment : balancing the risks and rewards / edited by Shrikaant Kulkarni, PhD, Iuliana Stoica, PhD, A.K. Haghi, PhD.
Names: Kulkarni, Shrikaant, editor. | Stoica, Iuliana, editor. | Haghi, A. K., editor.
Series: AAP research notes on nanoscience & nanotechnology.
Description: First edition. | Series statement: AAP research notes on nanoscience & nanotechnology | Includes bibliographical references and index.
Identifiers: Canadiana (print) 20220140685 | Canadiana (ebook) 20220140723 | ISBN 9781774638620 (hardcover) | ISBN 9781774638637 (softcover) | ISBN 9781003277200 (ebook)
Subjects: LCSH: Carbon nanotubes. | LCSH: Carbon nanotubes—Environmental aspects.
Classification: LCC TA455.C3 C37 2022 | DDC 620.1/93—dc23

Library of Congress Cataloging-in-Publication Data

....

CIP data on file with US Library of Congress

....

ISBN: 978-1-77463-862-0 (hbk)
ISBN: 978-1-77463-863-7 (pbk)
ISBN: 978-1-00327-720-0 (ebk)

ABOUT THE AAP RESEARCH NOTES ON NANOSCIENCE AND NANOTECHNOLOGY BOOK SERIES

AAP Research Notes on Nanoscience & Nanotechnology reports on research development in the field of nanoscience and nanotechnology for academic institutes and industrial sectors interested in advanced research.

BOOKS IN THE AAP RESEARCH NOTES ON NANOSCIENCE AND NANOTECHNOLOGY BOOK

- **Nanostructure, Nanosystems and Nanostructured Materials: Theory, Production, and Development**
  Editors: P. M. Sivakumar, PhD, Vladimir I. Kodolov, DSc,
  Gennady E. Zaikov, DSc, and A. K. Haghi, PhD
- **Nanostructures, Nanomaterials, and Nanotechnologies to Nanoindustry**
  Editors: Vladimir I. Kodolov, DSc, Gennady E. Zaikov, DSc,
  and A. K. Haghi, PhD
- **Foundations of Nanotechnology:**
  Volume 1: Pore Size in Carbon-Based Nano-Adsorbents
  A. K. Haghi, PhD, Sabu Thomas, PhD, and Moein MehdiPour MirMahaleh
- **Foundations of Nanotechnology: Volume 2: Nanoelements Formation and Interaction**
  Sabu Thomas, PhD, Saeedeh Rafiei, Shima Maghsoodlou, and Arezo Afzali
- **Foundations of Nanotechnology: Volume 3: Mechanics of Carbon Nanotubes**
  Saeedeh Rafiei
- **Engineered Carbon Nanotubes and Nanofibrous Material: Integrating Theory and Technique**
  Editors: A. K. Haghi, PhD, Praveen K. M., and Sabu Thomas, PhD
- **Carbon Nanotubes and Nanoparticles: Current and Potential Applications**
  Editors: Alexander V. Vakhrushev, DSc, V. I. Kodolov, DSc,
  A. K. Haghi, PhD, and Suresh C. Ameta, PhD
- **Advances in Nanotechnology and the Environmental Sciences: Applications, Innovations, and Visions for the Future**
  Editors: Alexander V. Vakhrushev, DSc, Suresh C. Ameta, PhD,
  Heru Susanto, PhD, and A. K. Haghi, PhD
- **Chemical Nanoscience and Nanotechnology: New Materials and Modern Techniques**
  Editors: Francisco Torrens, PhD, A. K. Haghi, PhD,
  and Tanmoy Chakraborty, PhD
- **Nanomechanics and Micromechanics: Generalized Models and Nonclassical Engineering Approaches**
  Editors: Satya Bir Singh, PhD, Alexander V. Vakhrushev, DSc,
  and A. K. Haghi, PhD

ABOUT THE EDITORS

Shrikaant Kulkarni, PhD

Adjunct Professor, Faculty of Science & Technology,
Vishwakarma University, Pune, India

Shrikaant Kulkarni, PhD, has 37 years of teaching and research experience at both undergraduate and postgraduate levels. He has been teaching subjects such as engineering chemistry, green chemistry, nanotechnology, analytical chemistry, catalysis, chemical engineering materials, industrial organization, and management, to name a few, over the years. He possesses master's degrees in chemistry, business management, economics, political science; an MPhil and PhD in chemistry; as well as other diplomas in HR, industrial psychology, higher education, population education, etc. He has published over 100 research papers in national and international journals and conferences. He has authored 18 book chapters in CRC, Springer, and Elsevier books. He has edited two books about green engineering and renewable materials, which are in the production stage to be published by Apple Academic Press/CRC Press in 2020–2021. Another three books—on carbon nanotubes for green environment and carbon-based nanomaterials for energy storage and artificial intelligence for chemical sciences—are in the process of development. He has coauthored four textbooks on chemistry as well. His areas of interests are analytical and green and sustainable chemistry. He is a reviewer and editorial board member of many journals in green and analytical chemistry of international repute. He has been invited by UNESCO to give a talk on *Green Chemistry Education for Sustainable Development* at the IUPAC International Conference on green chemistry held at Bangkok (Thailand) in September 2018, which was well received. He is an esteemed team member of UNSDG working for the attainment of sustainable development goals. He was appointed as an innovation summit judge in a Conrad challenge competition for teams from across the world, sponsored by NASA. He has been instrumental in formulating and coordinating RIO & COP programs dedicated to sustainable development at his institute by UNCSD. He has worked as a resource person for various national and international events.

Iuliana Stoica, PhD

Scientific Researcher, Department of Physical Chemistry of Polymers, "Petru Poni" Institute of Macromolecular Chemistry, Romania

Iuliana Stoica, PhD, is a scientific researcher in physics at the Romanian Academy, "Petru Poni" Institute of Macromolecular Chemistry, Department of Physical Chemistry of Polymers. She received her PhD from the Department of Polymer Physics and Structure of the Romanian Academy at the same institute. She joined a postdoctoral fellowship program at Politehnica University of Bucharest, Faculty of Applied Chemistry and Materials Science, Department of Bioresources and Polymer Science. Her area of scientific activity is focused on characterization of a wide range of polymers, copolymers, polymeric composites, and polymeric mixtures. She was a main or coauthor for over 100 papers in peer-reviewed ISI journals, and she has contributed several book chapters on polymer and materials science. She has edited two books in green polymer chemistry and biodegradable polymers and bio-based plastics from Apple Academic Press/CRC Press. She was a member of the organizing and program committees of several scientific conferences. She was also reviewer for a number of prestigious journals in the field of polymer science.

A. K. Haghi, PhD

Professor Emeritus of Engineering Sciences, Former Editor-in-Chief, International Journal of Chemoinformatics and Chemical Engineering and Polymers Research Journal; Member, Canadian Research and Development Center of Sciences and Culture

A. K. Haghi, PhD, is the author and editor of 200 books, as well as 1000 published papers in various journals and conference proceedings. Dr. Haghi has received several grants, consulted for a number of major corporations, and is a frequent speaker to national and international audiences. Since 1983, he served as professor at several universities. He is former Editor-in-Chief of the *International Journal of Chemoinformatics and Chemical Engineering* and *Polymers Research Journal* and is on the editorial boards of many international journals. He is also a member of the Canadian Research and Development Center of Sciences and Cultures (CRDCSC), Montreal, Quebec, Canada.

CONTENTS

Contributors .... *xiii*

Abbreviations .... *xv*

Preface .... *xix*

1. Carbon Nanotubes for Clean Water 1
    Gurcharn Kaur and Jatinder Singh Aulakh

2. Graphene Oxide-Carbon Nanotube Composites for
    Wastewater Treatment 19
    Priya Banerjee, Papita Das, Aniruddha Mukhopadhayay, and Asim Kumar Ghosh

3. Undoped and Doped Carbon Nanotubes for
    Remediation of Contaminants from Wastewater 51
    Hemlata Karne and Shrikaant Kulkarni

4. An Overview of the Application of Carbon Nanotubes for
    Enhanced Oil Recovery (EOR) and Carbon Sequestration 77
    Krishna Raghav Chaturvedi and Tushar Sharma

5. Carbon Nanomaterial Embedded Membranes for
    Heavy Metal Separation 97
    Pallavi Mahajan-Tatpate, Supriya Dhume, and Yogesh Chendake

6. Carbon Nanotubes for Biomedical Applications 123
    Shrikaant Kulkarni

7. Carbon Nanotubes for Greening the Environment 159
    Sukanchan Palit

8. Rheological Behavior of Carbon Nanotubes-Based
    Materials and Its Role in Processing into Various Products 185
    Andreea Irina Barzic

9. Thermal and Electrical Transport in Carbon
    Nanotubes Composites 209
    Andreea Irina Barzic

10. **FTIR Spectroscopy for Carbon Nanotube-Based Nanomaterials in Biomedical Applications**... 233

Mioara Drobota, Maria Andreea Lungan, and Iulian Radu

11. **Carbon Nanotubes-Based Composite Materials for Electromagnetic Shielding Applications** 257

Adrian Ghemes, Gabriel Ababei, George Stoian, Luiza Budeanu-Racila, Nicoleta Lupu, and Horia Chiriac

12. **Carbon Nanotube-Based Nanocomposites: Promising Materials for Advanced Biomedical Applications** 273

Simona Luminita Nica and Delia Mihaela Rata

Index.. *291*

CONTRIBUTORS

Gabriel Ababei
National Institute of Research and Development for Technical Physics, Iasi, Romania

Jatinder Singh Aulakh
Chemistry Department, Punjabi University Patiala, Patiala, Punjab,
India; E-mail: chemiaulakh@gmail.com

Priya Banerjee
Department of Environmental Studies, DDE, Rabindra Bharati University, Rabindra Bhavan,
Bidhannagar, Kolkata 700091, India; E-mail: prya_bnrje@yahoo.com

Andreea Irina Barzic
"Petru Poni" Institute of Macromolecular Chemistry, Laboratory of Physical Chemistry of Polymers,
700487, Iasi, Romania; E-mail: irina_cosutchi@yahoo.com

Krishna Raghav Chaturvedi
Enhanced Oil Recovery Laboratory, Rajiv Gandhi Institute of Petroleum Technology, Jais, India

Yogesh Chendake
Department of Chemical Engineering, Bharati Vidyapeeth (Deemed to be) University,
College of Engineering, Pune, India

Horia Chiriac
National Institute of Research and Development for Technical Physics, Iasi, Romania

Papita Das
Department of Chemical Engineering, Jadavpur University, Kolkata 700032, India
School of Advanced Studies in Industrial Pollution Control Engineering, Jadavpur University,
Kolkata 700032, India

Supriya Dhume
Department of Chemical Engineering, Bharati Vidyapeeth (Deemed to be) University,
College of Engineering, Pune, India

Adrian Ghemes
National Institute of Research and Development for Technical Physics, Iasi, Romania;
E-mail: aghemes@phys.iasi.ro

Asim Kumar Ghosh
Membrane Development Section, Chemical Engineering Group, Bhabha Atomic Research Centre,
Trombay, Mumbai 400085, India

Hemlata Karne
Department of Chemical Engineering, Vishwakarma Institute of Technology, Pune, India

Gurcharn Kaur
Chemistry Department, Punjabi University Patiala, Patiala, Punjab, India

Shrikaant Kulkarni
Faculty of Science & Technology, Vishwakarma University, Pune, India;
E-mail: srkulkarni21@gmail.com

Maria Andreea Lungan
Sara Pharm Solutions, Calea Rahovei 266268, Electromagnetica Business Park, Bucureşti

Nicoleta Lupu
National Institute of Research and Development for Technical Physics, Iasi, Romania

Pallavi Mahajan-Tatpate
School of Petroleum, Polymer and Chemical Engineering, Polymer Engineering Department, MIT, World Peace University, Pune, India

Mioara Drobota
"Petru Poni" Institute of Macromolecular Chemistry, Aleea Grigore Ghica Voda 41A, Iasi, 700487, Romania; E-mail: miamiara@icmpp.ro

Aniruddha Mukhopadhayay
Department of Environmental Science, University of Calcutta, Kolkata 700019, India

Simona Luminita Nica
"Petru Poni" Institute of Macromolecular Chemistry, Iasi, Romania; E-mail: nica.simona@icmpp.ro

Sukanchan Palit
Department of Chemical Engineering, University of Petroleum and Energy Studies, Energy Acres, Dehradun 248007, Uttarakhand, India; E-mail: sukanchan68@gmail.com, sukanchan92@gmail.com, sukanchanp@rediffmail.com

Luiza Budeanu-Racila
National Institute of Research and Development for Technical Physics, Iasi, Romania

Iulian Radu
Department of Surgery, University of Medicine and Pharmacy "Grigore T. Popa" Iasi, Romania; Regional Institute of Oncology, I-st Surgical Oncology, Iasi, Romania

Delia Mihaela Rata
"Apollonia" University of Iasi, Faculty of Medical Dentistry, Iasi, Romania

Tushar Sharma
Enhanced Oil Recovery Laboratory, Rajiv Gandhi Institute of Petroleum Technology, Jais, India; E-mail: Tushar Sharma: tsharma@rgipt.ac.in

George Stoian
National Institute of Research and Development for Technical Physics, Iasi, Romania

ABBREVIATIONS

AFM	atomic force microscopy
AOP	advanced oxidation process
BET	Brunauer, Emmett and Teller
BHT	butylated hydroxytoluene
BPA	bisphenol A
CA	cellulose acetate
CAM	chorioallantoic membrane
CB	carbon black
CBN	carbon-based nanomaterial
CBNM	carbon-based nanomaterial
CNH	nanohorn
cND	carboxylated nanodiamond
CNF	carbon nanofibers
CNS	central nervous system
CNT-comp	carbon nanotube composite
CNTs	carbon nanotubes
COD	chemical oxygen demand
CVD	chemical vapor deposition
DDS	drug delivery systems
DEP	diethyl phthalate
DEX	dexamethasone
DFT	density functional theory
DMA	dimethyl arsenic sodium
DMF	dimethylformamide
DOX	drug doxorubicin
DWCNT	double-walled carbon nanotubes
ECM	extracellular matrix
ED	electrodialysis
EDA	electrondonor–acceptor
EDA	ethylenediamine
EDX	energy-dispersive X-ray spectroscopy
EMI	electromagnetic interference
EOR	enhanced oil recovery

EPR	enhanced permeability and retention
EVA	ethylene vinyl acetate
FE-SEM	emission scanning electron microscopy
FITC-D	fluorescein isothiocyanate dextran
FLG	few-layer graphene
FMND	fluorescent magnetic nanodiamond
FO	forward osmosis
f-SWCNTs	functionalized single-walled carbon nanotubes
FT-IR	Fourier transform infrared spectroscopy
GA	graphene aerogel
GCN	graphene and carbon nanotube nanocomposite
GFP	green fluorescent protein
GO	graphene oxide
GQDs	graphene quantum dots
HiPco	high-pressure carbon monoxide
IFT	interfacial tension
IR	infrared
LAOS	large amplitude oscillatory shear
LbL	layer-by-layer
LDPE	low-density polyethylene
LVE	linear viscoelastic
MAP	mussel adhesive protein
MB	methyl blue
MF	microfiltration
MFCs	microbial fuel cells
MIR	mid-infrared region
MM	mixed matrix
MND	modified nanodiamond
MO	methyl orange
MPD	m-phenylenediamine
MRE	molecular recognition element
MRI	magnetic resonance imaging
MWCNTs	multiwalled carbon nanotubes
ND	nanodiamond
NF	nanofiltration
NIR	near infrared
NPs	nanoparticles
NSC	neural stem cell

OG	orange G
ORR	oxygen reduction reaction
OTC	oxytetracycline
PA	polyamide
PAN	polyacrylonitrile
PANGs	polyampholyte nanogels
p-ASA	p-arsanilic acid
PBS	phosphate buffered
PC	polycarbonate
PCL	polycaprolactone
PCL–NG	PCL–nanographite
PDT	photodynamic therapy
PE	polyethylene
PEG–DXR	polyethyleneglycol–doxorubicin
PEI	polyetherimide
PEK	polyetherketone
PEEK	polyetheretherketone
PES	polyethersulfone
PHT	photohyperthermia
PNG	nano-graphene oxide
PP	polypropylene
PS	polysulfone
PS	polystyrene
PTFE	polytetrafluoroethylene
PNS	peripheral nervous system
PS	persulfate
PVA/PVP	polyvinylalcohol-polyvinylpyrolidone
PVC	polyvinyl chloride
PVDF	polyvinylidene fluoride
RF	radiofrequency
rGO	reduced GO
RO	reverse osmosis
SAOS	small-amplitude oscillatory shear
SCs	Schwann cells
SDBS	sodium dodecylbenzene sulfonate
SDS	sodium dodecyl sulfate
SE	shielding effectiveness
SEM	scanning electron microscopy

SWCNTs	single-walled carbon nanotubes
SWCNH	single-walled carbon nanohorn
TEM	transmission electron microscopy
TERT-siRNA	telomerase reverse transcriptase RNA
TFC	thin-film composite
TFN	thin-film nanocomposite
TGA	thermogravimetric analysis
TMC	trimesoyl chloride
TOF	time of flight
TPU	thermoplastic polyurethane
UF	ultrafiltration
VA	vertically aligned
VNA	vector network analyzer
XPS	X-ray photoelectron spectroscopy
XRD	X-ray diffraction
WGA	wheat germ agglutinin
WHO	World Health Organization
τD	relaxation time
τR	relaxation time

PREFACE

Nanomaterials are materials that possess structural units at nanoscale in at least one of its dimensions, and its study is a rapidly evolving frontier area in materials science and engineering. Materials property profile changes in terms of uniqueness in the nanoregime, for example, the mechanical strength of materials, can be enhanced or quantum effects may arise on downsizing the materials in nanoscale or synthesizing quantum dots out of it.

Carbon plays a pivotal role in the field of nanotechnology as silicon or germanium play in electronics, and indeed carbon nanotubes as an allotrope of carbon and its applications in greening the environment deserve a separate volume. Carbon, graphite, diamond, and fullerenes provide only a fundamental form of conventional carbon materials. The advent of other forms of carbon nanomaterials, such as carbon nanotubes and fullerenes, cover all nanostructured carbons, namely, nanodiamonds, fullerenes, nanotubes, nanofibers, cones, and whiskers, which have evolved over the time.

Carbon nanotubes, whiskers, and nanofibers are excellent state-of-the-art materials used for exploring one-dimensional phenomena and are promising nanomaterials and nanostructures of pivotal importance. The role of nanomaterials in industries is on the rise continuously due to their tenability in properties in tandem with the requirements of real-world problems. Nanofibers can be used as insulations and reinforcement of composite materials, and various materials and structures employing carbon nanotubes are continuously under development. Research initiatives aimed at designing and fabricating such innovative and novel materials also are underway. Deriving such carbon nanotubes with diverse morphologies and surface area-to-volume ratio holds a lot of promises and potential and are precursors for advanced discoveries and novel applications. We may find ourselves in the midst of carbon age within a decade or so.

This book describes a host of carbon nanotubes, including single, multiwalled, doped, undoped, functionalized, etc. They cover the range of nanomaterials with a plethora of properties. There are semiconductors, metals, and dielectrics amongst carbons; further, the band gap of semiconducting carbon nanotubes can be controlled by regulating the tube diameter and surface curvature. They can be made either transparent or opaque, while their surfaces may be passive or chemically reactive. Thus, different permutations and combination of mechanical, electrical, or chemical properties can be obtained by controlling its structure and surface chemistry and topography of these carbon-based nanomaterials.

Some of the more conventional and very well-known carbon nanomaterials, such as soot, carbon black, or intercalated graphite, have not found a place in this book. The volume is designed specifically to provide an overview of carbon nanotubes pertaining to its applications in maintaining the environment green, clean, and sustainable. It means the use of carbon nanotubes in preserving the ecological balance by conserving the natural resources by optimally using them by current scientists, graduate students, and engineering professionals; this volume will treat the subject matter employing the terminology known to a materials scientist or engineer.

This new volume describes the synthesis, characterization, and unique applications of undoped/doped carbon nanotubes as well as hybrids of them with grapheme or nanocomposites.

This volume describes new approaches used for tapping the potential and promise of key materials in isolation or combined with other materials.

This research-oriented book represent a spectrum of application of carbon nanotubes as novel materials for energy storage, and environment remediation, water treatment, green health care products, etc.

This reference book also discusses the preparation; structures and properties of carbon nanotubes both doped and undoped grapheme-CNT hybrids, providing exciting examples of different types of carbon nanotubes and grapheme-CNT hybrids with varying nanostructures.

The new book consists of valuable chapters authored by the leading academicians and researchers working in the field. Providing coverage of the state-of-the-art nanomaterial design and development across the globe, it takes a review of both commercially available and emerging materials and more importantly their potential technologies that should be ecologically benign, reliable, efficient, and cost-effective.

In fact, this book is edited as a reference book for academicians and researchers specializing in this field and it also broadly and sufficiently covers spectrum of carbon nanotubes and can be made use of as a textbook for teaching a graduate, post graduate courses with carbon-based nanomaterials as an umbrella, their synthesis methods, characterization tools, and environmental applications on priority.

CHAPTER 1

CARBON NANOTUBES FOR CLEAN WATER

GURCHARN KAUR and JATINDER SINGH AULAKH*

Chemistry Department, Punjabi University Patiala, Patiala 147002, Punjab, India

Corresponding author. E-mail: chemiaulakh@gmail.com

ABSTRACT

Nowadays, water treatment is a requirement to cover water shortage due to industrialization and urbanization. Owing to their unique physical, chemical, and structural properties, carbon nanotubes are playing a promising role in cleaning the wastewater. Carbon nanotubes provide potential solutions to environmental problems by using them as adsorbent, photocatalysts, and membranes. Adsorption is a technique where chemical interaction between the adsorbate and adsorbent play a considerable role to remove toxic contaminants from wastewaters. There are different ranges in diameters are available, such as 10–20, 20–40, and 40–60 nm for the degradation of contaminants in wastewaters. 10–20 nm diameter size showed unique results in the degradation of contaminants. Further, smooth hydrophobic walls, specially aligned atoms and the diameter of carbon nanotubes, advanced membrane technologies for water treatments are developed and are of two types as vertically aligned and mixed matrix carbon nanotube membranes. Mixed carbon nanotube membranes are preferred to use due to tremendous separation performance, high water flux, low biofouling potential in water treatments. Scientists are still searching in this field to find an efficient and cost-effective catalyst for wastewater treatments.

1.1   GENERAL INTRODUCTION

Nowadays, the water crisis is one of the greatest challenges. Contaminated water is a ubiquitous problem around the world. Demand for water is growing rapidly because of the rapid increase in population and urbanization. However, resources are limited in arid and densely populated areas. So, shortage of water resources in the world makes scientists aware to find efficient technologies for the treatment of contaminated water and desalination of seawater. This is not only the problem of water shortage, pollution is also a big issue of shortage of water. The level of water pollution has reached at peak and society is also concerning about the detrimental effects of pollution on the environment, aquatic, and aerial. Due to the increase in industrialization, urbanization, domestic and agricultural waste, and deforestation, water resources as lakes, rivers, aquifers, oceans and groundwater are vulnerable to pollution. Human activities are producing many harmful pollutants that are very toxic to living organisms.[1] These pollutants can be divided into different types as organic (e.g., agricultural pesticides and herbicides, dyes, alcohols, carboxylic acid, aliphatic and aromatic compounds, and oils), inorganic (e.g., silver (Ag), lead (Pd), cadmium (Cd), nickel (Ni), arsenic (As), mercury (Hg) and salts), nutrients (e.g., ammonia (NH_3), phosphate (PO_4^{3-}), nitrite (NO_2^-) and nitrate (NO_3^-)), radioactive pollutants (e.g., iodine (^{131}I), sulfur (^{35}S), phosphorous (^{32}P), calcium (^{45}Ca), cobalt (^{60}Co), and carbon (^{14}C)) and pathogens (e.g., bacteria, fungi and viruses).[2] Due to the increase in the world population, the water demand is increased by 64 billion cubic meters per annum.[3,4] A report by United Nations (UN) depicted that almost 2 billion population in the world did not have clean water in 2018 and it would be nearly 1.8 billion by 2025.[5,6] So, treatments of water to get clean water has become an interesting topic of the public and government. There is a need to use effective and cost-effective technologies to disinfect the water, especially in rural areas. Several conventional techniques are used as filtration, sedimentation, flocculation, coagulation, centrifugal separation to clean the water. Despite this,other methods as evaporation, distillation, crystallization, precipitation, solvent extraction, ion exchange, adsorption, reverse osmosis (RO), forward osmosis (FO), microfiltration (MF), ultrafiltration (UF), electrolysis, and electrodialysis are used to clean water.[7] Among these, engineering expertise are required that is not cost-effective. From the above techniques, adsorption is considered as the best technique especially onto nanomaterials and this is a proficient technique to remove the heavy metals from the wastewater as compared with other methods that are not efficient in the treatment of water.[8-10] Several adsorbents as chelating minerals,[11,12] activated carbon,[13] biopolymers,[14,15] engineered nanomaterials[16,18] and nanocomposites[19-22] have been utilized for the removal of pollutants.

Nanomaterials have unique properties which are absent in bulk-sized materials. Nanomaterials are used due to their large surface area, small size, reactivity, and active aromaticity. The high surface area-to-volume ratio of the nanomaterial helps to increase the sorption efficiency which can be improved more by the surface modification through the use of functional groups on the surface of the material without the interaction with the bulk properties.[23-26] The surface modified materials have been explored immensely and resulted in enhanced performances in wastewater treatments.[27,28] Several nanomaterials like nanospheres,[29,30] nanorods[31,32] and nanotubes[33,35] are used for the removal of distinct pollutants in the wastewater. Carbon nanotubes (CNTs) have attracted enormous attention of the researchers and scientists since the first discovery in 1991 due to the unique physical, chemical, electrical, and structural properties of CNTs.[36] CNTs are utilized in several technologies to mitigate water scarcity by cleaning water in the world. Various carbon nanotube nanotechnologies have immense water treatment to clean water in many fields like catalysts, sorbents, and membranes.

1.2   FUNDAMENTAL ASPECTS OF CARBON NANOTUBES

Due to binding capabilities of carbon, it can form carbon-to-carbon long chains in both straight and complex branching's with the collection of atoms in different geometrical arrangements.[37] CNTs are one of the allotropic forms of carbon its structural form can be considered as elongated fullerenes into the tabular form.[38,39] Nanotubes have a length of several micrometers and a diameter of few nanometers. Due to the unique mechanical and stiffness properties of CNTs, it can be prepared with the highest length to diameter ratio of 132,000,000:1 which is the longest material compared with the others.[40] In CNTs, carbon atoms are sp^2 hybridized and are arranged in as hexagonal pattern same as that of graphite and exist as a closed cylindrical shape. CNTs demonstrate open and closed-end with fullerene caps.[41] CNTs can be classified based on the number of carbon layers into three categories as single-walled carbon nanotubes (SWCNTs), they consist of a single carbon layer with a diameter of few nanometers and exist in metallic and nonmetallic nature, useful in tiny electronic devices; double-walled carbon nanotubes (DWCNT), they consist of two carbon layers and have more chemical resistivity than the SWCNTs, and multiwalled carbon nanotubes (MWCNTs) have more than two carbon layers with their length of several (100–1000) nm, a width of 1–3 nm, and an outer diameter of 3–30 nm.[42] The morphological structural representation of single-walled and multiwalled CNTs as shown in Figure 1.1 The preparation of CNTs is relatively expensive, for the single-walled carbon nanotubes the current price is around 75–250 dollars/g and about 5–25 dollars/g for multiwalled carbon nanotubes.[43] However, CNTs are prepared by different methods as arc discharge, laser ablation, high-pressure carbon monoxide (HiPco) process, and chemical vapor deposition (CVD) with high quality at low prices. For the large-scale production of CNTs, CVD and HiPco methods are used extensively. A production rate of 595 kg/h can be achieved by using chemical vapor deposition in a fluidized bed reactor.[44] However, there is development and improvement still in progress to produce CNTs on the commercial scale with high purity.[45] The large-scale production of CNTs at a low price may pave to the wide applications like catalysis, biomedical devices, food industry, energy storage, electronics, water treatment, and agriculture sector due to unique physiochemical properties.[46] However, through the manipulation in the design of CNTs, CNTs can be incorporated more efficiently in water purification and disinfection processes. Finally, it can be depicted that CNTs are developed by the use of technology and can be used in water protection methods to remove contaminants from water.

CNTs as adsorbents for removal of organic and inorganic pollutants: Adsorption is an efficient method for the removal of organic and inorganic compounds to clean the water. This is mainly a surface process where hazardous contaminants are treated with a suitable adsorbent. Adsorption process is demonstrated in terms of adsorption isotherm and removal efficiency. adsorption depends on several factors, such as the nature of adsorbent and adsorbate, contact time between the adsorbent and adsorbate, pH of the solution and temperature. However, to remove pollutants from the wastewater, chemical interaction between the adsorbate and adsorbent played a considerable role. There are main five possible chemical interactions, such as л-л interaction, hydrophobic effect, electrostatic interaction, л-л electron donor–acceptor (EDA) and hydrogen bonding are reported which are responsible to remove the pollutants on the surface of adsorbents.[47] Several types of adsorbents like peat, biomass, activated carbon, and biomaterials to remove toxic contaminants are used, and they have low efficiency in removal process.[48] Thus, scientists are searching to develop new adsorbents which possess high adsorption capabilities and removal efficiencies for the treatment of wastewater. For this reason, CNTs and their composites have been reported[49–51] with strong adsorption capabilities that depend on their type, presence of various functional groups,[52,53] nature of pollutants and reaction conditions as pH, temperature, and contact time. Because of the hollow, layered structure and tunable surface chemistry, the use of CNTs has been increased several years later.[54] The removal of organics, heavy metals, and radioactive materials has been investigated as depicted in Table 1.1.

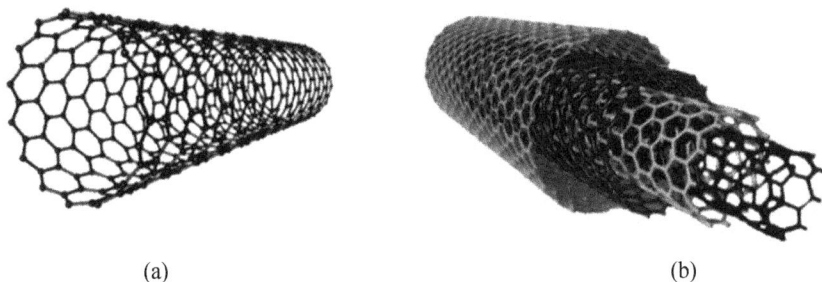

(a)   (b)

FIGURE 1.1 Morphological structure of (a) SWCNTs and (b) MWCNTs.

Catalytical degradation of contaminants by CNTs: Though adsorption is an effective way to clean the water, however, it discharges secondary waste products which cannot be treated again and are being dumped as it is. Toxic organic contaminants in the secondary waste product can be degraded catalytically by carbon materials by the advanced oxidation process (AOP). During this process, reactive hydroxyl radicals (OH) are generated which are quite oxidative toward contaminants in wastewaters and result in the decomposition of the hazardous organic pollutant.[79] This degradation process can encircle the decomposition of pollutants even at very low concentration (ppb level) without generating side products.[80] The mechanistic aspect of degradation of organic molecules is illustrated in Figure 1.2 which is illustrated by Ag–ZnO-MWCT composite. In this mechanism, it was proposed that the combination of ZnO photoexcitation and charge transfer to the Ag-CNT surface which creates superoxide radicals (O_2^-) by reacting with oxygen which then reacts

TABLE 1.1 Removal of Pollutants by CNTs.

Adsorbents type	Adsorbent dose	Pollutants	Operating conditions			Adsorption capabilities (mg/g)/removal efficiency (%)	References
			pH	T (°C)	t (min)		
As-received CNTs	500 mg/m³	Atrazine and trichloroethylene	5.7	25	1440	–	[55])
CNTs	10 mg	Microcystins (MC)	7	30	1440	14.8/–	[56]
MWCNTs	1 g/L	Bi(III)	0.1	20	30	–/ 98.91	[57]
MWCNTs	1 g/L	MB	–	21	360	465.5/76.3	[58]
MWCNTs	1 g/L	DR 80	–	21	360	380.7/96.1	[58]
MWCNTs-Cu₂O	0.25 g/L	MG	4–10	20	120	1495.46/99	[59]
MWCNTs-SH	1 g/L	MB	6	25	60	7.2/–	[60]
SWCNTs-COOH	2.5 g/L	BR	9	25	80	148/–	[61]
F-MWCNTs	0.01 g/L	Cu(II)	3	25	60	118.41/93	[62]
MWCNTs–polyamine	1.5 g/L	Pb(II)	6	25	180	3.6/–	[63]
MWCNTs–polyamine	1.5 g/L	Pb(II)	6	25	1	–/99	[64]
MWCNTs @SiO₂-NH₂	0.5 g	Pd (II)	5.2	30	240	147/–	[65]
CNT-PAMAM	0.03 g/L	Zn	8	25	240	–/88	[66]
CNT- PAMAM	0.03 g/L	Ni	8	25	240	–/98	[66]
CNTs activated by KOH etching	10–25 mg	Pharmaceutical	–	800	120	–	[67]
CNTs purified by HNO₃	20	Lead	5	140	60	17.5/87.8	[68]
MWCNTs	0.1 M	Natural organic matter	7	22	–	–	[69]

TABLE 1.1 *(Continued)*

Adsorbents type	Adsorbent dose	Pollutants	Operating conditions			Adsorption capabilities (mg/g)/removal efficiency (%)	References
			pH	T (°C)	t (min)		
As grown CNTs	0.05 g	1,2- Dichlorobenzene (DCB)	5.5	25	1440	–	[70]
MWCNTs grown by	50 mg	ciprofloxacin	5	25	300	112–231/–	[71]
CVD purified CNTs	1 mg	Polyaromatic hydrocarbons	7	40	1440	–	[72]
	1 mg	phenolic compounds	7	20	4320	–	[73]
	20 mg	Triton-X series	6	20	2820	330–740 mmol/kg/–	[74]
	100 mg/L	Uranium	7	20	1440	18–193 mmol/g/–	[75]
CNTs opened	1 g	Phenol	–	20	360	–	[76]
Amorphous Al$_2$O$_3$ supported on CNTs	0.2 g	Fluoride	6	25	720	13.2/–	[77]
Ceria nanoparticles supported on CNTs	0.025 g	Arsenate	3.1	25	1440	–/80	[78]

with water to produce OH$^{\cdot}$ radicals that decompose dyes by oxidative degradation. Interestingly, AOP can be used undersunlight rather than under UV or visible/mercury light lamps. AOP is divided into two types, such as photochemical and nonphotochemical AOP. Photochemical AOP is preferred over nonphotochemical AOP because it could not decompose the dyes completely.[81]

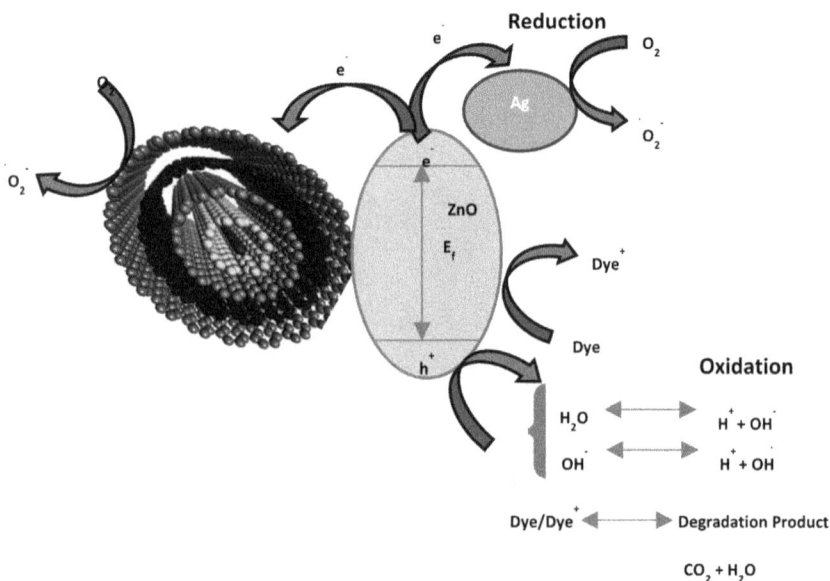

FIGURE 1.2   Mechanism of photocatalytic degradation of dye using CNTs.

There are different reports on the degradation of organic contaminants by conventional or AOP processes that are published which are presented in the following section. In a recent study, CNTs/TiO$_2$/AgNPS/surfactant nanocomposite was synthesized to degrade MB. It was demonstrated that the surface and electronic properties are being enhanced by incorporating the CNTs, while AgNPs improve the stability of this nanocomposite by the reduction of the reclamation of electron–hole pair which lower the bandgap energy (2.25 eV). Due to the large surface area of Ag NPs, it has high catalytic degradation of organic contaminants. Further, MWCNTs coated with TiO$_2$ are reported for the photocatalytic disinfection of bacterial endospores.[82] Recently, aluminum silicate multiwalled carbon nanotube nanocomposite coated with ferrocene groups (Si/Al@Fe/MWCNTs) was used for the decolorization and decomposition of methyl blue (MB), methyl orange (MO), acid orange 7 (AO7), and orange G (OG). It is observed that maximum removal efficiencies of this nanocomposite for MB, MO, AO7, and OG were 75.6%, 99.9%, 32.9% and 63.7%, respectively in 140 min without the addition of H_2O_2 and by adding H_2O_2 in the reaction, these dyes have been removed completely in 10, 6, 30, and 12 min for the above dyes. This could be due to the generation of 'OH in the reaction which helps to decompose organic dyes. In another study, microwave-induced carbon nanotubes were synthesized with 450 W microwave power and 2450 MHz frequency to remove organic pollutants from the aqueous solution. In this study, CNTs of different diameter ranges, such as 10–20, 20–40, and 40–60 nm are investigated for the degradation of contaminants in wastewaters. It is observed that there is 100% degradation of MO using CNTs with a diameter of 10–20 nm in just 7 min with 25 mg/L of the initial concentration of the solution and 1.2 g/L of catalyst amount. Furthermore, some more organic pollutants are tested under optimized conditions and it is observed that MWCNTs could be efficient to remove pollutants to clean wastewater. The rate constant (k) values for MO, MB, sodium dodecyl benzene sulfonate (SDBS), and bisphenol A (BPA) using 10–20 nm MWCNTs were 0.334, 0.679, 0.726, and 0.168 min^{-1}, respectively. It is predicted that smaller diameters of CNTs are more effective due to the large surface area.[83] In addition to this, a combination of MWCNTs and TiO_2 nanocomposite to degrade organic toxic molecules was reported and a very low amount of MWCNTs to TiO_2 nanoparticles degraded 4-chlorophenol completely in 2 h at 365 nm irradiation. This could be due to MWCNTs which act as an electron sink and facilitate charge separation.[84] A novel nanocomposite of SnO_2/ MWCNTs was fabricated by the sol–gel process in the autoclave followed by thermal annealing and used for the decomposition of MB dye and this nanocomposite degraded MB completely with higher reaction rate than pure SnO_2.[85] Different CNT-based nanocomposite can be prepared by the co-precipitation decomposition method to develop an efficient nano photocatalyst material. It was demonstrated that the addition of 5 wt.% MWCNTs to Ag-ZnO/CNT could produce efficient catalysts as it showed 100% of AO7 than Ag-ZnO alone which degraded 87% under UV light under optimized parameters.[86] A recent study, magnetic hybrid material was fabricated by polypyrrole/CNTs-CoFe$_2$O$_4$ (CNTs-CoFe$_2$O$_4$@PPy) to remove anionic and cationic dyes.[87] This catalyst showed good adsorption and catalytic performance due to the greater stability of this material and completely decompose MB in 30 min with an initial concentration of 50 mg/L and can be separated magnetically and reused. Recently, copper ferrite decorated multiwalled carbon nanotubes magnetic nanoparticles (CuFe$_2$O$_4$/MWCNTs MNPs) were prepared to degrade DEP using persulfate (PS). This catalytical performance of this catalyst is due to large surface area and can be used five times efficiently. It can be observed that different metal oxides and CNTs can be used as a single nanocomposite to remove the toxins from the wastewater to use for domestic use.

CNT-based membrane to clean water: Based on the high aspect ratios, smooth hydrophobic walls, specially aligned atoms and diameter of CNTs, this material is used in advanced membrane technologies to clean wastewaters, and better results are obtained using aligned CNT-based membranes than other conventional membranes.[88] CNT-based membranes can be classified into two types as vertically aligned (VA) CNT membrane and mixed matrix (MM) CNT membranes. VA membrane is fabricated by organizing CNTs perpendicularly with the support of filler material while the fabrication of MM membrane is simple and the MM membrane offers an economically viable process and showed enhanced water flux than the conventional membranes. Some of the limitations, such as irregular pore size, fouling, regeneration, and pore size demonstrated by the existing methods can be overcome using the MM-CNT membrane and its composition. The detailed literature of CNT-based membranes, their composites and their applications in wastewater treatment are summarized in Table 1.2. The nitrogen-doped CNT composite membrane was fabricated by a chemical vapor deposition process using nanoporous anodic aluminum membranes, the efficient composite was achieved in a ratio of 5 to 1 of carbon and nitrogen and is used in molecular separation, catalysis, and sensing application.[89] Apart from this, silver-doped CNT membranes were fabricated by a wet chemical process to remove the bacteria in wastewater. It is observed that nearly 100% of bacteria were removed by Ag-doped CNT-based membranes under optimized conditions.[90]

1.3 CONCLUSION

CNTs are used in various water treatment techniques. The use of CNTs is advantageous over conventional materials due to unique physical, chemical, and structural properties. By surface modification, CNTs show excellent properties and can be used in adsorption, degradation, and as membrane to purify water. At the end, it's good to say that CNTs are good materials to clean water.

TABLE 1.2   CNT-Based Membranes Used to Clean Water.

Membrane type	Type of pollutants	Membrane performance	Reference
N-doped CNTs–NAAMs	$(Ru (BPY)_3)^{2+}$	3.06 ± 0.10 (nmol/min/cm^2)	[89]
Ag-doped CNT	*E. coli*	100%	[90]
N- doped CNTs–NAAMs	$(RosB)^{2-}$	3.76 ± 0.17 (nmol/min/cm^2)	[91]
ZnO/MWCNTs	Direct Red 16	16.7 LMH of water flux and > 90% DR 16 rejection	[92]
Fe-Ag/ f-MWCNTs/ PES	CrO_4^{2-}	36.9 LMH water flux and 94.8% rejection of CrO_4^{2-}	[93]
Ar/O$_2$-plasma functionalized-MWCNTs	Zn^{2+}	90%	[94]
MWCNTs–biopolymer buckypaperCHT BP	trace organic	>95% rejection	[95]

KEYWORDS

- **carbon nanotubes**
- **CNTs**
- **water treatments**
- **adsorption**
- **degradation**
- **membranes**

REFERENCES

1. Xitong, L.; Wang, M.; Zhang, S.; Pan, B. Application Potential of Carbon Nanotubes in Water Treatment: A Review. *J. Environ. Sci.* **2013,** *25* (7), 1263–1280.
2. Shannon, M. A.; Bohn, P. W.; Elimelech, M.; Georgiadis, J.; Mariˉnas, G. B. J.; Mayes, A. M. Science and Technology for Water Purification in the Coming Decades. *Nature.* **2008,** *452* (7185), 301–310.

3. WWAP. *The United Nations World Water Development Report 3: Water in a Changing World*; UNESCO/Earthscan: Paris/London, 2009.

4. WHO. UNICEF: Progress on Sanitation and Drinking-Water-2013 Update: Joint Monitoring Programme for Water Supply and Sanitation, 2013.

5. UN. *The Report of the High-Level Panel of Eminent Persons on the Post-2015 Development Agenda*; UN: New York, 2013.

6. Wilson, J. *Water and Wastewater Treatment Technologies: Global Markets*, 2013.

7. Jawed, A.; Saxena, V.; Pandey, L. M. Engineered Nanomaterials and Their Surface Functionalization for the Removal of Heavy Metals: A Review. *J. Water Process. Eng.* **2020,** *33*, 101009.

8. Meepho, M.; Sirimongkol, W.; Ayawanna, J. Samaria-Doped Ceria Nanopowders for Heavy Metal Removal from Aqueous Solution. *Mater. Chem. Phys.* **2018,** *214*, 56–65.

9. Ren, G.; Wang, X.; Huang, P.; Zhong, B.; Zhang, Z.; Yang, L.; Yang, X. Chromium (VI) Adsorption from Wastewater Using Porous Magnetite Nanoparticles Prepared from Titanium Residue by a Novel Solid-Phase Reduction Method. *Sci. Total Environ.* **2017,** *607*, 900–910.

10. Saad, A. H. A.; Azzam, A. M.; El-Wakeel, S. T.; Mostafa, B. B.; El-latif, M. B. A. Removal of Toxic Metal Ions from Wastewater Using ZnO@ Chitosan Core-Shell Nanocomposite. *Environ. Nanotechnol. Monit. Manage.* **2018,** *9*, 67–75.

11. Brooks, C. S. *Metal Recovery from Industrial Waste*, 1st ed.; CRC Press, 2018.

12. Sun, S.; Wang, L.; Wang, A. Adsorption properties of crosslinked carboxymethylchitosan resin with Pb (II) as template ions. *J. Hazard. Mater.* **2006,** 136, 930–937.

13. Sager, U.; Däubner, E.; Bathen, D.; Schmidt, W.; Weidenthaler, C.; Tseng, J.-C.; Pommerin, A. Sorptive Abscheidung von Ammoniakimmissionen an Aktivkohle und deren Modifikationenbei Umgebungsbedingungen-Sorptive Removal of Low Concentrations of Ammonia on Activated Carbon and Its Modifications at Ambient Conditions. *Gefahrst. Reinhalt. Luft.* **2016,** *76*, 338–343.

14. Ge, F.; Li, M.-M.; Ye, H.; Zhao, B.-X. Effective Removal of Heavy Metal Ions Cd2+, Zn2+, Pb2+, Cu2+ from Aqueous Solution by Polymer-Modified Magnetic Nanoparticles. *J. Hazard. Mater.* **2012,** *211*, 366–372.

15. Abbaszadeh, S.; Alwi, S. R. W.; Webb, C.; Ghasemi, N.; Muhamad, I. I. Treatment of Lead-Contaminated Water Using Activated Carbon Adsorbent from Locally Available Papaya Peel Biowaste. *J. Clean. Prod.* **2016,** *118*, 210–222.

16. Yuan, J.; Liu, X.; Akbulut, O.; Hu, J.; Suib, S. L.; Kong, J.; Stellacci, F. Superwetting Nanowire Membranes for Selective Absorption. *Nat. Nanotechnol.* **2008,** *3*, 332.

17. Malakar, A.; Das, B.; Sengupta, S.; Acharya, S.; Ray, S. ZnS Nanorod as an Efficient Heavy Metal Ion Extractor from Water. *J. Water Process. Eng.* **2014,** *3*, 74–81.

18. Cao, C.-Y.; Qu, J.; Yan, W.-S.; Zhu, J.-F.; Wu, Z.-Y.; Song, W.-G. Low-Cost Synthesis of Flowerlike α-Fe$_2$O$_3$ Nanostructures for Heavy Metal Ion Removal: Adsorption Property and Mechanism. *Langmuir* **2012,** *28*, 4573–4579.

19. Nasiri, R.; Arsalani, N.; Panahian, Y. One-Pot Synthesis of Novel Magnetic Three Dimensional Graphene/Chitosan/Nickel Ferrite Nanocomposite for Lead Ions Removal from Aqueous Solution: RSM Modelling Design. *J. Clean. Prod.* **2018,** *201*, 507–515.

20. Aigbe, U. O.; Das, R.; Ho, W. H.; Srinivasu, V.; Maity, A. A Novel Method for Removal of Cr (VI) Using Polypyrrole Magnetic Nanocomposite in the Presence of Unsteady Magnetic Fields. *Sep. Purif. Technol.* **2018,** *194*, 377–387.

21. Ge, L.; Wang, W.; Peng, Z.; Tan, F.; Wang, X.; Chen, J.; Qiao, X. Facile Fabrication of Fe@ MgO Magnetic Nanocomposites for Efficient Removal of Heavy Metal Ion and Dye from Water. *Powder Technol.* **2018,** *326,* 393–401.

22. Ghosh, A.; Pal, M.; Biswas, K.; Ghosh, U. C.; Manna, B. Manganese Oxide Incorporated Ferric Oxide Nanocomposites (MIFN): A Novel Adsorbent for Effective Removal of Cr (VI) from Contaminated Water. *J. Water Process. Eng.* **2015,** *7,* 176–186.

23. Huang, Q.; Liu, M.; Chen, J.; Wan, Q.; Tian, J.; Huang, L.; Jiang, R.; Wen, Y.; Zhang, X.; Wei, Y. Facile Preparation of MoS$_2$ Based Polymer Composites via Mussel Inspired Chemistry and Their High Efficiency for Removal of Organic Dyes. *Appl. Surf. Sci.* **2017,** *419,* 35–44.

24. Pandey, L. M.; Pattanayek, S. K. Hybrid Surface from Self-Assembled Layer and Its Effect on Protein Adsorption. *Appl. Surf. Sci.* **2011,** *257,* 4731–4737.

25. Hasan, A.; Pandey, L. M. Polymers, Surface-Modified Polymers, and Self Assembled Monolayers as Surface-Modifying Agents for Biomaterials. *Polym. Technol. Eng.* **2015,** *54,* 1358–1378.

26. Pandey, L. M. Effect of Solid Surface with Self Assembled Monolayers on Adsorption of Proteins. *Diss.*, 2012.

27. Hasan, A.; Saxena, V.; Pandey, L. M. Surface Functionalization of Ti6Al4V via Self Assembled Monolayers for Improved Protein Adsorption and Fibroblast Adhesion. *Langmuir* **2018,** *34,* 3494–3506.

28. Huang, Q.; Liu, M.; Chen, J.; Wan, Q.; Tian, J.; Huang, L.; Jiang, R.; Wen, Y.; Zhang, X.; Wei, Y. Facile Preparation of MoS$_2$ Based Polymer Composites via Mussel Inspired Chemistry and Their High Efficiency for Removal of Organic Dyes. *Appl. Surf. Sci.* **2017,** *419,* 35–44.

29. De Velasco-Maldonado, P. S.; Hernández-Montoya, V.; Montes-Morán, M. A.; Vázquez, N. A.- R.; Pérez-Cruz, M. A. Surface Modification of a Natural Zeolite by Treatment with Cold Oxygen Plasma: Characterization and Application in Water Treatment. *Appl. Surf. Sci.* **2018,** *434,* 1193–1199.

30. Gokila, S.; Gomathi, T.; Sudha, P.; Anil, S. Removal of the Heavy Metal Ion Chromiuim (VI) Using Chitosan and Alginate Nanocomposites. *Int. J. Biol. Macromol.* **2017,** *104,* 1459–1468.

31. Zhou, J.; Gao, F.; Jiao, T.; Xing, R.; Zhang, L.; Zhang, Q.; Peng, Q. Selective Cu (II) Ion Removal from Wastewater via Surface Charged Self-Assembled Polystyrene-Schiff Base Nanocomposites. *Colloids Surf. A Physicochem. Eng. Asp.* **2018,** *545,* 60–67.

32. Ren, B.; Shen, W.; Li, L.; Wu, S.; Wang, W. 3D CoFe$_2$O$_4$ Nanorod/Flower-Like MoS$_2$ Nanosheet Heterojunctions as Recyclable Visible Light-Driven Photocatalysts for the Degradation of Organic Dyes. *Appl. Surf. Sci.* **2018,** *447,* 711–723.

33. Karami, H. Heavy Metal Removal from Water by Magnetite Nanorods. *Chem. Eng. J.* **2013,** *219,* 209–216.

34. Li, J.; Xing, X.; Li, J.; Shi, M.; Lin, A.; Xu, C.; Zheng, J.; Li, R. Preparation of Thiol Functionalized Activated Carbon from Sewage Sludge with Coal Blending for Heavy Metal Removal from Contaminated Water. *Environ. Pollut.* **2018,** *234,* 677–683.

35. Ainscough, T. J.; Alagappan, P.; Oatley-Radcliffe, D. L.; Barron, A. R. A Hybrid Super Hydrophilic Ceramic Membrane and Carbon Nanotube Adsorption Process for Clean Water Production and Heavy Metal Removal and Recovery in Remote Locations. *J. Water Process. Eng.* **2017,** *19,* 220–230.

36. Mubarak, N.; Sahu, J.; Abdullah, E.; Jayakumar, N.; Ganesan, P. Novel Microwave Assisted Multiwall Carbon Nanotubes Enhancing Cu (II) Adsorption Capacity in Water. *J. Taiwan Inst. Chem. Eng.* **2015,** *53,* 140–152.

37. Iijima, S. Helical Microtubules of Graphitic Carbon. *Nature.* **1991,** *354* (6348), 56–58.

38. Mubarak, N.; Sahu, J.; Abdullah, E.; Jayakumar, N.; Ganesan, P. Single Stage Production of Carbon Nanotubes Using Microwave Technology. *Diam. Relat. Mater.* **2014,** *48,* 52–59.

39. Albuerne, J.; Zenkel, C.; Munirasu, S. Functionalization and Polymerization on the CNT Surfaces. *Curr. Org. Chem.* **2013,** *17,* 1867–1879.

40. Tanaka, K.; Yamabe; Fukui, T. K. *The Science and Technology of Carbon Nanotubes,* 1st ed.; Elsevier, 1999.

41. Wang, X.; Li, Q.; Xie, J.; Jin, Z.; Wang, J.; Li, Y.; Jiang, K;. Fan, S. Fabrication of Ultralong and Electrically Uniform Single-Walled Carbon Nanotubes on Clean Ubstrates. *NanoLetters.* **2009,** *9,* 5.

42. Aqel, A.; El-Nour, K. M. M. A.; Ammar, R. A. A.; Al-Warthan, A. Carbon Nanotubes, Science and Technology Part (I) Structure, Synthesis and Characterisation. *Arab. J. Chem.* **2012,** *5,* 1–23.

43. Yu, G.; Lu, Y.; Guo, J.; Patel, M.; Bafana, A.; Wang, X.; Qiu, B.; Jeffryes, C.; Wei, S.; Guo, Z. Carbon Nanotubes, Graphene, and Their Derivatives for Heavy Metal Removal. *Adv. Compos. Hybrid Mater.* **2018,** *1,* 56–78.

44. Cheap Tubes Inc., 2012. http://www. cheaptubesinc. com/carbonnanotubes-prices. htm

45. Agboola, A. E.; Pike, R. W.; Hertwig, T. A.; Lou, H. H. Conceptual Design of Carbon Nanotube Processes. *Clean Technol. Environ. Policy.* **2007,** *9* (4), 289–311.

46. Rafique, M. M. A.; Iqbal, J. Production of Carbon Nanotubes by Different Routes—A Review. *J. Encapsulation Adsorpt. Sci.* **2011,** *1,* 29–34.

47. Daer, S.; Kharraz, J.; Giwa, A.; Hasan, S. W. Recent Applications of Nanomaterials in Water Desalination: A Critical Review and Future Opportunities. *Desalination.* **2015,** *367,* 37–48.

48. Yang, K.; Xing, B. Adsorption of Organic Compounds by Carbon Nanomaterials in Aqueous Phase: Polanyi Theory and Its Application. *Chem. Rev.* **2010,** *110,* 5989–6008.

49. Ihsanullah, A.; Abbas, A. M.; Al-Amer, T.; Laoui, M. J.; Al-Marri, M. S.; Nasser, M. K.; Atieh, M. A. Heavy Metal Removal from Aqueous Solution by Advanced Carbon Nanotubes: Critical Review of Adsorption Applications. *Sep. Purif. Technol.* **2016,** *157,* 141–161.

50. Malakootian, M.; Mansoorian, H. J.; Hosseini, A.; Khanjani, N. Evaluating the Efficacy of Alumina/Carbon Nanotube Hybrid Adsorbents in Removing Azo Reactive Red 198 and Blue 19 Dyes from Aqueous Solutions. *Process Saf. Environ.* **2015,** *96,* 125–137.

51. Elsehly, E. M.; Chechenin, N. G.; Makunin, A. V.; Pankratov, D. A.; Motaweh, H. A. Ozone Functionalized CNT-Based Filters for High Removal Efficiency of Benzene from Aqueous Solutions. *J. Water Process Eng.* **2018,** *25,* 81–87.

52. Shouman, M. A.; Fathy, N. A. Microporous Nanohybrids of Carbon Xerogels and Multi-Walled Carbon Nanotubes for Removal of Rhodamine B Dye. *J. Water Process Eng.* **2018,** *23,* 165–173.

53. Karimifard, S.; Moghaddam M. R. A. Enhancing the Adsorption Performance of Carbon Nanotubes with a Multistep Functionalization Method: Optimization of

Reactive Blue 19 Removal through Response Surface Methodology. *Process Saf. Environ.* **2016,** *99*, 20–29.

54. Pourzamani, H.; Parastar, S.; Hashemi, M. The Elimination of Xylene from Aqueous Solutions Using Single Wall Carbon Nanotube and Magnetic Nanoparticle Hybrid Adsorbent. *Process Saf. Environ.* **2017,** *109*, 688–696.

55. Mackie, E. B.; Wolfson, R. A.; Arnold, L. M.; Lafdi, K.; Migone, A. D. Adsorption Studies of Methane Films on Catalytic Carbon Nanotubes and on Carbon Filaments. *Langmuir* **1997,** *13* (26), 7197–7201.

56. Brooks, A. J.; Lim, H.; Kilduff, J. E. Adsorption Uptake of Synthetic Organic Chemicals by Carbon Nanotubes and Activated Carbons. *Nanotechnol.* **2012,** *23* (29), 294008.

57. Yan, H.; Gong, A.; He, H.; Zhou, J.; Wei, Y.; Lv, L. Adsorption of Microcystins by Carbon Nanotubes. *Chemosphere.* **2006,** *62* (1), 142–148.

58. Al-Saidi, H. M.; Abdel-Fadeel, M. A.; El-Sonbati, A. Z.; El-Bindary, A. A. Multi-Walled Carbon Nanotubes as an Adsorbent Material for the Solid Phase Extraction of Bismuth from Aqueous Media: Kinetic and Thermodynamic Studies and Analytical Applications. *J. Mol. Liq.* **2016,** *216*, 693–698.

59. Saber-Samandari, S.; Saber-Samandari, S.; Joneidi-Yekta, H.; Mohseni, M. Adsorption of Anionic and Cationic Dyes from Aqueous Solution Using Gelatin-Based Magnetic Nanocomposite Beads Comprising Carboxylic Acid Functionalized Carbon Nanotube. *Chem. Eng. J.* **2017,** *308*, 1133–1144.

60. Li, X.; Zhang, Y.; Jing, L.; He, X. Novel N-Doped CNTs Stabilized Cu$_2$O Nanoparticles as Adsorbent for Enhancing Removal of Malachite Green and Etrabromobisphenol A. *Chem. Eng. J.* **2016,** *292*, 326–339.

61. Robati, D.; Mirza, B.; Ghazisaeidi, R.; Rajabi, M.; Moradi, O.; Tyagi, I.; Agarwal, S.; Gupta, V. K. Adsorption Behavior of Methylene Blue Dye on Nanocomposite Multiwalled Carbon Nanotube Functionalized Thiol (MWCNT-SH) as New Adsorbent. *J. Mol. Liq.* **2016,** *216*, 830–835.

62. Elsagh, A.; Moradi, O.; Fakhri, A.; Najafi, F.; Alizadeh, R.; Haddadi, V. Evaluation of the Potential Cationic Dye Removal Using Adsorption by Graphene and Carbon Nanotubes as Adsorbents Surfaces. *Arab. J. Chem.* **2017,** *10*, S2862–S2869.

63. Gupta, V. K.; Agarwal, S.; Bharti, A. K.; Sadegh, H. Adsorption Mechanism of Functionalized Multi-Walled Carbon Nanotubes for Advanced Cu (II) Removal. *J. Mol. Liq.* **2017,** *230*, 667–673.

64. Albakri, M. A.; Abdelnaby, M. M.; Saleh, T. A.; Al Hamouz, O. C. S. New Series of Benzene-1, 3, 5-Triamine Based Cross-Linked Polyamines and Polyamine/CNT Composites for Lead Ion Removal from Aqueous Solutions. *Chem. Eng. J.* **2018,** *333*, 76–84.

65. Al Hamouz, O. C.; Adelabu, I. O.; Saleh, T. A. Novel Cross-Linked Melamine Based Polyamine/CNT Composites for Lead Ions Removal. *J. Environ. Manage.* **2017,** *192*, 163–170.

66. Yang, K.; Lou, Z.; Fu, R.; Zhou, J.; Xu, J.; Baig, S. A.; Xu, X. Multiwalled Carbon Nanotubes Incorporated with or Without Amino Groups for Aqueous Pb (II) Removal: Comparison and Mechanism Study. *J. Mol. Liq.* **2018,** *260*, 149–158.

67. Hayati, B.; Maleki, A.; Najafi, F.; Daraei, H.; Gharibi, F.; McKay, G. Synthesis and Characterization of PAMAM/CNT Nanocomposite as a Super-Capacity Adsorbent

for Heavy Metal (Ni^{2+}, Zn^{2+}, As^{3+}, Co^{2+}) Removal from Wastewater. *J. Mol. Liq.* **2016,** *224,* 1032–1040.

68. Ji, L. L.; Shao, Y.; Xu, Z. Y.; Zheng, S. R.; Zhu, D. Q. Adsorption of Monoaromatic Compounds and Pharmaceutical Antibiotics on Carbon Nanotubes Activated by KOH Etching. *Environ. Sci. Technol.* **2010,** *44* (16), 6429–6436.

69. Li, Y. H.; Wang, S. G.; Wei, J. Q.; Zhang, X. F.; Xu, C. L.; Luan, Z. K. Lead Adsorption on Carbon Nanotubes. *Chem. Phys. Lett.* **2002,** *357* (3–4), 263–266.

70. Hyung, H.; Kim, J. H. Natural Organic Matter (NOM) Adsorption to Multi-Walled Carbon Nanotubes: Effect of NOM Characteristics and Water Quality Parameters. *Environ. Sci. Technol.* **2008,** *42* (12), 4416–4421.

71. Peng, X.; Li, Y.; Luan, Z.; Di, Z.; Wang, H.; Tian, B.; Jia, Z. Adsorption of 1, 2-Dichlorobenzene from Water to Carbon Nanotubes. *Chem. Phys. Lett.* **2003,** *376* (1–2), 154–158.

72. Carabineiro, S. A. C.; Thavorn-amornsri, T.; Pereira, M. F. R.; Serp, P.; Figueiredo, J. L. Comparison between Activated Carbon, Carbon Xerogel and Carbon Nanotubes for the Adsorption of the Antibiotic Ciprofloxacin. *Catalysis Today* **2012,** *186* (1), 29–34.

73. Yang, K.; Zhu, L. Z.; Xing, B. S. Adsorption of Polycyclic Aromatic Hydrocarbons by Carbon Nanomaterials. *Environ. Sci. Technol.* **2006,** *40* (6), 1855–1861.

74. Kah, M.; Zhang, X.; Jonker, M. T. O.; Hofmann, T. Measuring and Modeling Adsorption of PAHs to Carbon Nanotubes Over a Six Order of Magnitude Wide Concentration Range. *Environ. Sci. Technol.* **2011,** *45* (14), 6011–6017.

75. Bai, Y.; Lin, D.; Wu, F.; Wang, Z.; Xing, B. Adsorption of Triton X-Series Surfactants and Its Role in Stabilizing Multi-Walled Carbon Nanotube Suspensions. *Chemosphere* **2010,** *79*(4), 362–367.

76. Schierz, A.; Z¨anker, H. Aqueous Suspensions of Carbon Nanotubes: Surface Oxidation, Colloidal Stability and Uranium Sorption. *Environ. Pollut.* **2009,** *157* (4), 1088–1094.

77. Wi´sniewski, M.; Terzyk, A. P. P.; Gauden, A.; Kaneko, K.; Hattori, Y. Removal of Internal Caps during Hydrothermal Treatment of Bamboo-Like Carbon Nanotubes and Application of Tubes in Phenol Adsorption. *J. Coll. Interf. Sci.* **2012,** *381* (1), 36–42.

78. Li, Y. H.; Wang, S. G.; Cao, A. Y.; Zhao, D.; Zhang, X. F.; Xu, C. L. Adsorption of Fluoride from Water by Amorphous Alumina Supported on Carbon Nanotubes. *Chem. Phys. Lett.* **2001,** *350* (5–6), 412–416.

79. Peng, X. J.; Luan, Z. K.; Ding, J.; Di, Z. H.; Li, Y. H.; Tian, B. H. Trace Organic Contaminants >95% Rejection. *Mater Lett.* **2005,** *59* (4), 399–403.

80. Moradi, M.; Haghighi, M.; Allahyari, S. Precipitation Dispersion of Ag–ZnO Nanocatalyst over Functionalized Multiwall Carbon Nanotube Used in Degradation of Acid Orange from Wastewater. *Process Saf. Environ.* **2017,** *107,* 414–427.

81. Kumar, A. A Review on the Factors Affecting the Photocatalytic Degradation of Hazardous Materials. *Mater. Sci. Eng. Int. J.* **2017,** *1.*

82. Oturan, M. A.; Aaron, J.-J. Advanced Oxidation Processes in Water/Wastewater Treatment: Principles and Applications. A Review. *Crit. Rev. Environ. Sci. Technol.* **2014,** *44,* 2577–2641.

83. Krishna, V.; Pumprueg, S.; Lee, S. H.; Zhao, J.; Sigmund, W.; Koopman, B.; Moudgil, B. M. Photocatalytic Disinfection with Titanium Dioxide Coated Multi-Wall Carbon Nanotubes. *Process Saf. Environ.* **2005,** *83,* 393–397.

84. Chen, J.; Xue, S.; Song, Y.; Shen, M.; Zhang, Z.; Yuan, T.; Tian, F.; Dionysiou, D. D. Microwave-Induced Carbon Nanotubes Catalytic Degradation of Organic Pollutants in Aqueous Solution. *J. Hazard. Mater.* **2016,** *310*, 226–234.

85. Zouzelka, R.; Kusumawati, Y.; Remzova, M.; Rathousky, J.; Pauporte, T. Photocatalytic Activity of Porous Multiwalled Carbon Nanotube-TiO_2 Composite Layers for Pollutant Degradation. *J. Hazard. Mater.* **2016,** *317*, 52–59.

86. Zaman, S.; Zhang, K.; Karim, A.; Xin, J.; Sun, T.; Gong, J. R. Sonocatalytic Degradation of Organic Pollutant by SnO_2/MWCNT Nanocomposite. *Diamond Relat. Mater.* **2017,** *76*, 177–183.

87. Li, X.; Lu, H.; Zhang, Y.; He, F. Efficient Removal of Organic Pollutants from Aqueous Media Using Newly Synthesized Polypyrrole/CNTs-$CoFe_2O_4$ Magnetic Nanocomposites. *Chem. Eng. J.* **2017,** *316*, 893–902.

88. Majumder, M.; Chopra, N.; Andrews, R.; Hinds, B. J. Enhanced Flow in Carbon Nanotubes. *Nature.* **2005,** *438*, 44.

89. Alsawat, M.; Altalhi, T.; Santos, A.; Losic, D. Facile and Controllable Route for Nitrogen Doping of Carbon Nanotubes Composite Membranes by Catalyst-Free Chemical Vapour Deposition. *Carbon.* **2016,** *106*, 295–305.

90. Wei, G.; Quan, X.; Fan, X.; Chen, S.; Zhang, Y. Carbon-Nanotube-Based Sandwich-Like Hollow Fiber Membranes for Expanded Microcystin-LR Removal Applications. *Chem. Eng. J.* **2017,** *319*, 212–218.

91. Ihsanullah; Laoui, T.; Al-Amer, A. M.; Khalil, A. B.; Abbas, A.; Khraisheh, M.; Atieh, M. A. Novel Anti-Microbial Membrane for Desalination Pretreatment: A Silver Nanoparticle-Doped Carbon Nanotube Membrane. *Desalination* **2015,** *376*, 82–93.

92. Zinadini, S.; Rostami, S.; Vatanpour, V.; Jalilian, E. Preparation of Antibiofouling Polyethersulfone Mixed Matrix NF Membrane Using Photocatalytic Activity of ZnO/MWCNTs Nanocomposite. *J. Membr. Sci.* **2017,** *529*, 133–141.

93. Masheane, M. L.; Nthunya, L. N.; Malinga, S. P.; Nxumalo, E. N.; Mamba, B. B.; Mhlanga, S. D. Synthesis of Fe-Ag/f-MWCNT/PES Nanostructured-Hybrid Membranes for Removal of Cr(VI) from Water. *Sep. Purif. Technol.* **2017,** *184*, 79–87.

94. Farid, M. U.; Luan, H.-Y.; Wang, Y.; Huang, H.; Khan, R. J. Increased Adsorption of Aqueous Zinc Species by Ar/O_2 Plasma-Treated Carbon Nanotubes Immobilized in Hollow-Fiber Ultrafiltration Membrane. *Chem. Eng. J.* **2017,** *325*, 239–248.

95. Rashid, M. H.-O.; Triani, G.; Scales, N; Panhuis, M.; Nghiem, L. D.; Ralph, S. F. Nanofiltration Applications of tough MWNT Buckypaper Membranes Containing Biopolymers. *J. Membr. Sci.* **2017,** *529*, 23–34.

CHAPTER 2

GRAPHENE OXIDE-CARBON NANOTUBE COMPOSITES FOR WASTEWATER TREATMENT

PRIYA BANERJEE[1*], PAPITA DAS[2,3],
ANIRUDDHA MUKHOPADHAYAY[4], and ASIM KUMAR GHOSH[5]

1Department of Environmental Studies, DDE, Rabindra Bharati University, Rabindra Bhavan, Bidhannagar, Kolkata 700091, India

2Department of Chemical Engineering, Jadavpur University, Kolkata 700032, India

3School of Advanced Studies in Industrial Pollution Control Engineering, Jadavpur University, Kolkata 700032, India

4Department of Environmental Science, University of Calcutta, Kolkata 700019, India

5Membrane Development Section, Chemical Engineering Group, Bhabha Atomic Research Centre, Trombay 400085, Mumbai, India

*Corresponding author. E-mail: prya_bnrje@yahoo.com

ABSTRACT

According to recent studies, approximately, four billion people (attributing to two-thirds of the world population) are presently facing severe water scarcity for at least a month every year. Contamination of water by different emerging pollutants is considered as one of the most reasons for global water scarcity. Due to an increasing global scarcity of potable water sources, laws regarding uses of water resources have become more stringent making it imperative to design new, efficient, cost-effective and time-saving methodologies for the treatment and reuse of wastewater. In recent studies, nanotechnology has received significant importance and a wide range of nanomaterials have been investigated for wastewater treatment. The present study reviews different graphene oxide (GO)–carbon nanotube (CNT)-based nanocomposites for their wastewater treatment potential. Special emphasis has been laid upon the mode of application, efficiency of pollutant removal, and antibacterial activity demonstrated by these novel nanocomposites. It also discusses the limitations as well as future challenges of using these nanocomposites for wastewater treatment. The different aspects of these nanocomposite-based water treatment discussed in this study will facilitate the potential readers including academicians, environmentalists, and industrialists trying to address issues of water scarcity.

2.1 INTRODUCTION

The most vital ingredient required for sustaining the existence of all living organisms is clean water. Nevertheless, rapid growth of population and industrialization recorded over recent years have severely contaminated global water resources.[1] Moreover, besides other needs, an 8%, 22%, and 70% increase in demand of fresh water has been recorded over the domestic, industrial, and agricultural sectors respectively. This in turn has led to the voluminous discharge of effluents bearing different types of pollutants.[1] Of all other pollutants, heavy metals and dyes appear in nearly all types of effluents. Recent studies have reported polyaromatic hydrocarbons, pharmaceuticals, surfactants, and other harmful chemicals as other aqueous pollutants of emerging concern.[2–4] Trace quantities of these pollutants render water unsuitable for consumption and is not removed efficiently by conventional processes of wastewater treatment. These pollutants are detrimental for living beings and ecosystems alike. Therefore, in order to protect the human health and the neighboring environment from the detrimental effects of these pollutants, it has become imperative to ensure the complete removal of the same from effluents prior to the discharge of the treated water into adjacent water bodies.

Adsorption, biodegradation, coagulation, distillation, electrodialysis, electrolysis, evaporation, flotation, gravity separation and sedimentation, ion exchange, membrane-based separation, oxidation, precipitation, solvent extraction, etc. are conventional processes for wastewater treatment. In comparison to all other processes, adsorption has been considered as the

best due to its cost-effectiveness, convenient procedure and the wide availability of a huge array of adsorbents.[5] Moreover, adsorption is applicable
for several types of pollutants, including the inorganic, organic (soluble and
insoluble) and biological ones. Adsorption has been widely implemented
for reclamation for water resources for potable, industrial as well as other
purposes. Nevertheless, owing to certain limitations, adsorption has not been
popular for wastewater treatment on a commercial scale. These limitations
include aggregation or loss of efficiency on addition to real effluents, lack
of suitable adsorbents with high adsorption capacity and compatibility with
all kind of pollutants. Nonetheless, the cost-effectiveness of all contemporary processes for wastewater treatment in decreasing order reportedly
stands as adsorption > evaporation> aerobic > anaerobic > ion exchange >
electrodialysis > micro- and ultrafiltration > reverse osmosis > precipitation
> distillation > oxidation > solvent extraction.[1] Therefore, irrespective of
few limitations, adsorption has the potential to be recognized as an efficient
technology for water treatment in the near future.

In recent investigations, nanomaterials and their composites have reportedly exhibited excellent efficiency in adsorptive removal of several types of
pollutants from both simulated and real effluents.[6] In view of the importance
of water quality and emerging utilities of nanotechnology, attempts have
been made to discuss various aspects of water treatment by adsorption using
nanomaterials. In this regard, promoting nanomaterials presents opportunities
to develop local and practical solutions for tackling global water pollution.
This review article presents a brief overview of the technical applicability of
different nanomaterials for removing various aquatic pollutants.

2.2   NANOMATERIALS AS ADSORBENTS FOR WATER TREATMENT

Nanoadsorbents are particles of nanodimensions synthesized from organic
or inorganic substrates having a high affinity for adsorption of pollutants.
The small size, high porosity, large surface area and a large "surface area
to volume" ratio enable nanoadsorbents to sequester pollutants of different
speciation behavior, hydrophobicity, and molecular size.[7-9] Nanoadsorbents act rapidly and are capable of considerable pollutant-binding. Their
catalytic potential and high reactivity render them as better adsorbents in
comparison to conventional ones. They are also conveniently regenerated
after exhaustion.[10] These properties of nanoadsorbents have attracted
considerable scientific attention for the same from all over the world.

2.2.1 CARBON-BASED MATERIALS

For any adsorbent, large pore volume and surface area as well as proper functionalities are crucial for success. Previous studies have reported various porous adsorbents, such as activated carbon, mesoporous oxides, polymers and metal-organic frameworks, pillared clays, zeolites, etc. that have exhibited varying effectiveness in the removal of toxic pollutants from air, water, and soil.[11–14] Among them, carbon-based adsorbents like activated carbon, carbon nanotubes (CNTs), fullerenes, and graphene have demonstrated high adsorption capacity and thermal stability.[15–17] So far, graphene has received significant global attention as a carbonaceous adsorbent due to its atomic thickness and 2D structure.[18] Graphene oxide (GO), an oxidized precursor of graphene, is decorated with hydroxyl and epoxy groups along its basal plane and carboxyl groups on the edges.[19,20] Its hydrophilic nature, along with a very high negative surface charge (occurring due to its oxygenated functional groups), has rendered GO highly suitable for the adsorption of environmental pollutants.[21] Besides, it has been reported that GO is able to form a monolayer in solution and intercalate water molecules.[21] This layered structure of GO and its oxygenated functional groups were considered highly appropriate for the application in adsorption-based water treatment processes.[22]

2.2.2 GRAPHENE OXIDE

The most widely accepted Lerf–Klinowski model of GO has two distinct regions, one comprised of lightly functionalized, mostly sp^2-hybridized carbon atoms and the other of highly oxygenated, predominately sp^3-hybridized carbon atoms.[23] According to this model, the hydroxyl and epoxide functional groups are localized in the basal planes and are clustered as islands amidst lightly functionalized, graphene-like domains while the pH-sensitive carboxylic groups are arranged along the edges of the sheets.[23] The strong oxidizing conditions created for treatment of graphite/any other precursor substrate render the resultant GO both oxidizing and acidic, which promotes the applicability of GO as a catalyst in diverse chemical reactions.[23]

GO is an oxidized form of graphene[19] bearing an assortment of hydroxyl and epoxy groups arranged in the basal plane with carboxyl groups bordering the edges.[20] GO, a precursor for graphene preparation, retains much of the properties of the highly valued super material, that is, pure graphene. It is generally obtained through the strong oxidation of graphite by modified Hummer's method.[24] It is much easier and cheaper to make in bulk quantities and easy to process. Large quantities of oxygen atoms are present on the surface of the resulting GO in the forms of epoxy, hydroxyl, and carboxyl groups.[25]

Notably, the oxygenated groups present on GO renders the same hydrophilic and allow it to be readily exfoliated in both water and polar organic solvents thereby resulting in the formation of stable suspensions. The extremely hydrophilic nature of GO rendered it suitable for application in the aquatic and biological environment. Its hydrophilic nature coupled with very high (−) ve charge density (resulting from its oxygen-bearing functional groups)[21] rendered GO highly suitable for adsorption purposes. In addition, GO was determined to exist as a single layer in solution and possess the ability of intercalating water molecules.[21] The layered nature of GO and assistance of its functional groups were effectively utilized for dye adsorption from aqueous solutions in previous studies. Though GO showed high adsorption properties owing to the presence of many oxygen-containing functional groups on its surfaces,[26] it exhibited a tendency of agglomeration and was not easily separated from treated water.[27] As a result, other kinds of adsorbents, including polymers, waste materials, and inorganic compounds have been recently investigated for nanocomposite[27,28] formation with GO. The oxygen functionalities of GO could be promoted as nucleation sites of clays/metallic compounds/polymers to yield different GO-composites with advanced properties.

2.2.3 SYNTHESIS AND FUNCTIONALIZATION OF GRAPHENE OXIDE NANOCOMPOSITES

The efficiency of graphene-based nanomaterials as adsorbents is strongly dependent on their ability to be homogenously dispersed in the aqueous phase and their affinity for different pollutants.[29] However, graphene in aqueous suspensions tends to agglomerate and revert to its graphitic structure by restacking.[30] Besides, GO weakly binds to anionic compounds owing to strong electrostatic repulsions operating between them.[29] Additionally, both graphene and GO are not easily separated from treated water and thereby lead to severe recontamination of the same.[29]

Chemical functionalization of GO reportedly facilitates dispersion and stabilization of the same by resisting aggregation of GO sheets in suspension[31] and also enhances the interactions between GO and various organic and inorganic contaminants.[29] Chemical functionalization of GO may be carried out with metal/metal oxide nanoparticles, organic polymers,[29] and even nanoclays.[32] Previous investigations have reported that various polymers and nanoparticles may be directly deposited on graphene sheets without cross-linkers,[29] bridged using cross-linkers,[32] or inserted amidst adjacent sheets for achieving both higher adsorptive surface area and prevention of sheet aggregation.[32–34]

The resulting nanocomposite widely differs from its parent components, with totally novel functionalities and properties.[29] These incorporated materials have been reported to increase the distance separating adjacent GO sheets, widen interlamellar capillaries, and ensure higher fluid percolation, rendering the nanocomposite highly efficient for adsorption purposes.[32,35] Depending on its structure, size, and crystalline nature, the inserted substrate is also found to facilitate high selectivity and stronger bond formation with the target pollutants.[29] The primary objective is the maximum utilization of the combined advantages offered by all components of the nanocomposite for improved adsorption efficiency.[29] The different strategies for the synthesis of different graphene-based nanocomposites have been reviewed in detail by Huang et al[36]

Recently, a new group of materials have been derived from the hybridization of CNTs and graphene or GO sheets.[37] These hybridized materials have better performances than their corresponding CNTs and GO sheets. They could be used as transparent conductive electrodes for various applications, such as supercapacitors, lithium-ion batteries, photocatalytic cells, and high strength polymeric composites. This short review was aimed to summarize the latest progress in processing, characterization, and application of this unique group of CNTs–GO hybrids.

2.3 GO FUNCTIONALIZATION USING CNTs

CNTs are one of the allotropes of carbon and these can be considered as various structural forms of carbon element. Basically, CNTs are of cylindrical shape. CNTs are of two types, single-walled CNTs (SWCNTs) and multiwalled CNTs (MWCNTs), composed of single and multiple rolled up graphene sheets, respectively.[1] CNTs, with their high surface active site-to-volume ratio and controlled pore size distribution, have an excep-tional sorption capability and high sorption efficiency. Different methods of preparing GO-CNTs composites are discussed as follows:

2.3.1 SOLUTION PROCESSING/CASTING

Solution casting is the simplest way to prepare CNTs–graphene (G)/ reduced GO (rGO) hybrid thin films.[38–40] For instance,[41] reported an inex-pensive method of forming CNTs–rGO hybrid consisting of chemically converted graphene and pretreated CNTs. In this case, both SWCNTs and GO were dispersed in anhydrous hydrazine, which resulted in a stable suspension while GO was reduced at the same time. The ratio of CNTs to rGO could be readily controlled. The stable suspensions could be used to deposit thin films by solution casting and spin coating. It is necessary to mention that this method did not require surfactant, which could be the reason that the intrinsic electronic and mechanical properties of both components were retained. Another water-based method was reported by King et al.,[42] in which small amount of G was used to prepare improve electrical conductivity of SWCTN films.

Multicomponent hybrids have also been reported in recent studies. Preparation of these hybrids include modified CNTs or G/rGO.[39,40] Inves-tigators may also consider the incorporation of a third component for further enhancement of the functionalities of the hybrid materials.[43–45] For example, a previous study had reported the synthesis of a GO–SWCNTs composite using noncovalently dispersed p-type doped GO.[39] Pan et al.[40] reported the usage of chitosan for grafting water soluble rGO. This also facilitated the efficient noncovalent dispersion of MWCNTs in acidic solutions. Inclusion of chitosan in these composites exhibited significant enhancement of mechanical properties.

2.3.2 LAYER-BY-LAYER (LBL) DEPOSITION

LbL is a process for depositing composite thin films having microstruc-tures that are controllable at nanoscales. This process has been used for effective synthesis of CNTs–GO composite films.[46,47] Kim and Min[48] had also implemented the same for the synthesis of a ultrathin, double-layered and transparent composite film made up of MWCNTs and rGO. In their study, MWCNTs were coated on a thin rGO layer deposited on SiO_2/Si substrates via electrostatic adsorption.

Hong et al.[46] reported a relatively simple and flexible process for LbL deposition of composite thin films (prepared from MWCNTs and rGO) on silicon and quartz substrates. Multilayered composite thin films were prepared by alternative depositions of negatively charged rGO and positively charged MWCNTs layers. Similar findings were reported by Yu and Dai.[47] In this study, positively charged stable aqueous dispersion prepared using polymer-bound G nanosheets and negatively charged MWCNTs suspension were deposited on various substrates by multistep sequential assembly procedure. The resultant films have interconnected network of well-defined nanopores which facilitate rapid diffusion of ions.

2.3.3   VACUUM FILTRATION

This process has been used for synthesizing independant CNTs–rGO composites.[49,50] Khan et al.[49] synthesized composite sheets using SWCNTs and G/nanographite suspended in N-methyl pyrrolidone. Thickness of the hybrid sheets prepared by vacuum filtration varied from 100 to 500 mm. Tang and Gou[50] implemented this same technique for making paper-like composites using MWCNTs and few-layer graphene (FLG). In another study, a graphene–CNTs–graphene sandwich composite was prepared using vacuum filtration.[51] CNTs were synthesized between graphene sheets and distributed sparsely (but uniformly) on the entire sheet surface. This composite demonstrated high electroactivity, good electron conductivity, and low resistance to diffusion of protons/cations.

2.3.4   CHEMICAL VAPOR DEPOSITION (CVD)

A number of studies have implemented CVD for the synthesis of CNTs–G/ GO/rGO composites.[51–54] One of the most prominent advantages of CVD is its ability to yield composites having well-defined hierarchical micro-structures.[51,55,56] Fan et al.[51] synthesized 3D CNTs–G sandwich composites by growing CNTs between G layers with CVD. Both pseudo and double-layer capacitors prepared with this composite exhibited excellent electro-chemical performances.

2.4 WASTEWATER TREATMENT USING GO-CNTs COMPOSITES

2.4.1 MECHANISMS OF WASTEWATER TREATMENT

2.4.1.1 GO-CNTS AEROGELS FOR ADSORPTION

Adsorptive nanotechnology solution is expected to fulfill next-generation demand of clean water supply.[57] However, removal efficiency of adsorbents is limited by their surface area, active sites, nonselectivity, and adsorption kinetics. The smart adsorbents are thus needed which can have excellent adsorption capacity to remove pollutants to the ppb level. Carbon-based nanoadsorbents with their high specific surface area and associated sorption sites and fast kinetics can enable technological innovation in advancing the treatment efficiency of polluted water to provide clean and affordable water.[58–60]

Graphene is thus a good candidate to be an effective adsorbent for removing contaminants in water.[22,29,32,61,62] However, it encounters difficulties in recycling because it is not easily separated from the treated water.[57] Recently, a new group of materials have been synthesized from the hybridization of CNTs and GO (GO−CNTs), which showed better properties than their individual counterparts.[63] Due to the noncorrosive property, tunable surface chemistry, high surface area, and presence of surface oxygen-containing functional groups, CNTs and graphene materials have been chosen as platforms to build new adsorbents with enhanced or more functions.[64]

The abundant oxygen-containing groups (-OH, -COOH and epoxy groups) on the basal planes and edges of GO can provide negative charge to combine positively charged ions through electrostatic interaction or constitute complexes with ions.[65,66] However, GO is prone to restack or agglomerate because of the strong van der Waals interactions and π-π stacking,[67,68] which can consume a large number of active sites. Fortunately, the assembly of GO nanosheets into 3D graphene aerogel (GA) is considered one of the most promising strategies for overcoming these problems. As a newly emerging porous material, GA exhibits special mechanical property, low density, and high internal surface area.[69,70] The integration of individual GO sheet into GA preserves the specific properties inherited from GO to the maximum. It also exhibits a peculiar three-dimensional porous network formed of interconnected macro and micropores.

This unique structure prevents aggregation of GO, promotes the free diffusion of ions and enhances the probability of contact between adsorbate moieties and active sites of the adsorbent.[71–73] Thus, 3D GAs reveal high adsorption capacity for the removal of aqueous pollutants. However, it is difficult to simultaneously maintain both high structural stability and effective active site exposure for GA.[60,74] The high structure stability of GA mostly depends on the high compactness. However, higher compactness reportedly limits the rate of ion diffusion in turn resulting in deactivation of adsorption sites.[75,76] On the contrary, decreasing the compactness of GA may result in the loss of structural stability.

Recent studies have reported hybrid materials by combining CNTs and graphene for widespread applications including water purification,[77] shape memory,[78] selective oil/water separation,[79] solar steam generation[80] and removal of formaldehyde.[81] Incorporating CNTs into GA improves its mechanical properties and structural stability.[66] Process of synthesis and architechtural details of CNTs–G aerogel prepared by Zhan et al.[66] has been shown in Figure 2.1.

The CNTs incorporated in graphene interlayers prevent restacking of graphene nanosheets and provide more accessible adsorptive surface area. CNTs have been considered as high-efficiency adsorbents owing to their large specific surface area, low density and strong interaction with pollutant moieties. The graphene and CNTs hybrid aerogels, which possess the outstanding properties of both CNTs and graphene along with the unique characteristics of aerogels, yield new porous materials with desired structures and peculiar adsorption properties. So far, CNTs/graphene hybrid aerogels are mainly prepared by CVD, chemical reduction and hydrothermal reduction process.[82–85] The CVD method requires expensive reagents and complicated instruments, which limits its applications. The other two methods seem more plausible, but the high-energy consumption and heavy use of chemical reagents virtually violate the concept of environmental protection. Application of GO–CNTs aerogel for wastewater treatment reported in recent studies have been enlisted in Table 2.1.

2.4.1.2 GO-CNTS MEMBRANES

In water and wastewater treatment by membrane, nanocomposite membrane composed of polymer and nanomaterials show great promise in terms of three main characteristics, namely, enhanced performance,

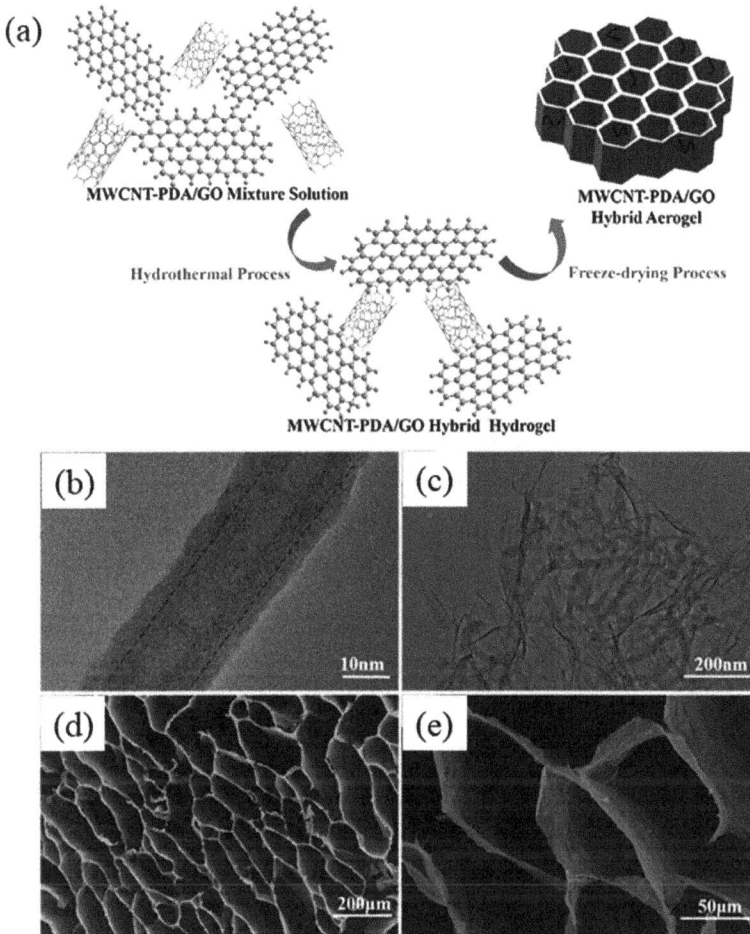

FIGURE 2.1   Carbon nanotube-graphene hybrid aerogels. (a) Schematic illustration of the synthesis process. (b) TEM images of MWCNTs–PDA samples. (c) TEM images of the hybrid aerogel. (d) Low-magnification and High-magnification SEM images of hybrid aerogel.

Source: Reprinted with permission from Ref. [66] ©2019 Elsevier.

improved property for cleaning and better fouling resistance. Among recently developed water treatment membranes, mixed-matrix membranes prepared using nanomaterials like CNTs, metal and metal oxides, graphene and GO, and zwitterionic materials show the maximum promise.[92] Being a carbon-based nanomaterial, CNTs, and GO have extra advantage of better

TABLE 2.1 Application of GO–CNTs Aerogel for Wastewater Treatment

Composition of aerogel	Type of CNTs	Organic pollutant	Experimental conditions	Adsorption capacity (mg/g)	References
Reduced GO and CNTs	Amino MWCNTs	Ethyl acetate, cyclohexane, acetone, dichloromethane, and sesame oil	—	152790, 122460, 219520, 242310, and 156350 respectively.	[86]
GO, CNTs, and a-FeOOH	Hydroxyl MWCNTs	Sodium arsenate (As(V)), dimethyl arsenic sodium (DMA), and p-arsanilic acid (p-ASA)	Initial sorbate conc. of 25 mg/L; 1.25 g/L adsorbent; Solution pH 3–9; Temperature 25°C; Time 60 min; Constant agitation of 160 rpm.	56.43, 24.43, and 102.11 for As(V), DMA, and p-ASA respectively	[87]
GO, CNTs, and Vitamin C	MWCNTs	Uranium (VI)	Initial sorbate conc. of 50 mg/L; 0.60 g/L adsorbent; Solution pH 5; Temperature 25°C; Time 24 h.	86.1	[88]
GO, CNTs, and a-FeOOH nanorods (applied as Fenton-like catalyst)	Carboxylated MWCNTs	Orange II (OII), rhodamine B (RhB), methylene blue (MB), phenol, and bisphenol A (BPA)	Initial sorbate conc. of 40 mg/L for OII; 1.25 g/L catalyst; 0.55 mM H_2O_2; neutral pH; temperature 20°C	-	[89]
GO and CNTs	1D CNTs	Cd^{2+}, Diethyl phthalate (DEP), MB, Oxytetracycline (OTC)	Initial sorbate conc. of 2000, 5, 5000, and 20 mg/L for Cd^{2+}, DEP, MB, and OTC, respectively; 0.004–0.02 g/L adsorbent; solution pH 3	235, 421, 680, 685, and 1729 respectively	[84]

TABLE 2.1 (*Continued*)

Composition of aerogel	Type of CNTs	Organic pollutant	Experimental conditions	Adsorption capacity (mg/g)	References
GO and CNTs	MWCNTs	NaCl, organic dyes (MB, RhB, acid fuchsin), and heavy metal ions (Pb^{2+}, Hg^{2+}, Ag^{+}, Cu^{2+})	Initial sorbate conc. of 35,000 mg/L for NaCl; 0.51, 0.46, 0.59, and 0.53 mmol/g for Pb^{2+}, Hg^{2+}, Ag^{+}, Cu^{2+} respectively	633.3 for NaCl; 190.9, 145.9, 66.4 for MB, RhB, and acid fuchsin respectively; 104.9, 93.3, 64, and 33.8 for Pb^{2+}, Hg^{2+}, Ag^{+}, Cu^{2+} respectively.	[77]
GO and CNTs	—	MB and methyl orange (MO)	Initial sorbate conc. of 10 mg/L; 2.5 g/L adsorbent; treatment period of 24 h	287.1	[85]
GO and CNTs	1D polydopamine (PDA) functionalized MWCNTs	Chloroform, n-hexane, n-heptane, n-dodecane, and acetone	Prefreezing temperature −50°C.	125,000–533,000	[90]
GO and CNTs	Polydopamine-modified MWCNTs	Cu(II) and Pb(II)	Initial sorbate conc. of 300 mg/L; 0.34 g/L adsorbent; Solution pH 6; Time 24 h; Temperature 25°C; Constant agitation of 200 rpm.	318.47 and 350.87 for Cu(II) and Pb(II) respectively.	[66]
GO, CNTs, and a-FeOOH nanorods	MWCNTs	Pb(NO$_3$)$_2$	500 mg/L sorbate; 0.2 g/L adsorbent	451	[91]

compatibility with organic polymers than the inorganic nanomaterials like metal and metal oxides, and also are believed to solve the trade-off issue between permeability and selectivity for membrane filtration applications. CNTs are used widely to make nanocomposite membrane for water purification due to their excellent mechanical, thermal, electrical, and chemical properties along with fast water transport and ability to mimic biological membranes.[93,94] Molecular modeling reveals that unlike pressure-driven membranes, the frictionless passage of water molecules takes place through highly hydrophobic CNTs. Through functionalization of tip and changing nanotube structural properties, the membrane manufacturer can manipulate easily the required membrane for the separation of pollutants in water purification technology. Organic functional group like amine or carboxylic acid functionalized CNTs are more useful not only for making the membrane more hydrophilic but also makes it more fouling resistant. In addition, the presence of functional groups on the surface of CNTs makes them more dispersible both in solvents as well as in a polymer matrix.[95] Similarly, a graphene derivative, GO is another carbon nanomaterial and its composite with polymers are used nowadays very effectively to enhance the performance of the nanocomposite membranes even at a very low loading.[96] In addition, GO-based membranes also show chlorine resistance and antifouling properties[97] which is an essential property for its use in wastewater treatment. Even though the applications of both CNTs and GO in membrane-based water and wastewater treatment appear very promising, but it is currently limited to bench-scale studies. Most of the CNTs- and GO-based mixed-matrix membranes are used in pressure-driven membrane processes, namely, ultrafiltration (UF), nanofiltration (NF), and reverse osmosis (RO).

Very limited studies are reported on nanocomposite membranes with aligned CNTs for practical water treatment process, but mixed-matrix membranes having dispersed CNTs as fillers in the polymer matrixes have been reported several times as a possible fouling resistant, low-energy replacement for conventional membranes. Among them, carboxylic acid functionalized CNTs-blended polysulfone[98] and polyethersulfone[99]-based UF membranes have been used for water treatment applications with enhanced membrane performance and fouling resistance due to increase in hydrophilicity. Similarly, CNTs along with silver nanoparticle are used to enhance antibacterial properties of polyacrylonitrile hollow fiber membranes where CNTs improve the dispersion of Ag nanoparticle.[100] A new invention where MWCNTs–polysulfone nanocomposite membranes were developed and used in a multifunctional membrane system by combining traditional pressure-driven membrane separation with sustainable solar thermal technology for efficient water treatment.[101] MWCNTs–polysulfone nanocomposite membrane exhibits water flux over 314 L/m^2/h at 0.10 Mpa pressure and it is 101.3% higher than that without solar irradiation. The scheme of such multifunctional separation system is shown in Figure 2.2.

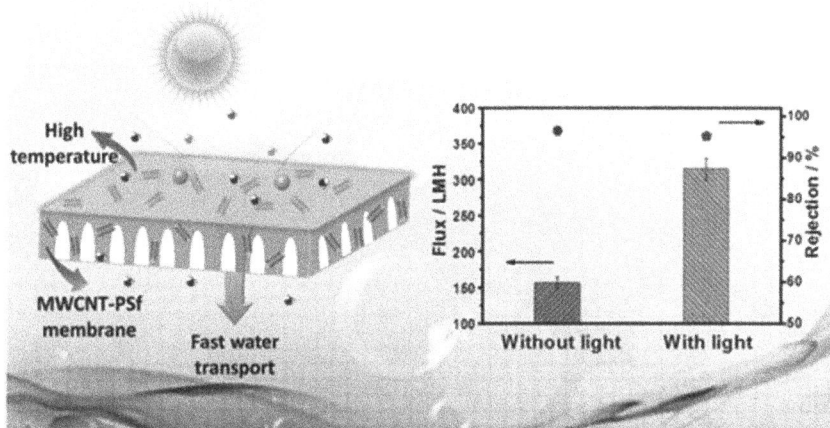

FIGURE 2.2   Schematic illustration of solar-intensified ultrafiltration system based on MWCNTs–PSf photothermal membrane for water treatment.

Source: Reprinted with permission from Ref. [101] © 2019, American Chemical Society.

NF and RO membranes are typically thin-film composite (TFC) type and the addition of a functionalized CNT layer improves the permeability of the membrane due to the increase of support hydrophilicity and porosity. Figure 2.3 illustrated the mechanism of functionalized CNTs-based mixed-matrix TFC membranes in improving membrane performances.[95] The TFC polyamide RO membranes prepared using interfacial polymerizations of trimesoyl chloride (TMC) and m-phenylenediamine (MPD) with the optimized carboxylated CNTs (only 0.002%) not only enhance the performances in terms of water flux with comparable salt rejection to those of common polyamide RO membranes without any CNTs but also enhance the chlorine resistance. In addition, membrane stability was found to be much better than that of the common polyamide RO membranes without any CNTs.[102]

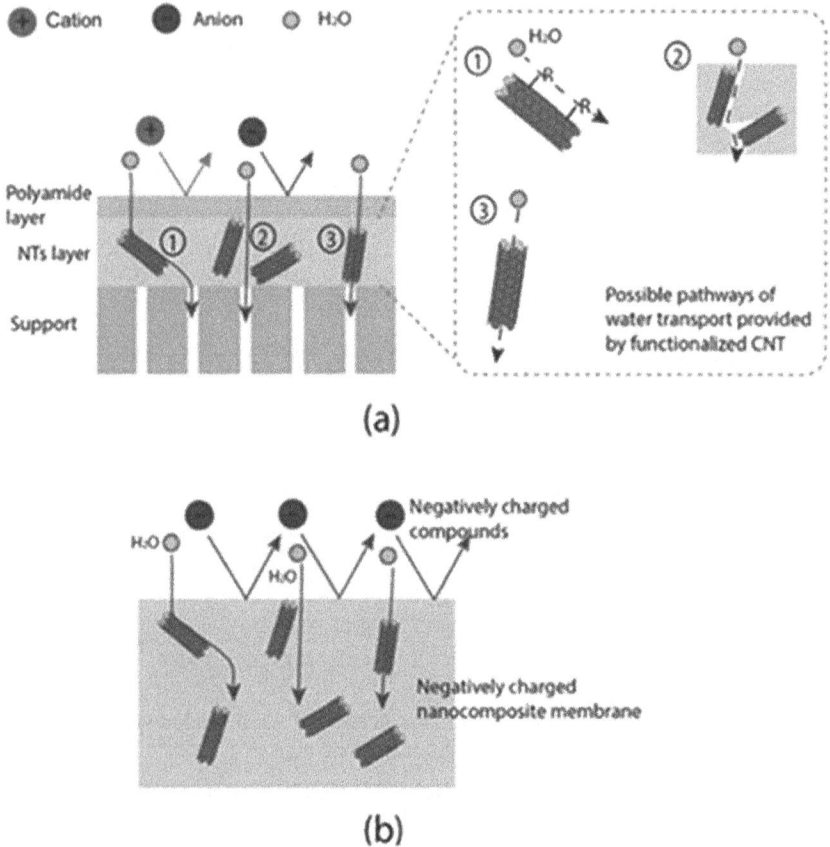

FIGURE 2.3 Mechanism of functionalized CNTs (mixed-matrix) in improving membrane performance. (a) Possible pathways of water transport provided by functionalized CNTs. (b) Rejection of negatively charged compounds.

Source: Reprinted with permission from Ref. [95]. © 2017, The Royal Society of Chemistry.

Very few attempts were made to developed membranes using both CNTs and GO. In the same line of common TFC-based RO polyamide membrane preparations by the interfacial polymerization of MPD and TMC, CNT with acidic groups (CNTa), GO, and both CNTa and GO (CNTa/GO) were used in aqueous solutions of MPD. In this study of CNTs/GO mixed filler-based RO polyamide membrane, 0.001–0.005

wt.% of CNTa and GO in aqueous solution was studied and around 27.2% water flux increase was observed maintaining high salt rejection.[103] It was found that the mixtures of CNTa and GO are much more dispersible than when CNTa or GO was used alone. As a new concept, GO/MWCNTs composite layer was introduced in TFC membrane between support layer and interracially formed polyamide layer to increase the water permeation by reducing the thickness of top polyamide payer ~60%.[104] Figure 2.4 illustrates a proposed model for the TFC membrane formation process with and without GO/MWCNTs composite layer.

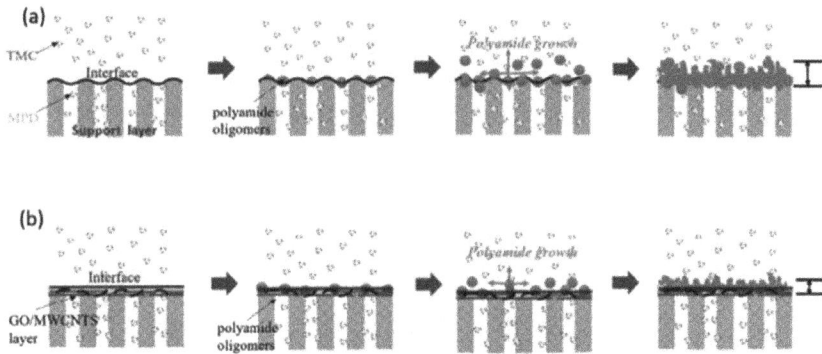

FIGURE 2.4   Proposed models for the polyamide formation process with the incorporation of GO/MWCNTs composite layer. (a) PA membrane formation process. (b) GPA membrane formation process.

Source: Reprinted with permission from Ref. [104]. © 2018, American Chemical Society.

Similar to CNTs-based membrane, GO nanocomposite membranes also exhibit higher water fluxes in water treatment. Novel GO-based composite membranes are of two types, one is free-standing membranes made only of GO and another is polymeric/ceramic membranes composite or modified with GO. A recent study revealed that the molecular separation of the GO membranes can occur through in-plane pores, interlayer spacing, and/or interactions with the oxygen-containing functional groups mechanisms.[105] An illustration of this mechanism is shown in Figure 2.5.

In order to improve the separation performance of GO membranes, physical approaches like GO nanosheet size, intercalating nanomaterials (such as carbon dots, single-walled CNTs, metal-organic frameworks), incorporating surfactant into laminated GO membrane, etc. was adopted along with some of the chemical approaches like changing GO structure by intercalating chemical groups like diamines monomer, dicarboxylic acids with different chain lengths etc.[106]. In order to prepare high performance membranes in water treatment, GO-embedded polyethersulfone nanocomposite membranes were synthesized to improve water flux, heavy metal and dye removal, hydrophilicity, and antifouling properties considerably.[107] Similarly, polysulfone/GO nanocomposite membranes were fabricated for removal of Bisphenol A (BPA) from water. The GO loading of 0.4% was found as optimum for further study on feasibility of scaling-up and comparative cost-efficiency of BPA removal from water.[108] GO-based RO membranes of thin-film nanocomposite (TFN) membranes were prepared by the in situ interfacial polymerization process with GO concentrations in the range of 0.01–0.02 wt.% to increase the hydrophilicity of the TFN membrane and the permeate water flux increased from 39.0 to 59.4 $L/m^2/h$, with slightly less rejections of NaCl and Na_2SO_4.[109]

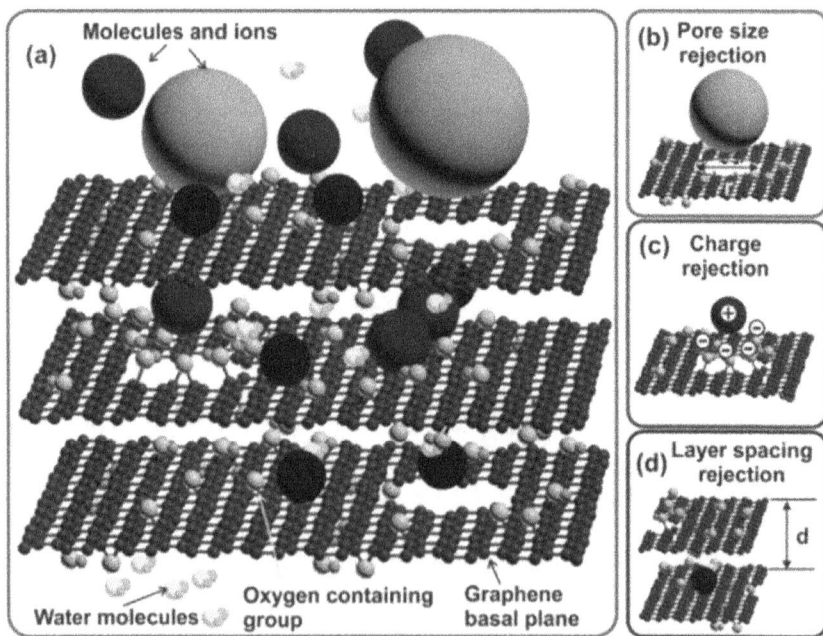

FIGURE 2.5   (a) Schematic illustration of molecules or ions passing through 2D-stacked GO and fundamental transport mechanisms of GO membranes. (b) In-plane pore size (defects). (c) Interactions (charge effect) with the oxygen-containing functional groups. (d) Interlayer spacing.

Source: Reprinted with permission from Ref. [105]. © 2019, American Chemical Society.

2.4.2 ANTIBACTERIAL ACTIVITY OF GO-CNTs COMPOSITES

Several studies have reported that GO exhibits strong antimicrobial properties against a different types of microorganisms, including both Gram-positive and Gram-negative pathogens, phytopathogens as well as biofilm-creating microorganisms.[110–114] The antimicrobial property of GO is reportedly mediated by physical and chemical interactions occurring when bacterial cells are directly exposed to GO.[115,116] The cytotoxicity of GO primarily affects the cell membrane.[110,117] Damages to the cell membrane are identified by altered morphology of the cell, changes in the transmembrane potential, uptake of membrane-impermeable dyes, and leakage of RNA and intracellular electrolytes.[110,112,118] GO may cause membrane damage by rupturing the cell membrane, disrupting cellular integrity, or GO-induced lipid peroxidation.[115] Oxidative stress is a major concern in bacterial cells exposed to GO.[113,114,119] The different mechanisms of cellular damage caused by graphene-family materials have been shown in Figure 2.6.

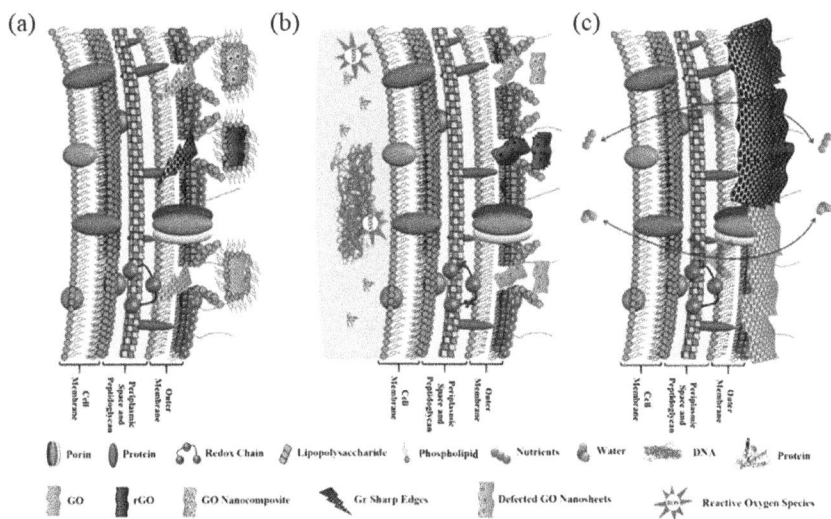

FIGURE 2.6 Main antimicrobial mechanisms of graphene-family materials. (a) Direct physical damage. (b) ROS-mediated oxidative stress. (c) Bacterial isolation via wrapping around the bacterial surface.

Source: Reprinted with permission from Ref. [120]. © 2020, Elsevier.

The contact-mediated antimicrobial properties of GO have inspired the development of several types of GO-based antimicrobial surfaces. These substrates have substituted biocide-releasing surfaces (containing antibiotics or silver) that are susceptible to eventual leaching or depletion.[115,121] Exposure to GO has been found to induce disruption of cellular integrity and loss of cell viability in turn reducing the bacterial build-up on the surface.[110,111,122] GO has successfully imparted antimicrobial properties to membranes applied for water treatment.[122-124]

However, detailed information regarding the effective antimicrobial activity of GO is yet to be determined. Studies have mostly reported antimicrobial properties of GO sheets in suspension whereby aggregation and cell wrapping takes place.[115] According to Liu et al.,[125] actual lateral dimension of GO sheets is strongly correlated with cell aggregation. However, aggregation of GO sheets reportedly enhances cell growth.[126] Nevertheless, incorporation of nanoparticles in graphene layers reportedly reduces the problem of aggregation.[32] Of all other nanoparticles, CNTs have exhibited strong antimicrobial activity and can be tailored to enhance their antimicrobial potential.[115,127] GO-CNTs composites have reportedly demonstrated stronger antibacterial activity against both Gram-positive (*Bacillus subtilis*) and Gram-negative (*Escherichia coli*) strains in comparison to Amoxicillin.[128] NF membranes prepared from rGO–CNTs hybrid materials have also demonstrated good antifouling properties.[129]

2.5 COMPUTATIONAL SIMULATION

Computational chemistry has been applied to simulate the heterogeneous reactions of pollutants by nanomaterials functionalized with metals, metal oxides, or carbon.[64] Insight into the removal reactions at the molecular-electronic level can be gained by computational simulation via theoretical calculations, including ab initio methods, the density functional theory (DFT), and their derivative models.[130,131] Thermodynamic properties, accurate bond energies, some experimental kinetic parameters that are difficult to deduce, and a better description of the reaction mechanism can be obtained from the models.[64,132] The results contribute to the design and modification of functional nanomaterials to improve their absorptive efficiencies for different types of pollutants.

Few studies have focused on the optimization of nanomaterial fabrication and their application for treatment of wastewater. Recent studies have investigated the structural and electronic properties of green chemically functionalized SWCNTs with ionic liquid functional groups via DFT.[133] Many studies have reported the application of soft computing techniques like response surface methodology and artificial neural networks for process optimization.[134–139] However, further studies on computer simulations of the interaction between pollutants and nanomaterials in aqueous phase are still needed to save valuable time and effort during basic lab experiments.

2.6   FUTURE RESEARCH

GO-CNTs composites have immense potential for different environmental applications, especially wastewater treatment. However, more studies are required for elucidating the fate, transport, and ecotoxicity of these composite materials in the environment. Application (e.g., adsorption capacity) and implication (e.g., toxicity) of GO-CNTs composites are dependant upon physicochemical properties of these materials including size, functional groups, surface charge, purity, etc. Process sustainability is dependent upon engineering of these composites in a method safe for both the environment and humans as well as the ability of the synthesized materials to yield the desired performance. Further studies should address the relationships between the structure design, obtained function, and environmental risk of engineered composites.

Toxicity analysis of CNTs and graphene nanomaterials should include the impact of the same on the entire life cycle of the target organism, rather than being restricted to exposure impacts. The impact of both production of CNTs and graphene nanomaterials and direct release of the same on the environment requires mitigation. Processes involving these materials essentially require a complete life cycle analysis for balancing the economics of commercial production, performance efficiency and environmental implication of the same. For ensuring wide scale commercial application of these composites, studies require to consider the following suggestions.

i.   The properties of nanomaterials primarily guided by the routes and conditions of synthesis. These parameters play a significant role in determining the size, structure, composition and reactivity of the synthesized nanomaterials. Few studies have reported the impact of changes in preparation conditions (like reaction time, reagent, temperature, and pH) on the performance efficiency of nanomaterials. Studies need to focus on these influencing factors both during composite synthesis and their application.

ii. Performance efficiency of these nanocomposites should be investigated with actual wastewater as simulated water contains only one pollutant in labscale batch experiments. Studies should also include more column and field studies.

iii. Most studies have focused on the removal efficiencies and capacities of heavy metals. More attention needs to be paid to the costs of using and supporting facilities required for the use of functionalized nanomaterials. Recycling of the nanomaterials also needs to be further explored. The resulting waste products will likely require further treatment due to the hazardous nature of pollutants and ecological risks of some nanomaterials.[64]

iv. Optimization of process parameters for both synthesis of the nanomaterials and their application for widescale wastewater treatment.

2.7 CONCLUSION

GO and CNTs hybrid materials have been investigated for wastewater treatment processes like catalysis, adsorption, and membrane filtration. Of these processes, membrane filtration has only been investigated on commercial scales. Some of the latest studies revealed that functionalized CNTs and GO-based membranes would be successful at greater extent both in laboratory and commercial scale, but implementation of these membranes in full scale for wastewater treatment requires feasibility study in pilot or full-scale studies. The performance assessments of these membranes in real feed system and the endurance test for sufficient time must be considered as one of the future works. In addition, the cost-effectiveness of both CNTs and GO-based membranes for full-scale wastewater treatment in contrast to other existing technologies need to be evaluated. For GO-based membranes, tuning the interlayer spacing for the enhancement of mechanical strength and improvement of ion permeability to achieve the consistent desirable separation performance are yet to be resolved. For the sustainable application of nanocomposite membranes in wastewater treatment, the membrane should be designed and developed using CNTs/GO in combination with other functional nanomaterials and the best compatible polymeric system. Other processes of wastewater treatment need to be evaluated in terms of the applicability of GO-CNTs hybrids in the same. Besides, studies are required on treatment of real effluents for understanding the applicability of the developed processes in actual situations.

KEYWORDS

- adsorption
- aerogel
- antimicrobial activity
- carbon nanotubes
- emerging pollutants
- graphene oxide
- membrane filtration
- nanocomposites
- wastewater treatment

REFERENCES

1. Santhosh, C.; Velmurugan, V.; Jacob, G.; Jeong, S. K.; Grace, A. N.; Bhatnagar, A. Role of Nanomaterials in Water Treatment Applications: A Review. *Chem. Eng. J.* **2016,** *306,* 1116–1137.
2. Banerjee, P.; Dey, T. K.; Sarkar, S.; Swarnakar, S.; Mukhopadhyay, A.; Ghosh, S. Treatment of Cosmetic Effluent in Different Configurations of Ceramic UF Membrane Based Bioreactor: Toxicity Evaluation of the Untreated and Treated Wastewater Using Catfish (*Heteropneustes fossilis*). *Chemosphere* **2016,** *146,* 133–144.
3. Barman, S. R.; Banerjee, P.; Das, P.; Mukhopadhayay, A. Urban Wood Waste as Precursor of Activated Carbon and Its Subsequent Application for Adsorption of Polyaromatic Hydrocarbons. *Int. J. Energy Water Resour.* **2018,** *2* (1–4), 1–13.
4. Bhattacharyya, S.; Banerjee, P.; Bhattacharya, S.; Rathour, R. K. S.; Majumder, S. K.; Das, P.; Datta, S. Comparative Assessment on the Removal of Ranitidine and

Prednisolone Present in Solution Using Graphene Oxide (GO) Nanoplatelets. *Desalin Water Treat* **2018**, *132*, 287–296.

5. Banerjee, P.; Sau, S.; Das, P.; Mukhopadhayay, A. Optimization and Modelling of Synthetic Azo Dye Wastewater Treatment Using Graphene Oxide Nanoplatelets: Characterization Toxicity Evaluation and Optimization Using Artificial Neural Network. *Ecotoxicol. Environ. Saf.* **2015**, *119*, 47–57.

6. Gautam, R. K.; Chattopadhyaya, M. C. *Nanomaterials for Wastewater Remediation*; Butterworth-Heinemann, 2016.

7. Hristovski, K.; Baumgardner, A.; Westerhoff, P. Selecting Metal Oxide Nanomaterials for Arsenic Removal in Fixed Bed Columns: From Nanopowders to Aggregated Nanoparticle Media. *J. Hazard. Mater.* **2007**, *147* (1–2), 265–274.

8. Li, L.; Fan, M.; Brown, R. C.; Van Leeuwen, J.; Wang, J.; Wang, W.; Song, Y.; Zhang, P. Synthesis, Properties, and Environmental Applications of Nanoscale Iron-Based Materials: A Review. *Crit. Rev. Environ. Sci. Technol.* **2006**, *36* (5), 405–431.

9. Li, X. Q.; Elliott, D. W.; Zhang, W. X. Zero-Valent Iron Nanoparticles for Abatement of Environmental Pollutants: Materials and Engineering Aspects. *Crit. Rev. Solid State Mater. Sci.* **2006**, *31* (4), 111–122.

10. Yang, K.; Xing, B. Desorption of Polycyclic Aromatic Hydrocarbons from Carbon Nanomaterials in Water. *Environ. Pollut.* **2007**, *145* (2), 529–537.

11. Adeyemo, A. A.; Adeoye, I. O.; Bello, O. S. Metal Organic Frameworks as Adsorbents for Dye Adsorption: Overview, Prospects and Future Challenges. *Toxicol. Environ. Chem.* **2012**, *94* (10), 1846–1863.

12. Gupta, S. S.; Bhattacharyya, K. G. Adsorption of Heavy Metals on Kaolinite and Montmorillonite: A Review. *Phys. Chem. Chem. Phys.* **2012**, *14* (19), 6698–6723.

13. Ngah, W. W.; Teong, L. C.; Hanafiah, M. M. Adsorption of Dyes and Heavy Metal Ions by Chitosan Composites: A Review. *Carbohydr. Polym.* **2011**, *83* (4), 1446–1456.

14. Wang, S.; Peng, Y. Natural Zeolites as Effective Adsorbents in Water and Wastewater Treatment. *Chem. Eng. J.* **2010**, *156* (1), 11–24.

15. Ren, X.; Chen, C.; Nagatsu, M.; Wang, X. Carbon Nanotubes as Adsorbents in Environmental Pollution Management: A Review. *Chem. Eng. J.* **2011**, *170* (2–3), 395–410.

16. Seymour, M. B.; Su, C.; Gao, Y.; Lu, Y.; Li, Y. Characterization of Carbon Nano-Onions for Heavy Metal Ion Remediation. *J. Nanopart. Res.* **2012**, *14* (9), 1087.

17. Wang, S.; Ng, C. W.; Wang, W.; Li, Q.; Li, L. A Comparative Study on the Adsorption of Acid and Reactive Dyes on Multiwall Carbon Nanotubes in Single and Binary Dye Systems. *J. Chem. Eng. Data* **2012**, *57* (5), 1563–1569.

18. Novoselov, K. S.; Geim, A. K.; Morozov, S. V.; Jiang, D.; Zhang, Y.; Dubonos, S. V.; Grigorieva, I.V.; Firsov, A. A. Electric Field Effect in Atomically Thin Carbon Films. *Science* **2004**, *306* (5696), 666–669.

19. Dreyer, D. R.; Park, S.; Bielawski, C. W.; Ruoff, R. S. The Chemistry of Graphene Oxide. *Chem. Soc. Rev.* **2010**, *39* (1), 228–240.

20. Mkhoyan, K. A.; Contryman, A. W.; Silcox, J.; Stewart, D. A.; Eda, G.; Mattevi, C.; Miller, S.; Chhowalla, M. Atomic and Electronic Structure of Graphene-Oxide. *Nano Lett.* **2009**, *9* (3), 1058–1063.

21. Ramesha, G. K.; Kumara, A. V.; Muralidhara, H. B.; Sampath, S. Graphene and Graphene Oxide as Effective Adsorbents toward Anionic and Cationic Dyes. *J. Colloid Interf. Sci.* **2011**, *361* (1), 270–277.

22. Banerjee, P.; Mukhopadhayay, A.; Das, P. Graphene Oxide Nanocomposites for Azo Dye Removal from Wastewater. In *Handbook of Textile Effluent Remediation*; Yusuf, M., Ed.; CRC Press, 2018; pp 303–341.

23. Dreyer, D. R.; Todd, A. D.; Bielawski, C. W. Harnessing the Chemistry of Graphene Oxide. *Chem. Soc. Rev.* **2014,** *43* (15), 5288–5301.

24. Hummers Jr, W. S.; Offeman, R. E. Preparation of Graphitic Oxide. *J. Am. Chem. Soc.* **1958,** *80* (6), 1339–1339.

25. Compton, O. C.; Nguyen, S. T. Graphene Oxide, Highly Reduced Graphene Oxide, and Graphene: Versatile Building Blocks for Carbon-Based Materials. *Small* **2010,** *6* (6), 711–723.

26. Cheng, Q.; Tang, J.; Ma, J.; Zhang, H.; Shinya, N.; Qin, L.C. Graphene and Nanostructured MnO_2 Composite Electrodes for Supercapacitors. *Carbon* **2011,** *49*, 2917–2925.

27. Zhao, J.; Ren, W.; Cheng, H.M. Graphene Sponge for Efficient and Repeatable Adsorption and Desorption of Water Contaminations. *J. Mater. Chem.* **2012,** *22* (38), 20197–20202.

28. Zhao, X.; Zhang, B.; Ai, K.; Zhang, G.; Cao, L.; Liu, X.; Sun, H.M.; Wang, H.S.; Lu, L. Monitoring Catalytic Degradation of Dye Molecules on Silver-Coated ZnO Nanowire Arrays by Surface-Enhanced Raman Spectroscopy. *J. Mater. Chem.* **2009,** *19* (31), 5547–5553.

29. Chowdhury, S.; Balasubramanian, R. Recent Advances in the Use of Graphene-Family Nanoadsorbents for Removal of Toxic Pollutants from Wastewater. *Adv. Colloid Interf. Sci.* **2014,** *204*, 35–56.

30. Cheng, J. S.; Du, J.; Zhu, W. Facile Synthesis of Three-Dimensional Chitosan–Graphene Mesostructures for Reactive Black 5 Removal. *Carbohydr. Polym.* **2012,** *88* (1), 61–67.

31. Kuilla, T.; Bhadra, S.; Yao, D.; Kim, N. H.; Bose, S.; Lee, J. H. Recent Advances in Graphene Based Polymer Composites. *Progress Polym. Sci.* **2010,** *35* (11), 1350–1375.

32. Banerjee, P.; Mukhopadhyay, A.; Das, P. Graphene Oxide–Nanobentonite Composite Sieves for Enhanced Desalination and Dye Removal. *Desalination* **2019,** *451*, 231–240.

33. Chang, H.; Wu, H. Graphene-Based Nanocomposites: Preparation, Functionalization, and Energy and Environmental Applications. *Energy Environ. Sci.* **2013,** *6* (12), 3483–3507.

34. Si, Y.; Samulski, E. T. Exfoliated Graphene Separated by Platinum Nanoparticles. *Chem. Mater.* **2008,** *20* (21), 6792–6797.

35. Xu, C.; Cui, A.; Xu, Y.; Fu, X. Graphene Oxide–TiO_2 Composite Filtration Membranes and Their Potential Application for Water Purification. *Carbon* **2013,** *62*, 465–471.

36. Huang, X.; Qi, X.; Boey, F.; Zhang, H. Graphene-Based Composites. *Chem. Soc. Rev.* **2012,** *41* (2), 666–686.

37. Kong, H. X. Hybrids of Carbon Nanotubes and Graphene/Graphene Oxide. *Curr. Opin. Solid State Mater. Sci.* **2013,** *17* (1), 31–37.

38. Huang, J. H.; Fang, J. H.; Liu, C. C.; Chu, C. W. Effective Work Function Modulation of Graphene/Carbon Nanotube Composite Films as Transparent Cathodes for Organic Optoelectronics. *Acs Nano* **2011,** *5* (8), 6262–6271.

39. Jang, W. S.; Chae, S. S.; Lee, S. J.; Song, K. M.; Baik, H. K. Improved Electrical Conductivity of a Non-Covalently Dispersed Graphene–Carbon Nanotube Film by Chemical p-Type Doping. *Carbon* **2012,** *50* (3), 943–951.

40. Pan, Y.; Bao, H.; Li, L. Noncovalently Functionalized Multiwalled Carbon Nanotubes by Chitosan-Grafted Reduced Graphene Oxide and Their Synergistic Reinforcing Effects in Chitosan Films. *ACS Appl. Mater. Interf.* **2011,** *3* (12), 4819–4830.

41. Tung, V. C.; Chen, L. M.; Allen, M. J.; Wassei, J. K.; Nelson, K.; Kaner, R. B.; Yang, Y. Low-Temperature Solution Processing of Graphene–Carbon Nanotube Hybrid Materials for High-Performance Transparent Conductors. *Nano Lett.* **2009,** *9* (5), 1949–1955.

42. King, P. J.; Khan, U.; Lotya, M.; De, S.; Coleman, J. N. Improvement of Transparent Conducting Nanotube Films by Addition of Small Quantities of Graphene. *ACS Nano* **2010,** *4* (7), 4238–4246.

43. Aravind, S. J.; Jafri, R. I.; Rajalakshmi, N.; Ramaprabhu, S. Solar Exfoliated Graphene–Carbon Nanotube Hybrid Nano Composites as Efficient Catalyst Supports for Proton Exchange Membrane Fuel Cells. *J. Mater. Chem.* **2011,** *21* (45), 18199–18204.

44. Velten, J.; Mozer, A. J.; Li, D.; Officer, D.; Wallace, G.; Baughman, R.; Zakhidov, A. Carbon Nanotube/Graphene Nanocomposite as Efficient Counter Electrodes in Dye-Sensitized Solar Cells. *Nanotechnology* **2012,** *23* (8), 085201.

45. Yen, M. Y.; Hsiao, M. C.; Liao, S. H.; Liu, P. I.; Tsai, H. M.; Ma, C. C. M.; Pu, N.W.; Ger, M. D. Preparation of Graphene/Multi-Walled Carbon Nanotube Hybrid and Its Use as Photoanodes of Dye-Sensitized Solar Cells. *Carbon* **2011,** *49* (11), 3597–3606.

46. Hong, T. K.; Lee, D. W.; Choi, H. J.; Shin, H. S.; Kim, B. S. Transparent, Flexible Conducting Hybrid Multilayer Thin Films of Multiwalled Carbon Nanotubes with Graphene Nanosheets. *ACS Nano* **2010,** *4* (7), 3861–3868.

47. Yu, D.; Dai, L. Self-Assembled Graphene/Carbon Nanotube Hybrid Films for Supercapacitors. *J. Phys. Chem. Lett.* **2010,** *1* (2), 467–470.

48. Kim, Y. K.; Min, D. H. Durable Large-Area Thin Films of Graphene/Carbon Nanotube Double Layers as a Transparent Electrode. *Langmuir* **2009,** *25* (19), 11302–11306.

49. Khan, U.; O'Connor, I.; Gun'ko, Y. K.; Coleman, J. N. The Preparation of Hybrid Films of Carbon Nanotubes and Nano-Graphite/Graphene with Excellent Mechanical and Electrical Properties. *Carbon* **2010,** *48* (10), 2825–2830.

50. Tang, Y.; Gou, J. Synergistic Effect on Electrical Conductivity of Few-Layer Graphene/Multi-Walled Carbon Nanotube Paper. *Mater. Lett.* **2010,** *64* (22), 2513–2516.

51. Du, F.; Yu, D.; Dai, L.; Ganguli, S.; Varshney, V.; Roy, A. K. Preparation of Tunable 3D Pillared Carbon Nanotube–Graphene Networks for High-Performance Capacitance. *Chem. Mater.* **2011,** *23* (21), 4810–4816.

52. Fan, Z.; Yan, J.; Zhi, L.; Zhang, Q.; Wei, T.; Feng, J.; Zhang, M.; Qian, W.; Wei, F. A Three-Dimensional Carbon Nanotube/Graphene Sandwich and Its Application as Electrode in Supercapacitors. *Adv. Mater.* **2010,** *22* (33), 3723–3728.

53. Dong, X.; Li, B.; Wei, A.; Cao, X.; Chan-Park, M. B.; Zhang, H.; Li, L.J.; Huang, W.; Chen, P. One-Step Growth of Graphene–Carbon Nanotube Hybrid Materials by Chemical Vapor Deposition. *Carbon* **2011,** *49* (9), 2944–2949.

54. Ma, Y.; Sun, L.; Huang, W.; Zhang, L.; Zhao, J.; Fan, Q.; Huang, W. Three-Dimensional Nitrogen-Doped Carbon Nanotubes/Graphene Structure Used as a Metal-Free Electrocatalyst for the Oxygen Reduction Reaction. *J. Phys. Chem. C* **2011,** *115* (50), 24592–24597.

55. Nguyen, D. D.; Tai, N. H.; Chen, S. Y.; Chueh, Y. L. Controlled Growth of Carbon Nanotube–Graphene Hybrid Materials for Flexible and Transparent Conductors and Electron Field Emitters. *Nanoscale* **2012,** *4* (2), 632–638.

56. Lee, D. H.; Kim, J. E.; Han, T. H.; Hwang, J. W.; Jeon, S.; Choi, S. Y.; Hong, S.H.; Lee, W.J.; Ruoff, R.S.; Kim, S. O. Versatile Carbon Hybrid Films Composed of Vertical Carbon Nanotubes Grown on Mechanically Compliant Graphene Films. *Adv. Mater.* **2010**, *22* (11), 1247–1252.

57. Sharma, V. K.; McDonald, T. J.; Kim, H.; Garg, V. K. Magnetic Graphene–Carbon Nanotube Iron Nanocomposites as Adsorbents and Antibacterial Agents for Water Purification. *Adv. Colloid Interf. Sci.* **2015**, *225*, 229–240.

58. Kim, H.; Hwang, Y. S.; Sharma, V. K. Adsorption of Antibiotics and Iopromide Onto Single-Walled and Multi-Walled Carbon Nanotubes. *Chem. Eng. J.* **2014**, *255*, 23–27.

59. Qu, X.; Alvarez, P. J.; Li, Q. Applications of Nanotechnology in Water and Wastewater Treatment. *Water Res.* **2013**, *47* (12), 3931–3946.

60. Shen, Y.; Fang, Q.; Chen, B. Environmental Applications of Three-Dimensional Graphene-Based Macrostructures: Adsorption, Transformation, and Detection. *Environ. Sci. Technol.* **2015**, *49* (1), 67–84.

61. Mahmoud, K. A.; Mansoor, B.; Mansour, A.; Khraisheh, M. Functional Graphene Nanosheets: The Next Generation Membranes for Water Desalination. *Desalination* **2015**, *356*, 208–225.

62. Pavagadhi, S.; Tang, A. L. L.; Sathishkumar, M.; Loh, K. P.; Balasubramanian, R. Removal of Microcystin-LR and Microcystin-RR by Graphene Oxide: Adsorption and Kinetic Experiments. *Water Res.* **2013**, *47* (13), 4621–4629.

63. Zhang, L.; Li, X.; Wang, M.; He, Y.; Chai, L.; Huang, J.; Wang, H.; Wu, X.; Lai, Y. Highly Flexible and Porous Nanoparticle-Loaded Films for Dye Removal by Graphene Oxide–Fungus Interaction. *ACS Appl. Mater. Interf.* **2016**, *8* (50), 34638–34647.

64. Sarkar, C.; Bora, C.; Dolui, S. K. Selective Dye Adsorption by pH Modulation on Amine-Functionalized Reduced Graphene Oxide–Carbon Nanotube Hybrid. *Indust. Eng. Chem. Res.* **2014**, *53* (42), 16148–16155.

65. Xu, J.; Cao, Z.; Zhang, Y.; Yuan, Z.; Lou, Z.; Xu, X.; Wang, X. A Review of Functionalized Carbon Nanotubes and Graphene for Heavy Metal Adsorption from Water: Preparation, Application, and Mechanism. *Chemosphere* **2018**, *195*, 351–364.

66. Fang, Q.; Zhou, X.; Deng, W.; Liu, Z. Hydroxyl-Containing Organic Molecule Induced Self-Assembly of Porous Graphene Monoliths with High Structural Stability and Recycle Performance for Heavy Metal Removal. *Chem. Eng. J.* **2017**, *308*, 1001–1009.

67. Zhan, W.; Gao, L.; Fu, X.; Siyal, S. H.; Sui, G.; Yang, X. Green Synthesis of Amino-Functionalized Carbon Nanotube-Graphene Hybrid Aerogels for High Performance Heavy Metal Ions Removal. *Appl. Surf. Sci.* **2019**, *467*, 1122–1133.

68. Chen, H.; Shao, D.; Li, J.; Wang, X. The Uptake of Radionuclides from Aqueous Solution by Poly (Amidoxime) Modified Reduced Graphene Oxide. *Chem. Eng. J.* **2014**, *254*, 623–634.

69. Hu, H.; Wang, X.; Wang, J.; Wan, L.; Liu, F.; Zheng, H.; Chen, R.; Xu, C. Preparation and Properties of Graphene Nanosheets–Polystyrene Nanocomposites via in Situ Emulsion Polymerization. *Chem. Phys. Lett.* **2010**, *484* (4–6), 247–253.

70. Si, Y.; Fu, Q.; Wang, X.; Zhu, J.; Yu, J.; Sun, G.; Ding, B. Superelastic and Superhydrophobic Nanofiber-Assembled Cellular Aerogels for Effective Separation of Oil/Water Emulsions. *ACS Nano* **2015**, *9* (4), 3791–3799.

71. Fang, Q.; Shen, Y.; Chen, B. Synthesis, Decoration and Properties of Three-Dimensional Graphene-Based Macrostructures: A Review. *Chem. Eng. J.* **2015**, *264*, 753–771.

72. Wang, H.; Yuan, X.; Zeng, G.; Wu, Y.; Liu, Y.; Jiang, Q.; Gu, S. Three Dimensional Graphene Based Materials: Synthesis and Applications from Energy Storage and Conversion to Electrochemical Sensor and Environmental Remediation. *Adv. Colloid Interf. Sci.* **2015**, *221*, 41–59.

73. Zhang, X.; Liu, D.; Yang, L.; Zhou, L.; You, T. Self-Assembled Three-Dimensional Graphene-Based Materials for Dye Adsorption and Catalysis. *J. Mater. Chem. A* **2015**, *3* (18), 10031–10037.

74. Shen, Y.; Zhu, X.; Chen, B. Size Effects of Graphene Oxide Nanosheets on the Construction of Three-Dimensional Graphene-Based Macrostructures as Adsorbents. *J. Mater. Chem. A* **2016**, *4* (31), 12106–12118.

75. Chen, Y.; Chen, L.; Bai, H.; Li, L. Graphene Oxide–Chitosan Composite Hydrogels as Broad-Spectrum Adsorbents for Water Purification. *J. Mater. Chem. A* **2013**, *1* (6), 1992–2001.

76. Yu, B.; Xu, J.; Liu, J. H.; Yang, S. T.; Luo, J.; Zhou, Q.; Wan, J.; Liao, R.; Wang, H.; Liu, Y. Adsorption Behavior of Copper Ions on Graphene Oxide–Chitosan Aerogel. *J. Environ. Chem. Eng.* **2013**, *1* (4), 1044–1050.

77. Sui, Z.; Meng, Q.; Zhang, X.; Ma, R.; Cao, B. Green Synthesis of Carbon Nanotube–Graphene Hybrid Aerogels and Their Use as Versatile Agents for Water Purification. *J. Mater. Chem.* **2012**, *22* (18), 8767–8771.

78. Liu, X.; Li, H.; Zeng, Q.; Zhang, Y.; Kang, H.; Duan, H.; Guo, Y.; Liu, H. Electro-Active Shape Memory Composites Enhanced by Flexible Carbon Nanotube/Graphene Aerogels. *J. Mater. Chem. A* **2015**, *3* (21), 11641–11649.

79. Ma, Q.; Cheng, H.; Fane, A. G.; Wang, R.; Zhang, H. Recent Development of Advanced Materials with Special Wettability for Selective Oil/Water Separation. *Small* **2016**, *12* (16), 2186–2202.

80. Hu, X.; Xu, W.; Zhou, L.; Tan, Y.; Wang, Y.; Zhu, S.; Zhu, J. Tailoring Graphene Oxide-Based Aerogels for Efficient Solar Steam Generation under One Sun. *Adv. Mater.* **2017**, *29* (5), 1604031.

81. Wu, L.; Qin, Z.; Zhang, L.; Meng, T.; Yu, F.; Ma, J. CNT-Enhanced Amino-Functionalized Graphene Aerogel Adsorbent for Highly Efficient Removal of Formaldehyde. *N. J. Chem.* **2017**, *41* (7), 2527–2533.

82. Lv, P.; Yu, K.; Tan, X.; Zheng, R.; Ni, Y.; Wang, Z.; Liu, C.; Wei, W. Super-Elastic Graphene/Carbon Nanotube Aerogels and Their Application as a Strain-Gauge Sensor. *RSC Adv.* **2016**, *6* (14), 11256–11261.

83. Shao, Q.; Tang, J.; Lin, Y.; Li, J.; Qin, F.; Yuan, J.; Qin, L. C. Carbon Nanotube Spaced Graphene Aerogels with Enhanced Capacitance in Aqueous and Ionic Liquid Electrolytes. *J. Power Sourc.* **2015**, *278*, 751–759.

84. Shen, Y.; Zhu, X.; Zhu, L.; Chen, B. Synergistic Effects of 2D Graphene Oxide Nanosheets and 1D Carbon Nanotubes in the Constructed 3D Carbon Aerogel for High Performance Pollutant Removal. *Chem. Eng. J.* **2017**, *314*, 336–346.

85. Wan, W.; Zhang, R.; Li, W.; Liu, H.; Lin, Y.; Li, L.; Zhou, Y. Graphene–Carbon Nanotube Aerogel as an Ultra-Light, Compressible and Recyclable Highly Efficient Absorbent for Oil and Dyes. *Environ. Sci.: Nano* **2016**, *3* (1), 107–113.

86. Cai, D.; Song, M.; Xu, C. Highly Conductive Carbon-Nanotube/Graphite-Oxide Hybrid Films. *Adv. Mater.* **2008**, *20* (9), 1706–1709.

87. Fu, D.; He, Z.; Su, S.; Xu, B.; Liu, Y.; Zhao, Y. Fabrication of α-FeOOH Decorated Graphene Oxide-Carbon Nanotubes Aerogel and Its Application in Adsorption of Arsenic Species. *J. Colloid Interf. Sci.* **2017**, *505*, 105–114.

88. Gu, Z.; Wang, Y.; Tang, J.; Yang, J.; Liao, J.; Yang, Y.; Liu, N. The Removal of Uranium (VI) from Aqueous Solution by Graphene Oxide–Carbon Nanotubes Hybrid Aerogels. *J. Radioanalyt. Nucl. Chem.* **2015**, *303* (3), 1835–1842.

89. Liu, Y.; Liu, X.; Zhao, Y.; Dionysiou, D. D. Aligned α-FeOOH Nanorods Anchored on a Graphene Oxide-Carbon Nanotubes Aerogel Can Serve as an Effective Fenton-Like Oxidation Catalyst. *Appl. Cataly. B: Environ.* **2017**, *213*, 74–86.

90. Zhan, W.; Yu, S.; Gao, L.; Wang, F.; Fu, X.; Sui, G.; Yang, X. Bioinspired Assembly of Carbon Nanotube into Graphene Aerogel with "cabbagelike" Hierarchical Porous Structure for Highly Efficient Organic Pollutants Cleanup. *ACS Appl. Mater. Interf.* **2018**, *10* (1), 1093–1103.

91. Zhang, M.; Gao, B.; Cao, X.; Yang, L. Synthesis of a Multifunctional Graphene–Carbon Nanotube Aerogel and Its Strong Adsorption of Lead from Aqueous Solution. *RSC Adv.* **2013**, *3* (43), 21099–21105.

92. Bassyouni, M.; Abdel-Aziz, M. H.; Zoromba, M. S.; Abdel-Hamid, S. M. S.; Drioli, E. A Review of Polymeric Nanocomposite Membranes for Water Purification. *J. Indust. Eng. Chem.* **2019**, *73*, 19–46.

93. Ihsanullah. Carbon Nanotube Membranes for Water Purification: Developments, Challenges, and Prospects for the Future. *Separation Purif. Technol.* **2019**, *209*, 307–337.

94. Holt, J. K.; Park, H. G.; Wang, Y.; Stadermann, M.; Artyukhin, A. B.; Grigoropoulos, C. P.; Noy, A.; Bakajin, O. Fast Mass Transport through Sub-2-Nanometer Carbon Nanotubes. *Sci.* **2006**, *312* (5776), 1034–1037.

95. Sianipar, M.; Kim, S. H.; Iskandar, F.; Wenten, I. G. Functionalized Carbon Nanotube (CNT) Membrane: Progress and Challenges. *RSC Adv.* **2017**, *7* (81), 51175–51198.

96. Sun, P.; Wang, K.; Zhu, H. Recent Developments in Graphene-Based Membranes: Structure, Mass-Transport Mechanism and Potential Applications. *Adv. Mater.* **2016**, *28* (12), 2287–2310.

97. Chae, H. R.; Lee, J.; Lee, C. H.; Kim, I. C.; Park, P. K. Graphene Oxide-Embedded Thin-Film Composite Reverse Osmosis Membrane with High Flux, Anti-Biofouling, and Chlorine Resistance. *J. Membr. Sci.* **2015**, *483*, 128–135.

98. Choi, J. H.; Jegal, J.; Kim, W. N. Fabrication and Characterization of Multi-Walled Carbon Nanotubes/Polymer Blend Membranes. *J. Membr. Sci.* **2006**, *284* (1–2), 406–415.

99. Celik, E.; Park, H.; Choi, H.; Choi, H. Carbon Nanotube Blended Polyethersulfone Membranes for Fouling Control in Water Treatment. *Water Res.* **2011**, *45* (1), 274–282.

100. Gunawan, P.; Guan, C.; Song, X.; Zhang, Q.; Leong, S. S. J.; Tang, C.; Chen, Y.; Chan-Park, M.B.; Chang, M.W.; Wang, K.; Xu, R. Hollow Fiber Membrane Decorated with Ag/MWNTs: Toward Effective Water Disinfection and Biofouling Control. *ACS Nano* **2011**, *5* (12), 10033–10040.

101. Song, X.; Wang, Y.; Wang, C.; Huang, M.; Gul, S.; Jiang, H. Solar-Intensified Ultrafiltration System Based on Porous Photothermal Membrane for Efficient Water Treatment. *ACS Sustain. Chem. Eng.* **2019**, *7* (5), 4889–4896.

102. Kim, H. J.; Choi, K.; Baek, Y.; Kim, D. G.; Shim, J.; Yoon, J.; Lee, J. C. High-Performance Reverse Osmosis CNT/Polyamide Nanocomposite Membrane by Controlled Interfacial Interactions. *ACS Appl. Mater. Interf. 2014, 6* (4), 2819–2829.

103. Kim, H. J.; Lim, M. Y.; Jung, K. H.; Kim, D. G.; Lee, J. C. High-Performance Reverse Osmosis Nanocomposite Membranes Containing the Mixture of Carbon Nanotubes and Graphene Oxides. *J. Mater. Chem. A 2015, 3* (13), 6798–6809.

104. Zhao, W.; Liu, H.; Liu, Y.; Jian, M.; Gao, L.; Wang, H.; Zhang, X. Thin-Film Nanocomposite Forward-Osmosis Membranes on Hydrophilic Microfiltration Support with an Intermediate Layer of Graphene Oxide and Multiwall Carbon Nanotube. *ACS Appl. Mater. Interf. 2018, 10* (40), 34464–34474.

105. Yang, T.; Lin, H.; Loh, K. P.; Jia, B. Fundamental Transport Mechanisms and Advancements of Graphene Oxide Membranes for Molecular Separation. *Chem. Mater. 2019, 31* (6), 1829–1846.

106. Ma, J.; Ping, D.; Dong, X. Recent Developments of Graphene Oxide-Based Membranes: A Review. *Membranes 2017, 7* (3), 52.

107. Marjani, A.; Nakhjiri, A. T.; Adimi, M.; Jirandehi, H. F.; Shirazian, S. Effect of Graphene Oxide on Modifying Polyethersulfone Membrane Performance and Its Application in Wastewater Treatment. *Sci. Rep. 2020, 10* (1), 1–11.

108. Nasseri, S.; Ebrahimi, S.; Abtahi, M.; Saeedi, R. Synthesis and Characterization of Polysulfone/Graphene Oxide Nano-Composite Membranes for Removal of Bisphenol A from Water. *J. Environ. Manage. 2018, 205,* 174–182.

109. Yin, J.; Zhu, G.; Deng, B. Graphene Oxide (GO) Enhanced Polyamide (PA) Thin-Film Nanocomposite (TFN) Membrane for Water Purification. *Desalination 2016, 379,* 93–101.

110. Akhavan, O.; Ghaderi, E. Toxicity of Graphene and Graphene Oxide Nanowalls against Bacteria. *ACS Nano 2010, 4* (10), 5731–5736.

111. Carpio, I. E. M.; Santos, C. M.; Wei, X.; Rodrigues, D. F. Toxicity of a Polymer–Graphene Oxide Composite against Bacterial Planktonic Cells, Biofilms, and Mammalian Cells. *Nanoscale 2012, 4* (15), 4746–4756.

112. Chen, J.; Peng, H.; Wang, X.; Shao, F.; Yuan, Z.; Han, H. Graphene Oxide Exhibits Broad-Spectrum Antimicrobial Activity against Bacterial Phytopathogens and Fungal Conidia by Intertwining and Membrane Perturbation. *Nanoscale 2014, 6* (3), 1879–1889.

113. Gurunathan, S.; Han, J. W.; Dayem, A. A.; Eppakayala, V.; Kim, J. H. Oxidative Stress-Mediated Antibacterial Activity of Graphene Oxide and Reduced Graphene Oxide in *Pseudomonas aeruginosa. Int. J. Nanomed. 2012, 7,* 5901.

114. Krishnamoorthy, K.; Veerapandian, M.; Zhang, L. H.; Yun, K.; Kim, S. J. Antibacterial Efficiency of Graphene Nanosheets against Pathogenic Bacteria via Lipid Peroxidation. *J. Phys. Chem. C 2012, 116* (32), 17280–17287.

115. Perreault, F.; De Faria, A. F.; Nejati, S.; Elimelech, M. Antimicrobial Properties of Graphene Oxide Nanosheets: Why Size Matters. *ACS Nano 2015, 9* (7), 7226–7236.

116. Sanchez, V. C.; Jachak, A.; Hurt, R. H.; Kane, A. B. Biological Interactions of Graphene-Family Nanomaterials: An Interdisciplinary Review. *Chem. Res. Toxicol. 2012, 25* (1), 15–34.

117. Tu, Y.; Lv, M.; Xiu, P.; Huynh, T.; Zhang, M.; Castelli, M.; Liu, Z.; Huang, Q.; Fan, C.; Fang, H.; Zhou, R. Destructive Extraction of Phospholipids from *Escherichia coli* Membranes by Graphene Nanosheets. *Nat. Nanotechnol. 2013, 8* (8), 594.

118. Hui, L.; Piao, J. G.; Auletta, J.; Hu, K.; Zhu, Y.; Meyer, T.; Liu, H.; Yang, L. Availability of the Basal Planes of Graphene Oxide Determines Whether It Is Antibacterial. *ACS Appl. Mater. Interf. 2014, 6* (15), 13183–13190.
119. Romero-Vargas Castrillón, S.; Perreault, F.; De Faria, A. F.; Elimelech, M. Interaction of Graphene Oxide with Bacterial Cell Membranes: Insights from Force Spectroscopy. *Environ. Sci. Technol. Lett. 2015, 2* (4), 112–117.
120. Wu, Y.; Xia, Y.; Jing, X.; Cai, P.; Igalavithana, A. D.; Tang, C.; Tsang, D.C.; Ok, Y. S. Recent Advances in Mitigating Membrane Biofouling Using Carbon-Based Materials. *J. Hazard. Mater. 2020, 382,* 120976.
121. Banerjee, I.; Pangule, R. C.; Kane, R. S. Antifouling Coatings: Recent Developments in the Design of Surfaces That Prevent Fouling by Proteins, Bacteria, and Marine Organisms. *Adv. Mater. 2011, 23* (6), 690–718.
122. Perreault, F.; Tousley, M. E.; Elimelech, M. Thin-Film Composite Polyamide Membranes Functionalized with Biocidal Graphene Oxide Nanosheets. *Environ. Sci. Technol. Lett. 2014, 1* (1), 71–76.
123. Lee, J.; Chae, H. R.; Won, Y. J.; Lee, K.; Lee, C. H.; Lee, H. H.; Kim, I.C.; Lee, J. M. Graphene Oxide Nanoplatelets Composite Membrane with Hydrophilic and Antifouling Properties for Wastewater Treatment. *J. Membr. Sci. 2013, 448,* 223–230.
124. Musico, Y. L. F.; Santos, C. M.; Dalida, M. L. P.; Rodrigues, D. F. Surface Modification of Membrane Filters Using Graphene and Graphene Oxide-Based Nanomaterials for Bacterial Inactivation and Removal. *ACS Sustain. Chem. Eng. 2014, 2* (7), 1559–1565.
125. Liu, S.; Hu, M.; Zeng, T. H.; Wu, R.; Jiang, R.; Wei, J.; Wang, L.; Kong, J.; Chen, Y. Lateral Dimension-Dependent Antibacterial Activity of Graphene Oxide Sheets. *Langmuir 2012, 28* (33), 12364–12372.
126. Ruiz, O. N.; Fernando, K. S.; Wang, B.; Brown, N. A.; Luo, P. G.; McNamara, N. D.; Vangsness, M.; Sun, Y.P.; Bunker, C. E. Graphene Oxide: A Nonspecific Enhancer of Cellular Growth. *ACS Nano 2011, 5* (10), 8100–8107.
127. Liu, X.; Wang, M.; Zhang, S.; Pan, B. Application Potential of Carbon Nanotubes in Water Treatment: A Review. *J. Environ. Sci. 2013, 25* (7), 1263–1280.
128. Khan, A. A. P.; Khan, A.; Rahman, M. M.; Asiri, A. M.; Oves, M. Lead Sensors Development and Antimicrobial Activities Based on Graphene Oxide/Carbon Nano-tube/Poly (O-toluidine) Nanocomposite. *Int. J. Biol. Macromol. 2016, 89,* 198–205.
129. Chen, X.; Qiu, M.; Ding, H.; Fu, K.; Fan, Y. A Reduced Graphene Oxide Nanofiltration Membrane Intercalated by Well-Dispersed Carbon Nanotubes for Drinking Water Purification. *Nanoscale 2016, 8* (10), 5696–5705.
130. Sun, L.; Zhang, A.; Su, S.; Wang, H.; Liu, J.; Xiang, J. A DFT Study of the Interaction of Elemental Mercury with Small Neutral and Charged Silver Clusters. *Chem. Phys. Lett. 2011, 517* (4–6), 227–233.
131. Wilcox, J.; Okano, T. Ab Initio-Based Mercury Oxidation Kinetics via Bromine at Postcombustion Flue Gas Conditions. *Energy Fuels 2011, 25* (4), 1348–1356.
132. Auzmendi-Murua, I.; Castillo, A.; Bozzelli, J. W. Mercury Oxidation via Chlorine, Bromine, and Iodine under Atmospheric Conditions: Thermochemistry and Kinetics. *J. Phys. Chem. A, 118* (16), 2959–2975.
133. Roohi, H.; Khyrkhah, S. Green Chemical Functionalization of Single-Wall Carbon Nanotube with Methylimidazolium Dicyanamid Ionic Liquid: A First Principle Computational Exploration. *J. Mol. Liquids 2015, 211,* 498–505.

134. Ali, I.; Alharbi, O. M.; ALOthman, Z. A.; Al-Mohaimeed, A. M.; Alwarthan, A. Modeling of Fenuron Pesticide Adsorption on CNTs for Mechanistic Insight and Removal in Water. *Environ. Res. 2019*, *170*, 389–397.
135. Dehghani, M. H.; Yetilmezsoy, K.; Salari, M.; Heidarinejad, Z.; Yousefi, M.; Sillanpää, M. Adsorptive Removal of Cobalt (II) from Aqueous Solutions Using Multi-Walled Carbon Nanotubes and γ-alumina as Novel Adsorbents: Modelling and Optimization Based on Response Surface Methodology and Artificial Neural Network. *J. Mol. Liquids 2020*, *299*, 112154.
136. Fiyadh, S. S.; AlSaadi, M. A.; AlOmar, M. K.; Fayaed, S. S.; Hama, A. R.; Bee, S.; El-Shafie, A. The Modelling of Lead Removal from Water by Deep Eutectic Solvents Functionalized CNTs: Artificial Neural Network (ANN) Approach. *Water Sci. Technol. 2017*, *76* (9), 2413–2426.
137. Hiew, B. Y. Z.; Lee, L. Y.; Lai, K. C.; Gan, S.; Thangalazhy-Gopakumar, S.; Pan, G. T.; Yang, T. C. K. Adsorptive Decontamination of Diclofenac by Three-Dimensional Graphene-Based Adsorbent: Response Surface Methodology, Adsorption Equilibrium, Kinetic and Thermodynamic Studies. *Environ. Res. 2019*, *168*, 241–253.
138. Ibrahim, R. K.; Fiyadh, S. S.; AlSaadi, M. A.; Hin, L. S.; Mohd, N. S.; Ibrahim, S.; Afan, H.A.; Fai, C.M.; Ahmed, A.N.; Elshafie, A. Feedforward Artificial Neural Network-Based Model for Predicting the Removal of Phenolic Compounds from Water by Using Deep Eutectic Solvent-Functionalized CNTs. *Molecules 2020*, *25* (7), 1511.
139. Nasiri, R.; Arsalani, N. Synthesis and Application of 3D Graphene Nanocomposite for the Removal of Cationic Dyes from Aqueous Solutions: Response Surface Methodology Design. *J. Clean. Prod. 2018*, *190*, 63–71.
140. Cai, J.; Tian, J.; Gu, H.; Guo, Z. Amino Carbon Nanotube Modified Reduced Graphene Oxide Aerogel for Oil/Water Separation. *ES Mater. Manuf. 2019*, *6*, 68–74.

CHAPTER 3

UNDOPED AND DOPED CARBON NANOTUBES FOR REMEDIATION OF CONTAMINANTS FROM WASTEWATER

HEMLATA KARNE[1] and SHRIKAANT KULKARNI[2*]

[1]Department of Chemical Engineering, Vishwakarma Institute of Technology, Pune, India

[2]Faculty of Science & Technology, Vishwakarma University, Pune, India

*Corresponding author. E-mail: shrikaant.kulkarni@vupune.ac.in

ABSTRACT

Water scarcity is a major problem in today's world. Day-by-day wastewater quantity goes on increasing as population goes on increasing. For the degradation of pollutants from water, various methods such as adsorption, catalysis, etc., were reported. Recently, carbon nanotubes are widely used for the removal of pollutants from water such as heavy metals, dyes, phenol, oil, and various organic and inorganic pollutants. To increase the capacity of nanotubes for the removal of pollutants, doping is carried out. Doped carbon nanotubes are used in several applications such as oxygen reduction reaction, photocatalytic activities, fuel cells, etc. Carbon nanotubes were generally doped with N_2, Ag, Al, Co, B, etc., for various applications. Carbon nanotubes combine other materials like grapheme; membranes are also used for the removal of harmful pollutants from wastewater. This work consists of introduction to undoped carbon nanotubes and doped carbon nanotubes doped with different elements, synthesis method, and their characterization. Further, it includes the applications of undoped and doped carbon nanotubes for removal of heavy metals, dyes, organic compounds, pathogens, etc., from wastewater. Doping of carbon nanotubes gives added advantage of better degradation and removal of pollutants from water. Composites of carbon nanotubes with various materials or membranes improve potential of carbon nanotubes for removal of pollutants from wastewater.

3.1 INTRODUCTION

The day-by-day generation of wastewater is going on increasing along with population increase. According to a World Bank report, 40% of the world's population did not have access to clean water or sanitation. Wastewater is being generated in huge amounts due to various activities, including domestic as well as industrial. Water pollution is caused mainly due to dumping of domestic wastewater as well as industrial wastewater directly in water resources. According to UN-Water statistics, every day, 2 million tons of human waste are disposed off in water resources in developing countries while 70% of industrial wastes are dumped untreated into water where they pollute the usable water supply.[1] Organic and inorganic pollutants present into water sources due to dumping of wastewater without treatment. These pollutants affect the health of human beings causing several diseases like diarrhea, cholera, etc., and ultimately result into health and economic losses of society.[2] Water demand is growing rapidly as a result of increasing population and rapid urbanization. However, water resources are limited in populated areas and arid regions. Due to which, there is an urgent need for treatment of wastewater and seawater desalination as freshwater resources are limited. In addition to this, the increase in pollution leads to water shortage problems.[3] Organic pollutants caused serious health and environmental effects. The main organic pollutants are heavy metals, humic substances, pharmaceuticals, industrial dyes, pesticides, oil, and other aromatic compounds etc.[4,5] The traditional methods used for water treatment are adsorption, activated carbon, activated sludge process, microfiltration, membrane filtration, and reverse osmosis. These traditional methods are not effective for the treatment of complex and complicated polluted water streams consisting of pharmaceutical, personal care products, surfactants, various industrial additives, and chemical purported. The traditional water treatment is also inadequate for the removal of toxic compounds such as heavy metals, dyes, and pathogens.[6] Therefore, various methods were developed consisting of several nanomaterials for treatment of water. Nanotechnology gives a promising option for the removal of such complex pollutants from wastewater. Commonly used nanomaterial for water treatment was carbon nanotubes (CNTs), TiO_2 nanotubes, graphene, zinc oxide nanotubes, etc.[7–9] However, widely used material for water treatment was CNTs-based material due to their efficiency and effectiveness for pollutants removal.

3.2 CARBON NANOTUBES FOR WASTE WATER TREATMENT

Traditionally, various methods are used for treatment of water such as adsorption, absorption, activated sludge process, anaerobic and aerobic ponds, and membrane technologies. But there are several limitations for traditional methods. Activated carbon was used as adsorbent in industrial treatment processes, but regeneration of it and deposition of it caused several problems. Membrane processes are efficient for removal of various pollutants from wastewater, but these processes are very costly. Therefore, there is an urgent need for an effective and low-cost technology for removal of pollutants from wastewater. Due to physical, chemical, electrical, and structural properties, CNTs are quite efficient for removal of pollutants from wastewater. CNTs are used as adsorbents, catalysts for degradation of pollutants, or blended with filters and membranes to remove various types of pollutants from wastewater.[1] Further, CNTs combined with metal oxides acted as the best catalyst, as CNTs have high electrical conductivity, structural flexibility, large surface area, and high mechanical strength. When metal oxides were doped on CNTs, it improved electronic conductivity of CNTs, which resulted in excellent electrocatalytic activities than the pristine CNTs. CNTs showed good electrocatalytic activities, due to different factors such as crystal phases, morphology, oxidation state, surface area, and electronic conductivity.[10]

Development of CNTs was started in 1991 and since then various researchers reported applications of CNTs in various fields. It was reported in literature that CNTs were used for drinking water treatment as well as for removal of several pollutants from wastewater due to its tunable physical, chemical, electrical, and structural properties. In water treatment, CNTs were used as sorbents, catalysts, filters, or membranes. Previously, manufacturing of CNTs on a large scale was a costly affair but, nowadays, due to the introduction of chemical vapor deposition method, mass-scale

production of CNTs was reported in literature. Due to large-scale production of CNTs, applications of CNTs in wastewater treatment increased.[3]

CNTs were used as adsorbents for removal of various organic as well
as inorganic pollutants from water. CNTs were used for removal dyes,
pharmaceutical products, phenols, heavy metals, etc. Removal of oil from
water with the help of CNTs was also reported by few researchers.[4]

3.3 DOPED CARBON NANOTUBES

3.3.1 NITROGEN DOPED

CNTs are widely used as adsorbent, photocatalyst, and oxygen reduction
reactions (ORRs). The efficiency of CNTs is further increased by doping
of it with various elements. One of the commonly used doping elements is
nitrogen. CNTs doped with nitrogen are used in several applications such as
photocatalytic degradation, ORR, etc. Doping of CNTs with nitrogen could
efficiently create the metal-free active sites for electrochemical reduction
of oxygen. Doping of nitrogen-induced intramolecular charge transfer and
intermolecular charge transfer induced by adsorption of polyelectrolyte
onto all carbon CNTs result in improved ORR catalytic activities.[13]

The modification of the intrinsic electronic structure of CNTs was
obtained by doping nitrogen on CNTs. Nitrogen-doped CNTs showed
enhanced catalytic performance and also showed the nonradical pathway
accompanied by the generation of radicals. The procedure of preparation
of CNTs is difficult and wasteful because preparing high-quality CNTs
require chemical vapor deposition equipment, mixed H_2/CH_4 atmosphere,
or the addition of some poisonous reagents like dicyandiamide. However,
nitrogen-doped CNTs were usually obtained through post-treatment of
CNTs with nitrogen containing reagents such as ammonia, melamine,
and urea. Nevertheless, these two-step methods always involve tedious
procedures and uniform distribution of N-atoms in the carbon matrix was
not confirmed. Recently, Wang et al. reported the growth of nitrogen-
doped CNTs under mild conditions by catalyzing it and with the assistance
of iron element and melamine. The traces of an amount of iron on the
surface of NCNT were also reported on active sites. However, this process
resulted in disordered CNTs with random size. Therefore, it still remains
a challenge to develop a method for preparation of nitrogen-doped CNTs
of uniform structure.[14]

Nitrogen-doped carbon materials have been one of the most popular families of materials showing promising applications in various fields. Nitrogen-doped CNTs framework was used for removal of Bisphenol A from wastewater. Bisphenol A is a commonly occurring pollutant, which is introduced in water bodies due to packaging material, pharmaceutical applications, etc. Nitrogen-doped CNTs showed higher Bisphenol A removal efficiency than porous carbon.[14]

Nitrogen-doped CNTs are also used in water desalination or in combination with membrane for water desalination. Researchers reported nitrogen-doped CNTs prepared by thermal conversion of polypyrrole nanotubes in nitrogen atmosphere. Their interconnected nanotube structure provided more accessible space for ion accommodation, shortened ion diffusion pathway for fast ion adsorption/desorption, and optimized nitrogen doping species, which resulted in improved electrical conductivity and increased sodium ion adsorption. Nitrogen-doped CNTs obtained by this method showed a maximum desalination capacity and a good cycling stability.[15]

Nitrogen-doped CNTs were also useful for solid-phase microextraction. Researchers reported metal organic framework-derived nitrogen-doped CNT cages with more active sites and better adsorption capacity. Compared with commercial fibers, these nitrogen-doped CNTs showed better extraction properties due to its π–π interactions, abundant active sites, and hollow cage structure, which was composed of interconnected crystalline N-doped CNTs. N-CNTC-coated fiber exhibited better extraction performance and shorter extraction equilibrium time than the solid N-doped C-coated fiber due to its hollow cage structure.[16]

Nitrogen-doped multi-walled CNTs combine with compound to produce hybrid, which one can also use in ORR. Researchers reported nitrogen-doped multi-walled CNTs (N-doped MWCNT)/$MnCo_2O_4$ hybrid preparation by hydrothermal method. This hybrid showed improved oxygen-reduction electrocatalytic activity in an aqueous alkaline medium when compared with pristine $MnCo_2O_4$.[10]

3.3.2   OTHER ELEMENTS DOPED

CNTs are not only doped with nitrogen but also with some other elements such as boron, sulfur, phosphorus, etc. Doping of CNTs are generally carried out to increase its efficiency by improving electron-donating properties, which result in improving ORR activities. Doping of CNTs is also carried out to increase photocatalytic activity. Due to the chemical nature of the heteroatoms and local structures, the heteroatom doping could either increase the oxygen reduction current or decrease the onset over potential. Improvement in ORR activities is due to facilitating the O_2 adsorption, increasing the total number of active sites, or improving the surface hydrophilicity.[13]

Yang et al. studied CNTs doped with boron and its effects on charged sites. They reported that doping boron could also turn CNTs into metal free ORR catalysts with positively shifted potential and reduction in current was also improved. This doping also showed good resistance toward methanol crossover and CO poisoning.[17] The larger electronegativity of carbon in comparison to the boron atom leads to the formation of positively charged boron due to the polarization in B–C hybridization bonds, which was essential for chemisorption of oxygen on boron dopant. In contrast, oxygen was adsorbed on the carbon atoms neighboring the nitrogen dopant in nitrogen-doped CNTs. These studies showed that the doping induced charge redistribution, regardless whether the dopants have lower or higher electronegativity than of carbon, could create charged sites that were favorable for oxygen adsorption as well as reduction process. While on pristine CNTs, these effects did not reported as there was no active sites and oxygen showed resistance to pristine CNTs due to orbital mismatch.[13]

Doped CNTs with aluminum to further increase adsorption capacity than that of pristine CNTs was also reported in literature. Aluminum-doped CNTs showed large surfaces as well as good mechanical properties, which resulted in high adsorption capability. Aluminum-doped CNTs were used for adsorption of various pollutants or removal of pollutants in water treatment.[5]

Boron doped on multiwalled CNTs was also used in sensors for detection on Bisphenol A in water samples. These types of biosensors were reported as highly sensitive electrochemical activities due to electroactivity toward o-quinone, a product of this enzymatic reaction of Bisphenol A (BPA) oxidation catalyzed by tyrosinace. This type of biosensor was reported by Zehani et al. who tested it on various river water samples for detection of Bisphenol A in wastewater. They reported that boron doped on multi-walled CNTs was promising and reliable analytical tool for onsite monitoring of BPA in wastewater.[18]

CNTs grown over Fe- and Ni-doped activated alumina were used for removal of heavy metals from industrial effluents. CNTs grown over

Fe- and Ni-doped activated alumina was prepared by chemical vapor deposition with acid wash and reported as nanofloral clusters. These clusters were used to remove Cr(VI) and Cd(II) from industrial waste water. These nanofloral clusters showed good adsorption for Cr(VI) and Cd(II), which was varied by both Langmuir and Freundlich models.[19]

3.3.3 METHODS OF SYNTHESIS FOR DOPED CNTs

Various methods were used for the production of CNTs such as laser ablation, arc discharge, electrolysis, sono-chemical or hydrothermal method, and various vapor deposition methods. The thermal chemical vapor deposition method has been used widely due to its simplicity and cost-effective nature, which is easily scalable for commercial level. It was also reported that CNTs produced by this method showed purity and higher yield as compared with other methods. The main parameters thath affect the growth of CNTs were carbon source, carrier gas, growth time, growth temperature, and properties of catalyst in chemical vapor deposition method. Commonly used catalyst for CNTs synthesis were Fe, Co, Ni, and their combinations, deposited on different support material such as SiO_2, MgO, and Al_2O_3.[20]

Doping of various species on CNTs were carried out by various methods such as chemical vapor deposition, electrodeposition, spray pyrolysis, wet impregnation method, facil hydrothermal method, arc discharge method, and aerosol pyrolysis method. Among all these methods, a commonly used method is vapor deposition method.[21–24]

3.3.4 CHARACTERIZATION OF DOPED CNTs

Characterization is the important parameter to indicate the successful synthesis of CNTs. There are various methods used for characterization of CNTs, but most commonly used characterization methods are X Ray diffraction (XRD), Fourier-transform infrared spectroscopy (FT-IR), Raman spectroscopy, transmission electron microscopy (TEM), X ray photoelectron spectroscopy (XPS), and scanning electron microscopy (SEM).[25] The structure and morphology of CNTs are mainly characterized by SEM, TEM, XRD, Raman spectroscopy, XPS, and UV-vis spectroscopy.[26] Doping of CNTs by different species such as nitrogen, boron, Li,

Al, etc., are carried out by various methods. The characterization of doped CNTs were carried out by XRD, emission SEM (FE-SEM), high-resolution TEM, Brunauer, Emmett and Teller (BET) technique, and thermogravimetric analysis (TGA).[4] These characterization techniques showed the presence of doped species on the surface of CNTs as well as active sites on the surface of CNTs.[5] Generally, optical properties of doped CNTs are determined by UV-vis spectroscopy, mineralogical properties are observed by Raman spectroscopy, crystallographic properties are determined by TEM, and its microstructure is observed by SEM.[27]

3.4   USE OF UNDOPED CNTs FOR HEAVY METAL REMOVAL

The group of race elements such as metals and metalloids with an atomic density greater than 4+_ 1 g/cm3, for example, Hg, Cu, Cr, Fe, Co, Cd, Pb, etc., is known as heavy metals. These metals ions are generally considered toxic mineral pollutants of soil and water. These heavy metals in wastewater effluent are observed due to natural sources such as volcanic activities, weathering of rocks and mineral, soil erosion, etc. While anthropogenic sources of heavy metals are fuel combustion, mineral processing, street run offs, agricultural activities, landfill,s and industrial activities. Untreated or inadequately treated metal-contaminated wastewater cause various health and environmental impacts due to stability, migration activity, and highly solubility of heavy metals in aqueous medium. Heavy metals are adsorbed by plants, thereby entering the animals and human bodied through food chains and causes damage to the body and various diseases. Therefore, removal of heavy metals from wastewater is an urgent need. Various traditional methods were used for removal of heavy water from wastewater such as membrane filtration, solvent extraction, chemical precipitation, ion exchange, coagulation, electrochemical removal, etc. However, these methods were not efficient for the complete removal of heavy metals from wastewater due to low efficiency, sensitive operating conditions, toxic sludge, and costly disposal. Therefore, it is necessary to develop new efficient method or material for removal of heavy metals from wastewater.[29]

Carbon materials are widely used as adsorbents in many engineering applications. As a new member of family, CNTs have greater potential in various applications such as field emitters for flat panel display, composite reinforcement, sensors, energy storage and energy conversion devices. and catalytic support phases, due to their extraordinary electrical, mechanical, thermal, and structural properties.[28] Due to chemical, mechanical, and physical properties CNTs replaced many traditional adsorbents for removal various pollutants from wastewater. In addition to this, CNTs also show chemical stability, large surface area, and availability of well-developed mesopores, which make CNTs suitable adsorbent for heavy metal removal from wastewater. As CNTs can be covalently or non-covalently functionalized with various species, they are able to effectively interact with adsorbates and have greater maximum adsorption capacity after modification. Therefore, CNTs replaced traditional adsorbents such as activated carbon in removal of heavy metals from wastewater.[29] CNTs have exhibited excellent adsorption properties in the removal heavy metals like lead, cadmium, chromium, copper, mercury, etc., being a promising candidate in wastewater treatment. The morphologies of the CNTs can affect their adsorption capabilities greatly. CNTs with poor crystallinity and morphology can be easily be introduced with much more functional groups, interestingly leading to better adsorption capabilities.[28] Tofighy and Mohammadi studied removal of some divalent heavy metal ions (Cu^{2+}, Zn^{2+}, Pb^{2+}, Cd^{2+}, and Co^{2+}) from water by using CNTs sheets and reported oxidized CNTs sheets was shown effective adsorbent for heavy metal ions removal from water. The kinetics of adsorption varied with initial concentration of heavy metal ions. They reported that usages of CNTs sheets for removal of heavy metals from wastewater without CNTs leakage into water is economically feasible.[30]

Lead is one of the heavy metal pollutant entering human life through industrial wastewater and causes different health damage and illnesses. As non-degradable nature of lead, many methods are developed for removal of lead from wastewater such as precipitation, membrane filtration, liquid-liquid extraction, ion exchange, and adsorption. Out of all these methods, adsorption is a commonly used method for removal of lead from wastewater due to its efficiency and simplicity. Many adsorbents have been developed for higher adsorption capacity and efficiency. CNTs is the most-used dadsorbent due to its large surface area and high efficiency for removal of lead from wastewater.[31] Lia et al. studied adsorption thermodynamics of removal of Pb2+ by using CNTs from wastewater and reported that removal of Pb from wastewater was spontaneous and endothermic reaction. It was also reported that Pb desorbed from CNTs by adjusting pH value, thus CNTs shown the potential for removal of Pb from wastewater.[32] Aliyu et al. studied preparation of multi-walled CNTs with catalytic chemical vapor deposition method. They studied removal of turbidity, iron (Fe), and lead (Pb) by using multi-walled CNTs and reported that multi-walled CNTs with higher length (58.17 μm) showed high removal capacity for turbidity and iron. They also reported that lower length (38.87 μm) multi-walled CNTs showed high removal of lead from waste water. The growth temperature during chemical vapor deposition showed a great effluence on the aspect ratio of the multi-walled CNTs.[33]

Tehrani et al. prepared multi-walled CNTs functionalized with tris (2-aminoethyl) amine for removal of lead ions from wastewater and reported that removal of lead from wastewater was strongly depend on pH. Under optimal conditions, maximum adsorption capacity obtained was 43 mg/g for lead. Langmuir isotherm showed best fit for reported data of removal of lead from wastewater by functionalized multi-walled CNTs.[31] Sahmetlioglu et al. prepared nanocomposite of CNTs with polypyrrole conducting polymer for the separation of lead at trace levels in wastewater. They reported the adsorption capacity of nanocomposite for lead was 25 mg lead/g composite.[34]

Chromium is commonly occurred pollutants in wastewater. Chromium produced dangerous effects on human body such as headache, diarrhea, and nausea. Chromium is a carcinogenic agent, which caused different types of cancers when accumulated in body.[60] Several researchers reported removal of chromium from wastewater with the help of CNTs-based material. A zero-valent iron multi-walled CNTs nanocomposites was reported for removal of Cr(VI) from wastewater with 36% higher efficiency on Cr removal as compared to bare zero valent iron. The removal of Cr followed pseudo first-order model under different pHs and initial concentrations of Cr.[35] Magnetic multi-walled CNTs were also reported for Cr(IV) removal from waste water. The adsorption capacity of magnetic multi-walled CNTs increased with initial Cr concentration while decreased with the increase of adsorbent dosage. The adsorption process is well explained by pseudo second-order kinetic model and Langmuir isotherm provide best fit for data of adsorption.[36] Activated carbon coated with CNTs were also reported for removal of chromium from wastewater. Kabbashi et al. studied removal of chromium by using CNTs from water with optimized parameters. They reported that activated carbon coated with CNTs showed highest adsorption capacity of 11.57 mg/g as compared to normal activated carbon. It was reported that activated carbon coated with CNTs adsorbed

23.7% more chromium from water as compared with normal-activated carbon. It was also reported that Langmuir and Freundlich isotherms fitted good for experimental data.[37] XU et al. studied adsorption of anionic chromate CrO_4^{2-} by functionalized CNTs from wastewater and reported that functionalized CNTs had a superior adsorption capacity for typical toxic heavy metals like anionic chromate from wastewater as compared to unmodified CNTs. The good adsorption capacity of functionalized CNTs was due to interaction of anionic chromate with surface oxygen containing functional groups on modified CNTs.[38]

Copper and cadmium are commonly occurred pollutants in wastewater. Cadmium produced dangerous effects on human body such as renal disorders and renal damage while copper causes liver damage, insomnia, and Wilson's disease. Cadmium is a carcinogenic agent, which caused different types of cancers when accumulated in body.[60] CNTs-based materials are reported to remove copper and cadmium from wastewater. Mohajeri et al. prepared a composite membrane of highly ordered CNTs and studied removal of copper (II) and cadmium (II) from simulated industrial wastewater. They reported that composite membrane with CNTs showed effective removal of copper and cadmium from wastewater due to CNTs hydrophobic inner walls, which had capability to adsorb heavy metals ions. They also reported that with increasing pH of the solution-enhanced heavy metal removal from wastewater.[39]

Mercury is one of the toxic pollutants, which causes contamination of groundwater in some developing countries like Korea, Bangladesh, and India. Mercury is carcinogenic in nature, which causes cancer when it enters in the human body. Other non-carcinogenic health effects of mercury on human body are cardiovascular, neurological, and pulmonary disorders as well as tremor and behavior disturbances. Traditional methods used for mercury removal are ion exchange, reverse osmosis, adsorption, coagulation, precipitation, etc.[60] Disadvantages of these methods are less effectiveness and high cost, toxic waste, etc. CNTs-based material was reported for removal of mercury from wastewater by researches. Deb et al. prepared amidoamine functionalized multi-walled CNTs and studied for separation of mercury from water. They reported that CNTs with amidoamine showed superior selectivity with high adsorption capacity for mercury as compared to other metal ions present in industrial wastewater. The adsorption followed pseudo second-order path and Redlich-Peterson isotherm showed good fit for experimental data.[40]

3.5   USE OF DOPED CNTs FOR HEAVY METAL REMOVAL

MFCs were also reported for removal of hexavalent chromium (Cr(VI)) and nontoxic chromium (Cr(III)) from waste water by use of modified CNTs. Surface of CNTs was modified by introduction of melamine and Fe and used as a catalyst in MFCs for removal of aquatic Cr and nontoxic Cr(III). Zhou et al. reported 55% chromium reduction efficiency in MFCs with use of modified CNTs with melamine and Fe, which was higher than commercial catalyst, with low-cost environmentally friendly solution.[41] N-doped CNTs encapsulating Ni nanoparticles was used for removal of Cr (IV) from polluted wastewater. N-doped CNTs showed excellent catalytic activity for Cr removal at mild conditions. Removal of Cr from wastewater by N-doped CNTs with Ni was significantly affected by the mass of nickel salt and synthesis temperature. N-doped CNTs composite showed excellent catalytic activity and recyclable capability for Cr removal from wastewater (Table 3.1).[61]

3.6   USE OF UNDOPED CNTs FOR DYE REMEDIATION

Every day huge amount of effluents containing dyes are released by industries such as paint formulation, carpet manufacture, printing houses, and paper and pulp production. Therefore, there is an urgent need for suitable technology for removal of harmful dyes from wastewater. These dyes are classified as anionic, cationic, non-ionic, and zwitterionic dyes and cationic dyes are more toxic than the anionic ones.[42] Various dyes, which cause water pollution, are methyl orange, rhodamine B, Coomassie brilliant blue R-250, orange G, etc., due to disposal of waste coming out of the textile industry, paper industry, etc. Main adsorption technique were reported to be used for removal of such dyestuff from wastewater and new adsorbents were introduced with improved surface area and increased number of active sites.[35]

Hu et al. prepared three-dimensional graphene oxide and CNTs nanostructures by method freeze-drying method. They reported that nanostructure of graphene oxide and CNTs showed good organic dye removal for both cationic (Rhodamine) and anionic (Methyle orange) in aqueous solution with adsorption capacities 248.48 mg/g for Rhodamine and 66.96 mg/g for methyl orange. They reported that the adsorption of both dyes follow pseudo second-order model while the adsorption equilibrium was well explained by the Langmuir and Freundlich isotherm theory.[45]

TABLE 3.1 Heavy Metal Removal from Wastewater by CNTs.

Adsorbent	Target pollutants	Highlights
CNTs sheets	Cu^{2+}, Zn^{2+}, Pb^{2+}, Cd^{2+}, Co^{2+}	Oxidized CNTs sheets shown effective adsorption
CNTs	lead	Adsorption was spontaneous and endothermic
Multi-walled CNTs	Iron & lead	Higher removal of iron by higher length CNTs, higher removal of lead by lower length CNTs
CNTs/Tris amine	lead	Adsorption capacity of functionalized CNTs was 43 mg/g
CNTs/ polypyrrole	lead	Adsorption capacity of nanocomposites was 25 mg/g
CNTs/Iron	Cr(VI)	Nanocomposites shown 36% higher efficiency on Cr removal as compared to bare zero valent iron.
Multi-walled CNTs	Cr(VI)	The adsorption capacity of magnetic multi walled CNTs increased with initial Cr concentration
CNTs/ activated carbon	Cr	Activated carbon coated with CNTs showed highest adsorption capacity of 11.57 mg/g
CNTs	CrO_4^{2-}	Functionalized CNTs had a superior adsorption capacity for anionic chromate from wastewater
CNTs/ membrane	Copper (II) and cadmium (II)	Composite membrane with CNTs showed effective removal of copper and cadmium
CNTs/ amidoamine	Mercury	CNTs with amidoamine showed superior selectivity with high adsorption capacity for mercury
CNTs/Fe/Ni	Cr(VI), Cd(II)	Good adsorption for Cr(VI) and Cd(II)
CNTs/ Melamine/Fe	Aquatic Cr, nontoxic Cr(III)	55% chromium reduction efficiency in MFCs, Low-cost environmentally friendly solution

3.7 USE OF DOPED CNTs FOR DYE REMEDIATION

All over the world due to serious environmental consequences of dyestuff usage and disposal, very strict regulations and measures are being adopted by governments. Various dyes, which cause water pollution are methyl orange, rhodamine B, Coomassie brilliant blue R-250, orange G, etc., due to disposal of waste coming out of the textile industry, paper industry etc.[35] Doped CNTs with Na and S were reported for photocatalytic degradation of p-chlorophenol and rhodamine B from wastewater. The enhanced photocatalytic activity of co-doped Na and S on CNTs was due to large surface area, narrowed bandgap, and enhanced visible light absorption with down-shifted valence band. Co-doped Na and S on CNTs was prepared by thermal polymerization using $NaHCO_3$ and thiourea as Na and S source, respectively.[36]

For removal of dyes like Rhodamine B from wastewater nitrogen-doped CNTs, $FeSO_4$ composite was also reported in literature. Phosphate residue is one of the hazardous wastes. $FeSO_4$ can be separated for it and used for formation of composite with CNTs for removal of dyes from wastewater. Wei et al. reported successful development of a method for separation of $FeSO_4$ from waste and used it for formation of nitrogen doped CNTs $FeSO_4$ composite. This composite was also reported for removal of Rhodamine B by photo degradation at concentration 15 mg/L and 98.9% degradation of dye occurred in 60 min with pseudo-first-order kinetics. After repeated cycles also this composite showed good catalytic capacity.[37]

Doped CNTs with metal can also directly act as a photocatalyst. Researchers reported that loading of silver on CNTs enhanced the photo-catalytic activity of CNTs. Doped Ag on CNTs was used for removal of rhodamine B from water. Doped Ag on CNTs exhibited photocatalytic degradation activity toward dyes like rhodamine B. First the rhodamine B molecule was adsorbed to CNTs, and then rhodamine B was excited upon visible light illumination. The photo-generated electrons could be transferred along the CNT surface and trapped by Ag particles. These trapped electrons reduced the adsorbed oxygen to superoxide anion radicals, resulting into further degradation of rhodamine B.[1]

Doped Au on CNTs were also reported for removal of dyes from wastewater by photocatalysis. In photocatalytic degradation of rhodamine B, applications of CNTs were beneficial as CNTs used as pillars of reduced graphene oxide platelets for dye degradation and the preparation of Au-CNT hybrid photocatalyst. Although Au nanoparticles are visible light photo-sensitizers, they don't hold photocatalytic activity, due to the fast rate of charge recombination. The excellent electro conductivity ability of CNTs made Au CNTs hybrid an effective visible light photocatalyst.[1]

Al-doped CNTs were reported as a multifunctional, adsorbent, and coagulant aid for organic pollutant removal such as methyl orange from aqueoussolution. Aluminum species were dispersed homogeneously on the surface of CNTs, and the structure of CNTs was modified for better adsorption of methyl orange from wastewater. Doping of aluminum efficiently improved adsorption ability for methyl orange onto the CNTs. Kang et al. reported the maximum adsorption capacity 69.7 mg/g of Al-doped CNTs for methyl orange removal based on the Langmuir isotherm model (Table 3.2).[5]

TABLE 3.2 Dye Removal from Wastewater by Different Doped CNTs.

Doped species on CNTs	Target pollutants	Highlights
CNTs/graphene oxide	Rhodamine and methyl orange	Nanostructure of graphene oxide and CNTs showed adsorption capacities 248.48 mg/g for Rhodamine and 66.96 mg/g for methyl orange
CNTs/Ag	Rhodamine B	Good photocatalytic degradation activity towards dyes like rhodamine B
CNTs/Al	Methyl orange	Good adsorption capacity for removal of dye
CNTs/Na/S	p-chlorophenol and rhodamine B	Enhancement in photocatalytic activity, Enhancement in visible light absorption
CNTs/ N/ FeSO$_4$	Rhodamine B	Good removal by photo degradation, pseudo-first-order kinetics in process

Organic Compound Removal from Wastewater
Undoped CNTs

For removal of various organic pollutants from wastewater, traditionally used method was adsorption due to its simplicity and efficiency over other methods. Commonly used adsorbents for adsorption were activated carbon, zeolites, and resins. Among all these, activated carbon is one of the most widely used adsorbents in water treatment because of its several advantages such as broad-spectrum removal capability toward pollutants, thermal stability, and chemical inertness. However, the use of activated carbon in water treatment also has some limitations like slow adsorption kinetics and difficulty for regeneration. To overcome these limitations, activated carbon fibers were developed as the second generation of carbonaceous adsorbents. The pores in activated carbon fiber are directly opening on the surface of the carbon matrix, which reduces the diffusion distance of pollutants to adsorption sites.[1] To further increase efficiency of adsorbent CNTs were developed which showed increase in surface area as well as increase in number of activated sites. Like activated carbon fiber, CNTs have one-dimensional structure. All adsorption sites are located on the inner and outer layer surface of CNTs. CNTs are promising third-generation carbonaceous adsorbents due to hollow and layered structures and tunable surface chemistry. In literature, a variety of organic compounds or heavy metals removal by CNTs was reported. The effect of various parameters on the removal process such as pH, ionic strength, and coexisting matter was also reported.[3]

Gethard et al. studied CNTs enhanced membrane distillation for waste concentration, which generated pure water. They reported that in a CNTs immobilized membrane, CNTs adsorbed waste material and provide additional pathway for enhanced water transport. CNTs immobilized membrane showed effective removal of waste from water than membrane without the nanotubes. The purified water contained less than 10% of residual organics as compared to the original water.[49]

Kanel et al. reported that CNTs yarn was a promising material for the removal of organic contaminants from aqueous waste streams due to thermal and chemical stability and high-surface area. They studied removal of model nitroaromatic compound, 2,4-dinitrotoluene from waste water. They reported that the adsorption follows pseudo-second-order model while the adsorption equilibrium was well explained by the Freundlich isotherm theory.[50]

Bisphenol A is commonly occurring pollutants in wastewater, which produce dangerous effects on environment and human health by entering in food chain. It shows endocrine disrupting property and high toxicity even at very low concentrations. Bisphenol A is widely used as intermediates in various processes such as production of polycarbonate plastic and epoxy resins lining of food containers. The traditional methods used for removal of Bisphenol A were chemical oxidation, adsorption, ultrasound digestion, and biodegradation but failed for complete separation due to unsatisfactory results and complicated procedures.[14] Nanocomposites of multi-walled CNTs and TiO_2/SiO_2 was reported for photo catalytic removal of bisphenol A and carbamazepine from wastewater. The nanocomposites showed high potential for removal of bisphenol A and carbamazepine from wastewater. The decomposition of carbamazepine over nanocomposites followed different mechanism than traditional photocatalyst P25 and the kinetics of removal of both pollutants followed a pseudo-first-order path.[43] Multi-walled CNTs were reported for removal of bisphenol F from wastewater. The multi-walled CNTs shown the great potential for removal of bisphenol F from wastewater. Fang et al. studied multi-walled CNTs for removal bisphenol F from wastewater and reported 96.2% removal of bisphenol F from wastewater was possible within 4 min under pH 4–10.

Bisphenol removal was very fast with multi-walled CNTs and followed pseudo-second-order model.[44]

3.8   DOPED CNTs IN MICROBIAL FUEL CELLS

In literature, a variety of organic compounds or heavy metals removal by CNTs was reported. The effect of various parameters on the removal process such as pH, ionic strength, and coexisting matter was also reported. The efficiency of CNTs is further increased by doping of it with various elements. Nanotubes were generally doped with N_2, Ag, Al, Co, B, etc., for various applications. Doping of CNTs gave added advantage of better degradation and removal of pollutants from water.

Microbial fuel cell is a very useful and environmentally friendly method for removal of pollutants from wastewater with additional benefit of electricity generation using microorganisms. The performance of microbial fuel cells (MFCs) is judged by chemical oxygen demand (COD) removal and electricity generation, which depend on several factors, such as reactor design, pH, microbial species, and electrode material. Microorganisms transfer electrons to electrodes and in anode, the organic matter in wastewater is oxidized by mediation of exoelectrogens. The anode material requires a large specific area for microbes colonization, stability, high conductivity, and catalytic activity. Most-used material for MFCs anode are carbon cloth, carbon paper, and carbon foam. However, these carbon materials have low electro-catalytic activity for the electrode microbial reactions. CNTs is a promising option for anode due to high conductivity and high surface to volume ratio, but the major herald is that CNT have cellular toxicity, which lead to proliferation inhibition and cell death and could not be used directly. Therefore, CNTs were reported to be used by putting coating on it or by doping with other species.[3]

In past decades, MFCs were reported as an evolving wastewater treatment technology for removal of harmful pollutants from wastewater. In typical MFCs, organic energy stored in organic compounds present in wastewater is directly converted to electricity by taking advantage of microbial extracellular respiration with an electrode. Due to which organic removal and electricity generation can be accomplished simultaneously at low energy consumption. Oxygen is widely used as an electron acceptor in MFCs due to its availability and high-reduction potential. But ORRs are complex and always require the presence of a catalyst. Platinum and its alloys are commonly used ORR catalysts because of their superior ORR catalytic activity but not an economically viable option in practice and therefore the major hindrance for the large-scale application of MFCs.[51]

The most important thing for the successful deployment of MFCs is the cost-effective catalysts for electricity generation form organic waste from wastewater. Doped CNTs with nitrogen were also reported as one of the good alternatives for effective catalysts in electricity generation from organic wastes in MFCs. Yang et al. synthesized N doped bamboo like CNT in which Co and Ni alloy was encapsulated at the end and/or middle section of the tubes. They reported the composites showed excellent electrocatalytic activity with competitive price due to increase in specific surface area and proved a promising candidate for ORR catalysts in MFCs for waste decomposition and energy recovery from wastes.[51]

Alkaline fuel cells are also a major class in MFCs. For alkaline fuel cells, development of highly efficient catalysts based on non-noble metal for ORR is of high significance. Nitrogen-doped CNTs were reported as catalysts used in alkaline fuel cells for organic decomposition from wastewater. Hanif et al. synthesized Ni and Co co-doped nitrogen CNTs and studied their catalytic performance in ORR for alkaline fuel cells. They reported that co-doped NCNTs with Co/Ni showed excellent ORR performance in alkaline fuel cells. The improved performance and stability of this composite was due to the synergetic effect of the nitrogen doped CNTs and Ni/Co active sites. The performance of this composite was higher than commercial Pt cathode which proved co-doped CNTs were good electrocatalysts for ORR in alkaline fuel cells.[52]

Nitrogen-doped CNTs frameworks were reported for catalytic degradation of Bisphenol A in the presence of peroxymonosulfate. Catalyst dose and operating temperature were main operating parameters, which hamper removal of bisphenol A by nitrogen-doped CNTs from wastewater. Wenjie et al. report efficient removal of Bisphenol A from wastewater by nitrogen-doped CNTs.[14]

In solid phase microextraction, an efficient and stable adsorbent is very important. It was reported that N-doped CNTs cage was efficiently used for solid phase microextraction of polychlorinated biphenyls. Compared with CNTs-coated fibers and commercial fibers, N-doped CNTs cage shown higher extraction property due to abundant active sites, hollow cage structure, and π-π interactions (Table 3.3).[16]

TABLE 3.3 Organic Pollutants Removal from Wastewater by CNTs.

Adsorbent	Target pollutants	Highlights
CNTs/ membrane	Organic pollutants	In a CNTs immobilized membrane, CNTs adsorbed waste material and provide additional pathway for enhanced water transport
CNTs yarn	Organic pollutants	CNTs yarn good capacity for the removal of organic contaminants
Multi-walled CNTs/ TiO$_2$/ SiO$_2$	Bisphenol A and carbamazepine	The nanocomposites showed high potential for removal of bisphenol A and carbamazepine
Multi-walled CNTs	Bisphenol F	Multi-walled CNTs shown 96.2% removal of bisphenol F
CNTs/ Co/ Ni/N	Organic pollutants	Excellent electrocatalytic activity, promising candidate for ORR catalysts in MFCs
CNTs/Co/Ni/N	Organic pollutants	Were good electrocatalysts for ORR in alkaline fuel cells
CNTs/N	Bisphenol A	Higher Bisphenol A removal efficiency than porous carbon
CNTs/ N	Polychlorinated biphenyls	Better extraction performance, shorter extraction equilibrium time

3.9   DOPED CARBON NANOTUBES FOR REMOVAL OF PATHOGENS

Many deaths occurred worldwide, but mostly in developing countries, due to waterborne diseases, which are caused due to contamination of water. Removal of pathogens is thus an extremely important factor for better health and hygiene. Direct contact with single-walled carbon tubes reported severe membrane damage and followed by cell death. Single-walled CNTs filters showed high bacterial retention while multi-walled CNTs showed high viral removal at low pressure due to exclusion effect. The combined effect of single-walled and multi-walled CNTs hybrid filter showed efficient bacterial inactivation and viral retention at low pressure.[3,53]

CNTs show unique characteristics, such as large surface areas, electrical properties, high stabilities, and strong adsorption properties. It was reported in literature that multi-walled CNTs can remove harmful viruses from wastewater and also be used as a disinfectant.[54] Mamba et al. reported that

25-nanometer-sized polio viruses can be removed by multi-walled CNTs from water as well as large pathogens such as Escherichia coli (E. Coli) and Staphylococcus aureus bacteria. They also reported about the good adsorption properties of β-cyclodextrin and shown that this material had a greater binding to water contaminants of up to 100,000.[55] But Lam et al. reported that CNTs introduced in the body interfere or damage DNA and that could produce harmful effects to organs.[56]

Multi-walled CNTs doped with silver were reported as anti-bacterial activity for removal of bacteria from wastewater. Rananga and Magadzu synthesized Al doped multi walled CNTs and tested for bacteria removal like *Escherichia coli* (*E. coli*). They reported that the nanocomposites caused a growth delay in the growth rate of E. coli bacteria. It was also reported that nanocomposite damage the outer membrane of E. coli, which were shown by formation of "pits."

3.10   USE OF UNDOPED CNTs WITH OTHER MATERIALS POLLUTANTS REMOVAL

Other compounds generally observed in wastewater are oil, some drugs, etc. CNTs were also used for removal of oil from wastewater. Oil refineries and water transport were responsible of leakage of huge amount of oil in water bodies and sea. These oil contaminations caused dangerous effects on aquatic life and ecosystem. Gui et al. reported a new sorbent material, magnetic CNTs, for removal of oil from wastewater. The sorption capacity of magnetic CNTs sponges was 56 g/g. Zhao et al. prepared a sponge-like hybrid of CNTs and exfoliated vermiculite with different CNT percentages by introducing aligned CNT arrays into natural vermiculite layers for oil adsorption. For diesel oil, the highest adsorption capacity reported was 26.7 g/g using hybrid CNTs and exfoliated vermiculite.

Toński et al. studied multi-walled CNTs for removal of anticancer drugs such as cyclophosphamide, ifosfamide, and 5-fluorouracil from wastewater mainly observed in hospital wastewater. They reported that CNTs showed highest surface area and had greatest adsorption capacity and cyclophosphamide showed sorption potential for all CNTs. Freundlich isotherm provided good fit for adsorption of anticancer drugs on multi-walled CNTs from wastewater.[57]

Yin et al. prepared CNTs mixed with organic binder for strips, which were used for removal of o-cresol from water. Th strips showed lower adsorption capacity of organics in water compared to the pristine CNTs but could be used as a reversible adsorbent. They also reported that the strips showed higher strength under compression and much higher resilience.[58]

3.11 USE OF DOPED CNTs WITH OTHER MATERIALS POLLUTANTS REMOVAL

Cobalt nanoparticles encapsulated in nitrogen-doped CNTs were reported for wastewater treatment for water recovery as well as hydrogen production. Yu et al. prepared cobalt nanoparticles encapsulated in nitrogen-doped CNTs by pyrolysis and showed that calcination temperature significantly affected the morphology and catalytic performance. They reported that wastewater was successfully recovered by codoped CNTs with cobalt and nitrogen and hydrogen production, which result into solving environmental issues as well as energy generation (Table 3.4).[59]

TABLE 3.4   Pollutants Removed from Wastewater by Different Undoped and Doped CNTs.

Doped species on CNTs	Target pollutants	Highlights
CNTs/sponges	Oil	Good adsorbent for spilled oil
CNTs/vermiculite	Oil	Oil adsorption from water
Multi-walled CNTs	Cyclophosphamide, ifosfamide, and 5-fluorouracil	CNTs shown greatest adsorption capacity for cyclophosphamide
CNTs/organic binder	o-cresol	CNTs mixed with organic binder strips showed higher strength under compression and much higher resilience
CNTs/Co/N	Pollutants	Water recovery was achieved with codoped CNTs with Co and nitrogen

3.12 CONCLUSION

The quantity of wastewater is getting increase over the time due to rampant industrialization population growth and widespread use of agrochemicals in agriculture. Numerous contaminants found in wastewater are namely heavy metals, dyes, and organic compounds, which have detrimental effects on both health and environment. Various materials have so far been used for the remediation of these pollutants present in waste water.

CNTs, by virtue of their host of chemical, mechanical properties, are used effectively and efficiently for the removal of such harmful chemicals from wastewater. Both undoped and doped CNTs are used for the remediation of heavy metals, dyes, organic compounds, pathogen, etc., from wastewater and are quite useful for it. Doping of CNTs with varying degrees of materials and composites of CNTs are found to be productive in their use. CNTs therefore either in isolation or doped with graphene offer an opportunity because of their unique properties and are ideal candidates in designing advanced and rapid water treatment technology-based solutions.

KEYWORDS

- **wastewater**
- **carbon nanotubes**
- **doped CNTs**
- **heavy metals**
- **dyes**
- **organic pollutants**

REFERENCES

1. Liu, X.; Zhang, S.; Pan, B. Potential of Carbon Nanotubes in Water Treatment. *J. Environ. Sci.* IntechOpen **2012**, 1–30.
2. Karne, H.; Bhatkhande, D.; Jabade, S. Mesophilic and Thermophilic Digestion of Faecal Sludge in a Pilot Plant Digester. *Int. J. Environ. Stud.* **2018**, *75*, 484–495.
3. Liu, X.; Wang, M.; Zhang, S.; Pan, B. Application Potential of Carbon in Water Treatment: A Review. *J. Environ. Sci.* **2013**, *25* (7), 1263–1280.
4. Ahmad, K.; Tarik, F.; Gordon, R.; Mohammed, M.; Ahmed Abdala, A-M.; Hilal, N.; Hussien, M. A. Enhancing Oil Removal from Water Using Ferric Oxide Nanoparticles Doped Carbon Nanotubes Adsorbents. *Chem. Eng. J.* **2016**, *293*, 90–101.
5. Kang, D.; Yu, X.; Ge, M.; Xiao, F.; Xu, H. Novel Al-Doped Carbon Nanotubes with Adsorption and Coagulation Promotion for Organic Pollutant Removal. *J. Environ. Sci.* **2017**, *54*, 1–12.
6. Amin, M. T.; Manzoor, A. A.; Alazba, U. A Review of Removal of Pollutants from Water/Wastewater Using Different Types of Nanomaterials. *Adv. Mater. Sci. Eng.* **2014**, 1–24. doi:10.1155/2014/825910

7. Hsieh, C. T.; Fan, W. S.; Chen, W-Y.; Lin, J-Y. Adsorption and Visible-Light-Derived Photocatalytic Kinetics of Organic Dye on Co-Doped Titania Nanotubes Prepared by Hydrothermal Synthesis. *Separation Purif. Technol.* **2009,** *67,* 312–318.

8. Varghese, J.; Varghese, K. T. Graphene/CuS/ZnO Hybrid Nanocomposites for High Performance Photocatalytic Applications. *Mater. Chem. Phys.* **2015,** *1671,* 258–264.

9. Subramaniam, M. N.; Goh, P. S.; Lau, W. J.; Ismail, A. F.; Karaman, M. Enhanced Visible Light Photocatalytic Degradation of Organic Pollutants by Iron Doped Titania Nanotubes Synthesized via Facile One-Pot Hydrothermal. *Powder Technol.* **2020,** *366,* 96–106.

10. Yuvaraj, S.; Vignesh, A.; Shanmugam, S. R.; Kalai, S. Nitrogen-Doped Multi-Walled Carbon Nanotubes- $MnCo_2O_4$ Microsphere as Electrocatalyst for Efficient Oxygen Reduction Reaction. *Int. J. Hydrogen Energy* **2016,** *41* (34), 15199–15207.

11. Gui, X.; Li, H.; Wang, K.; Wei, J.; Jia, Y.; Li, Z.; Fan, L.; Cao, A.; Zhu, H.; Wu, D. Recyclable Carbon Nanotube Sponges for Oil Absorption. *Acta Mater.* **2011,** *59,* 4798–4804.

12. Zhao, M.-Q.; Huang, J.-Q.; Zhang, Q.; Luo, W.-L.; Wei, F. Improvement of Oil Adsorption Performance by a Sponge-Like Natural Vermiculite-Carbon Nanotube Hybrid. *Appl. Clay Sci.* **2011,** *53,* 1–7.

13. Zhang, J.; Zhang, S.; Dai, Q.; Zhang, Q.; Dai, L. *Heteroatom-Doped Carbon Nanotubes as Advanced Electrocatalysts for Oxygen Reduction Reaction Nanocarbons for Advanced Energy Conversion*; Feng, X.;, 1st ed., 2015; pp 43–70.

14. Wenjie, M.; Wang, N.; Fan, Y.; Tong, T.; Han, X.; Du, Y. Non-Radical-Dominated Catalytic Degradation of Bisphenol A by ZIF-67 Derived Nitrogen-Doped Carbon Nanotubes Frameworks in the Presence of Peroxymonosulfate. *Chem. Eng. J.* **2018,** *336,* 721–731.

15. Pengfei, S.; Chen, W.; Jiayue, S.; Peng, L.; Tao, Y. Thermal Conversion of Polypyrrole Nanotubes to Nitrogen-Doped Carbon Nanotubes for Efficient Water Desalination Using Membrane Capacitive Deionization. *Separation Purif. Technol.* **2020,** *235,* 116–196.

16. Yuheng, G.; Xue, H.; Chuanhui, H.; Hui, C.; Lan, Z. Metal–Organic Framework-Derived Nitrogen-Doped Carbon Nanotube Cages as Efficient Adsorbents for Solid-Phase Microextraction of Polychlorinated Biphenyls. *Analyt. Chim. Acta* **2020,** *1095,* 99–108.

17. Yang, L.; Jiang, S.; Zhao, Y.; Zhu, L.; Chen, S.; Wang, X.; Wu, Q.; Ma, J.; Ma, Y.; Hu, Z. *Angewan. Chem.* **2011,** *50,* 7132–7135.

18. Nedjla, Z.; Philippe, F.; Mohamed, S.; Lachgar, A.; Baraket, N.; Jaffrezic, R. Highly Sensitive Electrochemical Biosensor for Bisphenol A Detection Based on a Diazonium-Functionalized Boron-Doped Diamond Electrode Modified with a Multi-Walled Carbon Nanotube-Tyrosinase Hybrid Film. *Biosens. Bioelectron.* **2015,** *7415,* 830–835.

19. Nalini, S.; Meha, J.; Nishith, V. Composite Nanofloral Clusters of Carbon Nanotubes and Activated Alumina: An Efficient Sorbent for Heavy Metal Removal. *Chem. Eng. J.* **2014,** *2351,* 1–9.

20. Is'haq, A.; Mohammed, M.; Bankole, T.; Ambali, S.; Abdulkareem, S.; Ochigbo, S.; Ayo, S.; Afolabi, O.; Abubakre, K. Full Factorial Design Approach to Carbon Nanotubes Synthesis by CVD Method in Argon Environment. *SA J. Chem. Eng.* **2017,** *24,* 17–42.

21. Larrude, D. G.; Maia da Costa, M. E. H.; Monteiro, F. H.; Pinto, A. L.; Freire Jr., F. L. Characterization of Phosphorus-Doped Multiwalled Carbon Nanotubes. *J. Appl. Phys.* **2012**, *111*, 315.

22. Garcia, G.; Juliano, B.; Cardoso, C.; Valnice, M.; Zanoni, B. Enhanced Photoelectrocatalytic Degradation of an Acid Dye with Boron-Doped TiO_2 Nanotube Anodes. *Cataly. Today* **2015**, *240*, 100–106.

23. Müller, C.; Al-Hamry, A.; Kanoun, O.; Rahaman, M.; Dietrich, R. T.; Zahn, E.; Yoshiko, M.; José, M.; Rosolen, H. Sensing Behavior of Endohedral Li-Doped and Undoped SWCNT/SDBS Composite Films. *Sensors* **2019**, *19*, 171.

24. Edgar, R.; Alvizo, P.; Jose, M.; Romo, H.; Humberto, T.; Mauricio, T.; Jaime Ruiz, G.; Jose, L.; Hernandez, L. Soft Purification of N-Doped and Undoped Multi-Wall Carbon Nanotubes. *Nanotechnology* **2008**, *19*, 155701.

25. Türkay, M.; Aytekin Aydın, H.; Levent, H.; Alper, D.; Kıymet, G. Synthesis, Characterization and Antibacterial Activity of Silver-Doped TiO_2 Nanotubes. *Spectrochim. Acta Part A: Mol. Biomol. Spectrosc.* **2018**, *205*, 503–507.

26. Lijun, J.; Yuheng, Z.; Shiyong, M.; Mindong, G., Xi, L. In Situ Synthesis of Carbon Doped TiO_2 Nanotubes with an Enhanced Photocatalytic Performance under UV and Visible Light. *Carbon* **2017**, *125*, 544–550.

27. Chen, W-F. Photocatalytic Performance of Undoped and Transition-Metal-Doped/Codoped TiO_2 Thin Films, A Thesis by School of Materials Science and Engineering, Faculty of Science, UNSW Australia, 2016.

28. Li, Y. H.; Zhao, Y. M.; Hu, W. B.; Ahmad, I.; Y. Zhu, Q.; Peng, X. J.; Luan, Z. K. Carbon Nanotubes—the Promising Adsorbent in Wastewater Treatment Journal of Physics. *Conf. Ser.* **2007**, *61*, 698–702.

29. Alexander, E.; Burakova, E.; Galunina, V.; Irina, V.; Burakovaa, A.; Kucherovaa, E.; Shilpi, A.; Alexey, G.; Tkacheva, V.; Guptab, K. Adsorption of Heavy Metals on Conventional and Nanostructured Materials for Wastewater Treatment Purposes: A Review. *Ecotoxicol. Environ. Saf.* **2018**, *148*, 702–712.

30. Maryam, A. T.; Toraj, M. Adsorption of Divalent Heavy Metal Ions from Water Using Carbon Nanotube Sheets. *J. Hazard. Mater.* **2011**, *185* (1), 140–147.

31. Mohammad, S.; Tehrani, P.; Abroomand, A.; Parvin, E.; Namin, S.; Moradi, D. Removal of Lead Ions from Wastewater Using Functionalized Multiwalled Carbon Nanotubes with Tris (2-Aminoethyl) Amine. *J. Environ. Protect.* **2013**, *4*, 529–536.

32. Yan-Hui, L.; b,, Zechao, D.; Jun, D.; Dehai, W.; Zhaokun, L.; Yanqiu, Z. Adsorption Thermodynamic, Kinetic and Desorption Studies of Pb^{2+} on Carbon Nanotubes. *Water Res.* **2005**, *39*, 605–609.

33. Ahmed, A.; Ishaq, K.; Saka, A.; Abdul, K. Effects of Aspect Ratio of Multi- Walled Carbon Nanotubes on Coal Washery Waste Water Treatment. *J. Environ. Manage.* **2017**, *202* (1), 84–93.

34. Polypyrrole/Multi-Walled Carbon Nanotube Composite for the Solid Phase Extraction of Lead(II) in Water Samples. *Talanta* **2014**, *119*, 447–451.

35. Xiao, s.; Jiang, X.; Guangming, J.; Xinhua, X. Removal of Chromium(VI) from Wastewater by Nanoscale Zero-Valent Iron Particles Supported on Multiwalled Carbon Nanotubes. *Chemosphere* **2011**, *85* (7), 1204–1209.

36. Zhuo-nan, H.; Xiao-ling, W.; De-suo, Y. Adsorption of Cr(VI) in Wastewater Using Magnetic Multi-Wall Carbon Nanotubes. *Water Sci. Eng.* **2015**, *8* (3), 226–232.

37. Nassereldeen, A.; Kabbashi, A.; Nour, Ma'an, H.; Al-Khatib, Md. A. M. *Removal of Chromium With CNT Coated Activated Carbon for Waste Water Treatment, Encyclopedia of Renewable and Sustainable Materials*; Elsevier Chapter 10785 2019; pp 536–547.

38. Xu, Y-J.; Rosa, A.; Liu, X.; SU, D-s. Characterization and Use of Functionalized Carbon Nanotubes for the Adsorption of Heavy Metal Anions. *N. Carbon Mater.* **2011**, *26* (1), 57–62.

39. Mahdi, M.; Hamed, A.; Vahid, K. Synthesis of Highly Ordered Carbon Nanotubes /Nanoporous Anodic Alumina Composite Membrane and Potential Application in Heavy Metal Ions Removal from Industrial Wastewater. *Mater. Today Proc.* **2017**, *4* (3), 4906–4911.

40. Singha Deb, A. K.; Vidushi, D.; Dasgupta, K.; Musharaf Ali, Sk.; Shenoy, K. T. Novel Amidoamine Functionalized Multi-Walled Carbon Nanotubes for Removal of Mercury(II) Ions from Wastewater: Combined Experimental and Density Functional Theoretical Approach. *Chem. Eng. J.* **2017**, *313* (1), 899–911.

41. Shaofeng, Z.; Beiping, Z.; Zhiyang, L.; Lihua, Z.; Yong, Y. Autochthonous N-Doped Carbon Nanotube/Activated Carbon Composites Derived from Industrial Paper Sludge for Chromate (VI) Reduction in Microbial Fuel Cells. *Sci. Total Environ.* **2020**, *712*, 136513.

42. Ghaedia, M.; Shokrollahia, A.; Tavallalib, H.; Shojaiepoora, F.; Keshavarzb, B.; Hossainiana, H.; Soylakc, M.; Purkaitd, M.K. Activated Carbon and Multiwalled Carbon Nanotubes as Efficient Adsorbents for Removal of Arsenazo(III) and Methyl Red From Waste Water. *Toxicol. Environ. Chem.* **2011**, *93* (3–4), 438–449.

43. Bożena, Czech; Buda, Waldemar M. Photocatalytic Treatment of Pharmaceutical Wastewater Using New Multiwall-Carbon Nanotubes/TiO_2/SiO_2 Nanocomposites. *Environ. Res.* **2015**, *137*, 176–184.

44. Zhang, L.; Pan, F.; Liu, X.; Yang, L.; Jiang, X.; Yang, J.; Shi, W. Multi-Walled Carbon Nanotubes as Sorbent for Recovery of Endocrine Disrupting Compound- Bisphenol F from Wastewater. *Chem. Eng. J.* **2013**, *218* (15), 238–246.

45. Chenxi, H.; David, G.; Xianghui, H.; Fang, X. High Rhodamine B and Methyl Orange Removal Performance of Graphene Oxide/Carbon Nanotube Nanostructures. *Front. Chem.* **2020**, *5*, 1–10.

46. Ufana, R.; Ashraf, S. M. Semi-Conducting Poly(1-Naphthylamine) Nanotubes: A pH Independent Adsorbent of Sulphonate Dyes. *Chem. Eng. J.* **2011**, *174* (2–31), 546–555.

47. Cai, J.; Zhou, M.; Xuedong, X. X.; Du, S. Boron and Cobalt Co-Doped TiO_2 Nanotubes Anode for Efficient Degradation of Organic Pollutants. *J. Hazard. Mate.* **2020**, *396*, 122723.

48. Lianmei, W.; Zhang, Y.; Shengwen, C.; Luping, Z.; Lijun, W. Synthesis of Nitrogen-Doped Carbon Nanotubes-$FePO_4$ Composite from Phosphate Residue and Its Application as Effective Fenton-Like Catalyst for Dye Degradation. *J. Environ. Sci.* **2019**, *7*, 188–198.

49. Ken, G.; Ornthida, S-K.; Somenath, M. Carbon Nanotube Enhanced Membrane Distillation for Simultaneous Generation of Pure Water and Concentrating Pharmaceutical Waste. *Separation Purif. Technol.* **2012**, *90*, 239–245.

50. Kanel, S. R.; Heath, M.; Dhriti, N.; Shankar, M.; Goltz, M. N. The Use of Carbon Nanotube Yarn as a Filter Medium to Treat Nitroaromatic-Contaminated Water. *N. Carbon Mater.* **2016**, *31* (4), 415–423.

51. Yang, H.; Heyang, Y.; Zhenhai, W.; Shumao, C.; Xiaoru, G.; Zhen, H.; Junhong, C. Nitrogen-Doped Graphene/CoNi Alloy Encased within Bamboo-Like Carbon Nanotube Hybrids as Cathode Catalysts in Microbial Fuel Cells. *J. Power Sourc.* **2016**, *307*, 561–568.

52. Saadia, H.; Naseem, I.; Xuan, S.; Tayyaba, N.; Kannan, A. M. NiCo–N-Doped Carbon Nanotubes Based Cathode Catalyst for Alkaline Membrane Fuel Cell. *Renew. Energy* **2020**, *15*, 508–516.

53. Alizae, S. A.; Mariya, I.; Ali, Q.; Savita, A.; Ghazala, Y. Applications of Carbon Nanotubes (CNTs) for the Treatment of Drinking and Waste Water-a Brief Review. *Int. J. Adv. Res. Dev.* **2016**, *1* (12), 11–16.

54. Rananga, L. E.; Magadzu, T. Interaction of Silver Doped Carbon Nanotubes-Cyclodextrin Nanocomposites with *Escherichia coli* Bacteria during Water Purification. *Water Sci. Technol.* **2014**, *14* (3), 367–375.

55. Mamba, B.; Krause, R. W.; Malefetse, T. J. Monofunctionalised Cyclodextrin Polymers for the Removal of Organic Pollutants from Water. *Environ. Chem. Lett.* **2007**, *5*, 79–84.

56. Lam, C.; James, J. T.; McCluskey, R.; Arepalli, S.; Hunter, R. L. A Review of Carbon Nanotube Toxicity and Assessment of Potential Occupational and Environmental Health Risks. *Crit. Rev. Toxicol.* **2006**, *36* (3), 189–217.

57. Michał, T.; Joanna, D.; Monika, P.; Jerzy, W.; Anna, B-B. Preliminary Evaluation of the Application of Carbon Nanotubes as Potential Adsorbents for the Elimination of Selected Anticancer Drugs from Water Matrices. *Chemosphere* **2018**, *201*, 32–40.

58. Zefang, Y.; Duoni, H.; Chen, J.; Wang, F. W. Resilient, Mesoporous Carbon Nanotube-Based Strips as Adsorbents of Dilute Organics in Water. June **2018**, *132*, 329–334.

59. Yu, J.; Li, G.; Liu, H.; Wang, A.; Yang, L.; Zhou, W.; Hu, Y.; Chu, B. Simultaneous Water Recovery and Hydrogen Production by Bifunctional Electrocatalyst of Nitrogen-Doped Carbon Nanotubes Protected Cobalt Nanoparticles. *Int. J. Hydrogen Energy* **2018**, *43*, 12110–12118.

CHAPTER 4

AN OVERVIEW OF THE APPLICATION OF CARBON NANOTUBES FOR ENHANCED OIL RECOVERY (EOR) AND CARBON SEQUESTRATION

KRISHNA RAGHAV CHATURVEDI and TUSHAR SHARMA*

Enhanced Oil Recovery Laboratory, Rajiv Gandhi Institute of Petroleum Technology, Jais, India

Corresponding author. E-mail: tsharma@rgipt.ac.in

ABSTRACT

While global energy demand is increasing at an exponential pace, oil-alternative fuel sources have been unable to match the supply needs. This renders us dependent on hydrocarbons to fulfill our energy needs for the immediately foreseeable future. However, the burning of hydrocarbons releases copious amounts of CO_2, which in the atmosphere, causes global warming. These twin challenges can be tackled by implementing combined enhanced oil recovery (EOR) and carbon sequestration projects. Any process that alters the property of the reservoir fluids or rock to further improve hydrocarbon recovery is known as EOR while carbon sequestration involves the injection of atmospheric carbon down into the subsurface, where it would be unable to play any role in further aggravating global warming. However, the combined implementation of EOR and carbon sequestration is plagued with unique challenges that can only be countered by synthesizing and improving the properties of carbon-based nanomaterials like carbon nanotubes (CNTs).

Given their superior properties like high conductivity, mechanical strength, and superior aspect ratios (length/diameter ≥ 1000), which are highly relevant in subsurface applications, CNTs have elicited a great interest amongst researchers who have explored the use of single-walled CNTs (SWCNTs) and multi-walled CNTs(MWCNTs) for oilfield applications. These CNTs are dispersed in varying base-fluids to form colloidal suspensions (nanofluids), which are then used for field application. Past studies have proven that CNTs reduce the interfacial tension (IFT) between oil/water while mixing CNTs with acidic oil forms an in-situ surfactant, enabling the lowering of the capillary pressure within the small pore increasing crude oil recovery. It is hypothesized that the in-situ surfactant would also form foam with injected CO_2, reducing its mobility leading to the formation of a uniform front and increasing the carbon storage potential of the reservoir. However, for the application of CNTs in the combined processes (EOR and carbon sequestration) to succeed, various factors like reservoir temperature, pressure, salinity, and heterogeneity would have to be accounted for and understood. Hence, the chapter would aim to elaborate and explain, in detail the application of CNTs in EOR and carbon sequestration under prevalent reservoir conditions to promote a carbon-negative future.

4.1   INTRODUCTION

While hydrocarbons are and will remain as the main source of energy for the human technical progress, various market disruptions in the recent past (Shale fracking boon, credit crisis, COVID19 pandemic) have majorly impacted oil exploration and production, causing a fall in the finding of new fields.[1,2] Since most existing oilfields are getting depleted at an accelerated pace, it is essential that new methods to improve the productivity of the oil be explored. Simultaneously, the rise in CO_2 emissions due to the high burning of fossil fuels is causing an increase in global temperatures and researchers are investigating various methods to sequester this CO_2 in the sub-surface.[3] In solving these issues together, nanotechnology can play a major role.[4] It has the potential to radically change the various processes in the upstream and downstream segments of the oil industry. However, in this study, the focus will remain on the upstream applications of nanotechnology. In particular, the focus will be on the novel suspensions, nanofluids, which will be the highlight of this study.

Nanofluids are colloidal suspensions of nanoparticles (NPs) under 100 nm, which are dispersed in an aqueous media for flow-related applications.[5] The selection of the aqueous media is highly important in this, as it directly influences the background properties like rheology and stability of the nanofluids.[6] Compared to other fluids, nanofluids exhibit properties that are highly conducive for oilfield applications like viscosity enhancement, polymer stabilization, surfactant adsorption mitigation, IFT reduction, and wettability alteration.[7–9] While several NPs are used to formulate nanofluids like silica, titania, ZnO, etc., this study aims to investigate the use of novel single-walled carbon nanotubes (SWCNTs) and multi-walled carbon nanotubes (MWCNTs) for oilfield applications. Given their superior properties like superior mechanical strength, high conductivity, and desirable aspect ratios (length/diameter ≥ 1000), CNT nanofluids have found several applications in various upstream oilfield applications (Fig. 4.1). In the exploration of oil and gas, the role of coated carbon-nano structures for real-time evaluation and 2-D detection technology was proposed in a US patent application.[10] Researchers also proposed the application of CNT-based nanofluids for applications in drilling fluids and cementing to improve their properties in HPHT conditions.[11,12] The oil industry is also investigating the use of CNTs (given their high absorption potential, unique physical, and chemical properties) for waste water treatment by removing the heavy metals from the produced formation brines.[13] Additionally, the role of CNT nanofluids for EOR injection, CO_2 absorption and IFT reduction has also been explored in past literature, which has been provided in the following sections. The upcoming sections of this chapter are fully dedicated to the formulation of a stable CNT nanofluid for flow applications. Due care will be taken to ensure that the desired properties like initiating wettability alteration, causing IFT reduction, enhancing oil recovery, and maximizing CO_2 absorption are not affected by the formulation methodology.

4.2 SYNTHESIS &DISPERSION STABILITY OF CNT NANOFLUIDS FOR OILFIELD APPLICATION

There are two main methods to formulate CNT nanofluids for flow applications. One method involves the synthesis of CNT NPs in the base fluid itself using various precursor chemicals and catalysts. This method is known as the single-step synthesis method, which yields highly stable and homogenous nanofluids, greatly viable for flow applications.[14] However, it is highly unconventional that single-step synthesized nanofluids would be economically viable for synthesis in large volumes.[14] Also, it is highly doubtful that the entire precursor compounds would be converted to CNTs, causing major concerns of impurity in major applications.[15] An alternative approach to synthesize CNT nanofluids can the two-step method. In this approach, commercially obtained nanopowder is dispersed into a base fluid using industrial mixers and homogenizers.[16] The CNT nanopowder can be prepared using the chemical vapor deposition (CVD) method, which is a very simple and highly economical technique. CVD is used for the synthesis of CNTs at ambient temperature and pressure conditions. Previously, Hassan and co-workers used the CVD method to synthesize nanopowder for oil recovery applications.[17] They prepared a solution of ferrocene and toluene, which was then kept in a two-stage furnace (first stage temperature = 200°C, reaction stage temperature = 760°C). The lower temperature (200°C) ensured the vaporization of the solution upon its injection while the CNTs were found to have grown in the quartz reaction tube. The center of the tube was cooled to room temperature and then scrapped out to obtain the CNTs.[18] The obtained nanopowder is then dispersed in a base fluid (primarily, water) using an industrial mixer by stirring at a high speed (≈6000 RPM) for 10–15 min followed by homogenizing the solution using a sonicator (25 kHz, 150 W). This prepares the CNT nanofluids, which can then be used for flow behavior studies in porous media.

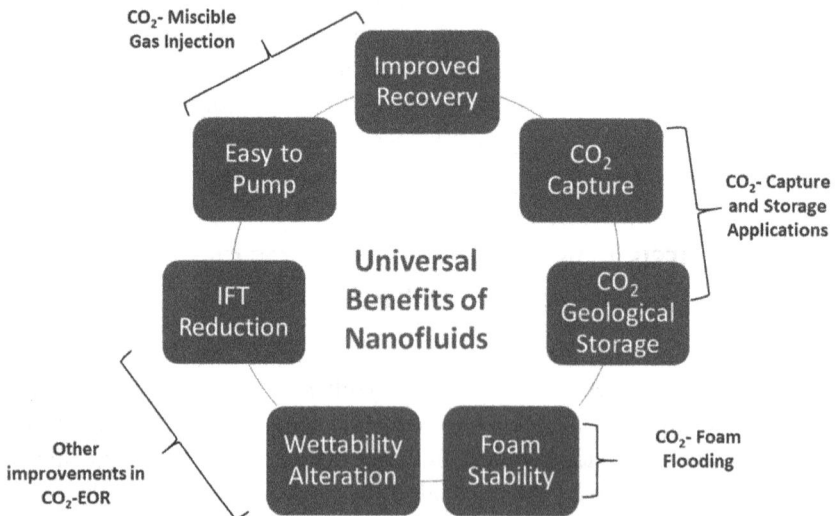

FIGURE 4.1   Application of CNT-nanofluids in oilfield applications.

However, a major challenge in the wider application of CNTs nano-fluids is their tendency to agglomerate and settle. A similar property is also exhibited in other nanofluids as the smaller NPsare bound together by attractive van der Waals interactions when dispersed in an aqueous media. In this study, commercially obtained CNTs nanopowder: MWCNTs (Sigma-Aldrich, India) of size O.D. 6–13 nm × L 2.5–20 μm nanopowder of >98% carbon basis of desired wt.% (0.05–0.1 wt.%) was initially dispersed in distilled water, which acts as the base fluid and the solution was mechanically stirred for 1 h followed by homog-enization for 30 min to form CNT nanofluid dispersion. However, the CNT nanofluids prepared in water were highly unstable and all nano-fluids settled within 4 h (Table 4.1). The agglomeration and settlement in nanofluids were also understood by the zeta-potential values of the nanofluids (measured by a Nano ZS, Malvern, UK). Conventionally, a zeta-potential value between −30 and 30 eV indicates the instability of nanofluids.[19] Hence, water-based nanofluids of CNTs are highly inadequate for wider industrial applications requiring long-term disper-sion stability. To improve the dispersion stability, a polymer solution of 1000 ppm polyacrylamide (PAM, a widely used oilfield polymer) was formulated. To this solution, an anionic surfactant, sodium dodecyl sulfate (SDS) was added at measured CMC value determined using the electrical conductivity method and the solution was stirred for only 15 min (Fig. 4.2).[9]

While the added polymer provides steric stabilization and improved rheological attributes to the nanofluids, the SDS is used for surface modi-fication and formulation of a stable dispersion of CNTs. In other literature too, the role of surfactants to formulate stable nanofluids has been widely reported.[9,20,21] This can be attributed to the ability of the surfactants to deposit and accumulate on the surface of the NPs. Hence, it is highly essen-tial that the researchers fully understand the presence of surface charge on the CNTs in different dispersion media. This will improve the dispers-ibility and stability of CNTs. Conventionally, SDS is preferred because of its cheap cost, easily understandable cheap nature and interaction with CNTs, and its tendency to reduce agglomeration in CNTs by penetrating the CNT aggregates.[22,23] However, past studies have also investigated the use of nonionic surfactants like TX-100 or anionic surfactant sodium dodecylbenzenesulfonate in this role.[24,25] Compared to other methods, surfactant treatment is a much better technique of nanofluids stabilization as it exhibits an insignificant effect on the electrical conducting behavior of the nanofluids (unlike the chemical functionalization technique). The role of ultra-sonication is essential as due to its help, SDS molecules can end the initial gap between single tubes inside the bundle, further separating them and due to this increased accumulation of surfactants in the gap site, a stable dispersion is formed.[26] The use of PAM and SDS yielded highly stable nanofluids with dispersion stability of 18–22 d with stable zeta-potential values (Table 4.1). After 18 d, the 0.1 wt.% CNT nanofluids were found to sediment, which proves that using a higher concentration of NPs is more prone to agglomeration.[27]

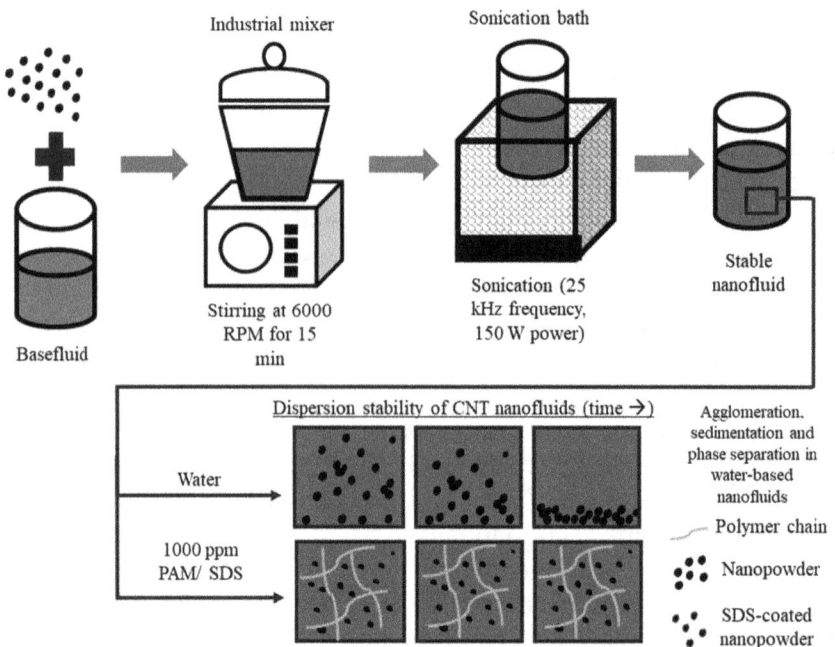

FIGURE 4.2   Formulation of CNT nanofluids via the two-step method and their dispersion in different base fluids.

From the presented results, it is evident that for wider flow applications and studies, a PAM/SDS-based nanofluid should be formulated. Hence, in this study, all further investigations were performed with a 1000 ppm PAM-based nanofluid with 0.05 and 0.1 wt.% CNT composition.

TABLE 4.1 Dispersion Stability of CNT Nanofluids.

Base fluid	Nanopowder (wt.%)	Dispersion stability	Zeta-potential (eV)	
			After preparation	After agglomeration
DI water	0.05	≈4 h	−22	−13
	0.1	≈3.5 h	−21	−9
1000 ppm PAM/SDS	0.05	22 d	−36	−27
	0.1	18 d	−35	−24

4.3 INTERFACIAL TENSION

For any flow behavior-based investigation, it is highly essential to investigate the IFT behavior of a fluid. The IFT is an attractive force that exists between the molecules of two fluids existing at the interface of their molecules. Given the wider relevance of nanofluids for carbon-based applications, it thus becomes highly essential to investigate the IFT of the CO_2/CNT nanofluids system. One of the easiest methods to analyze the IFT, the pendant drop analysis method was used for the measurement of the extent surface tension of CNT nanofluids in the presence of CO_2. For this application, a custom-built set-up was utilized, which consisted of an imaging system (camera and light source); an experimental system, which was a syringe pump injection CNT nanofluids using a syringe pump via a microscopic syringe; and a CO_2 cylinder, which was used to maintain the CO_2 environment (Fig. 4.3).

The detailed procedure to obtain the IFT has been provided in past literature.[17,28] For the sake of brevity, it can be stated that the IFT is determined by analyzing the shape of the droplet using the following equation,

$$\gamma = \frac{\Delta \rho g D^2}{H} \tag{4.1}$$

where $\Delta \rho$ is the contrast in density between the nanofluids, CO_2, D is the diameter of the bubble, and H is the shape factor (obtained using correlations). Initially, the IFT between CO_2 and crude oil (Tarapur oilfield, Ahmedabad, India) was investigated using the pendant drop method and the value was found to be 19.42 mN/m. The IFT between water and oil was found to be 16.36 mN/m. From the given results, it is evident that water has a greater potential to mobilize crude oil due to its lower IFT value. The ST values for water-CO_2 and PAM-CO_2 were also determined to ensure comparability between fluids. The ST between water and CO_2 was found to be 72 mN/m and the ST between PAM/CO_2 was 44 mN/m. Then, the ST values were obtained for each of the nanofluid mixtures with CO_2. While the ST between 0.05 wt.% CNT nanofluids was found to be 31.27 mN/m, the ST between 0.1 wt.% CNT nanofluids was found to be 34.25 mN/m, which indicates that the inclusion of NP_S is more conducive for CO_2 application and EOR. Even at lower concentrations (0.05 and 0.1 wt.%), CNT nanofluids were able to sufficiently reduce the ST. This behavior has been universally observed in other nanomaterials due to the smaller size, which allows for a greater number of molecules being present on their surface and able to participate in chemical interactions.[29,30] MWCNTs have the advantage of a highly unique shape, that is, greater length/diameter ratio, which is essential for interaction with gas molecules and, thus, they stand as viable candidates for carbon utilization in subsurface reservoirs.[17]

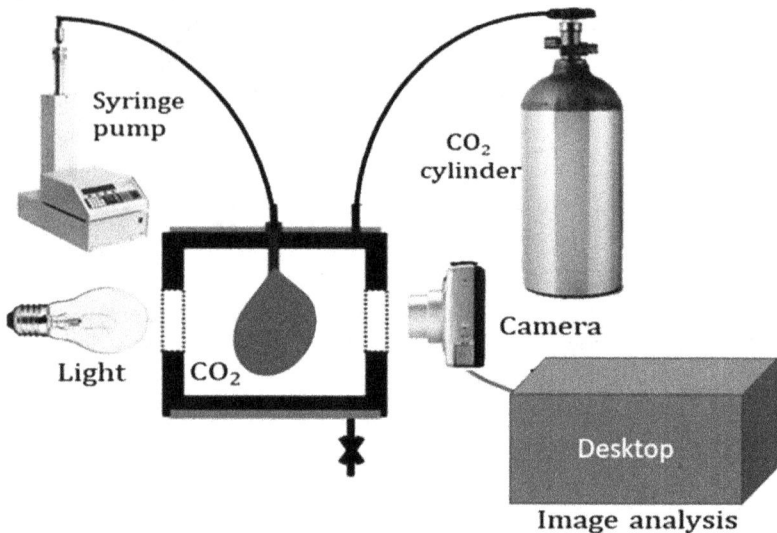

FIGURE 4.3   Formulation of CNT nanofluids via the two-step method and their dispersion in different base fluids.

4.4   SURFACE CARBON CAPTURE USING CNTs

The carbon retention potential of CNT nanofluids was then explored next for a variety of pressure (0–8 bar) and temperature conditions (30°C–50°C).

Similar experiments have been performed in past studies.[16,31,32] The experimental setup has been provided in Figure 4.4.

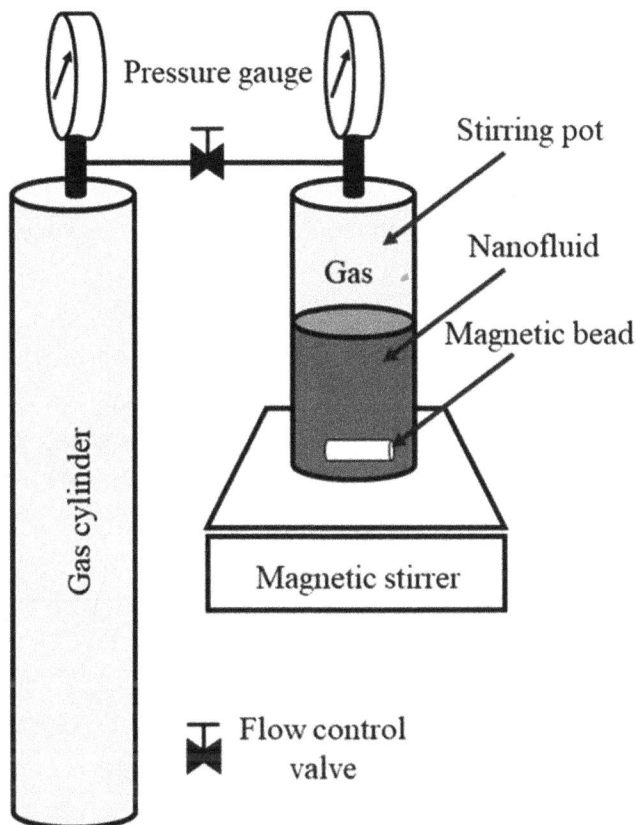

FIGURE 4.4 Schematic of set-up utilized for CO_2 absorption studies.

The CO_2 absorption setup comprised of a stirring pot (volume= 25 mL) mounted on a magnetic-stirrer cum hot plate. An uniform rpm of 300 was maintained in the pot using a magnetic bead to induce agitation, which is conducive for higher CO_2 absorption.[33] A CO_2 cylinder with a flow control valve was used to introduce the CO_2 in the stirring pot, which had been earlier evacuated using a vacuum pump. A pressure gauge was used to monitor the pressure inside the stirring pot continuously and the amount of CO_2 absorbed in the CNT nanofluids was determined by the well-established pressure decay method. Immediately after the introduction of the CO_2 inside the stirring pot, the initial number of moles of CO_2 inside the pot was calculated by

$$n^i_{CO_2} = \frac{P_i(V_{res})}{Z^i_{CO_2} RT_{eq}}$$

(4.2)

where $n^i_{CO_2}$ was the initial moles, P_i initial pressure, V_{res} is the interstitial space (10 mL), R stands for the gas constant, T_{eq} is the experimental temperature, and $Z^i_{CO_2}$ stands for the compressibility factor. After some time, the pressure in the stirring pot stopped to reduce and the equilibrium pressure was reached at which the number of moles of CO_2 left unabsorbed in the cell were determined

$$n^{eq}_{CO_2} = \frac{P_{eq}(V_{Cell} - V_{NF})}{Z^f_{CO_2} RT_{eq}}$$

(4.3)

where $n^{eq}_{CO_2}$ is the unabsorbed CO_2 moles, P_{eq} equilibrium pressure, and $Z^f_{CO_2}$ stands for the compressibility factor. By using these two values, the number of CO_2 moles solvated in the nanofluids was calculated by taking their difference. The values of CO_2 molality, that is, moles absorbed per kilogram of solvent have been provided in Figure 4.5.

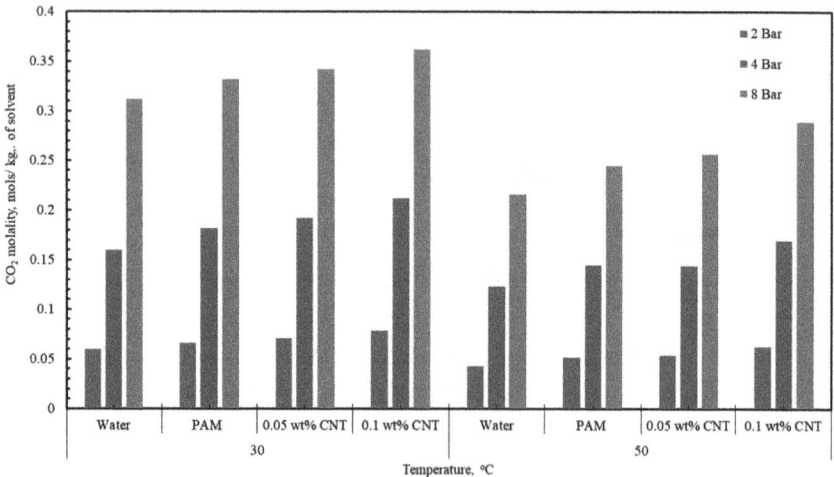

FIGURE 4.5 CO_2 absorption as a function of pressure and temperature for water, PAM, and CNT nanofluids.

From the presented results, it was observed that compared to both water and PAM, CNT nanofluids display enhanced CO_2 absorption, which is highly favorable for their use in carbon utilization. Increasing pressure was found to increase CO_2 solvation in all fluids as higher pressure tends to force a greater amount of gas inside the body of the fluid.[32] Conversely, increasing the temperature of solutions reduced their CO_2 solvation capacity as the increasing temperature induces lower solvation of gas in the fluid. The CNT nanofluids displayed an enhanced CO_2 absorption, which was in the range of 20–36% for the various test conditions. The enhanced CO_2 absorption by MWCNTs can be attributed to their tendency to adhere to the edges of the CO_2 bubbles and increasing their strength, which allows them to resist deformation and, ultimately, escape from the fluid. The increased mass transfer of CO_2 inside the fluid can also be attributed to the presence of the Brownian motion and the Grazing effect, which influences the CO_2 transfer positively.[34] Compared to other nanofluids like SiO_2, ZnO, or TiO_2, CNT nanofluids display an increased CO_2 absorption due to their hydrophobicity and increased absorption capacity.[16,34] The retention of CO_2 inside the body of the nanofluids was established by taking a drop of the CO_2-solvated fluid under an optical microscope (Fig. 4.6).

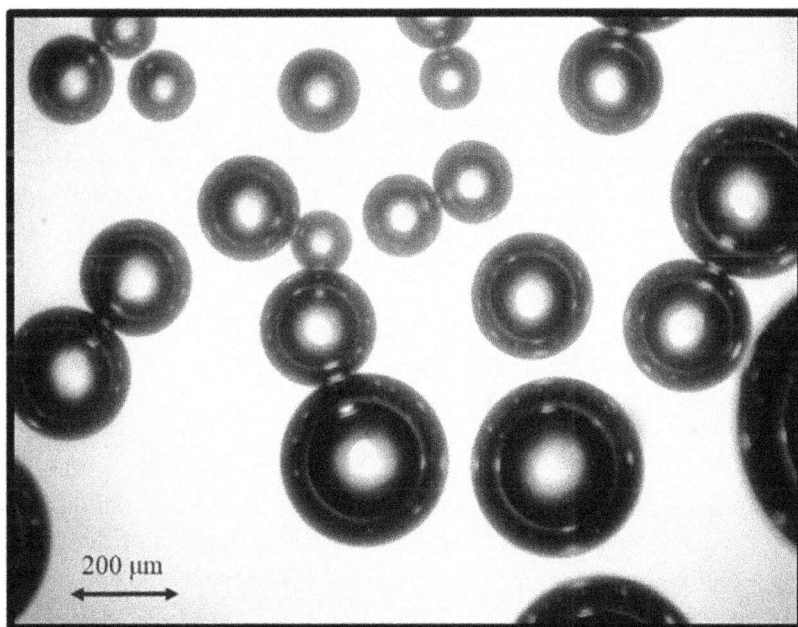

FIGURE 4.6   Microscopic image of CO_2 bubbles stabilized in 0.05 wt.% CNT nanofluids (8 bar 30°C)

The retention of the CO_2 inside the fluids was observed by carefully observed the bubbles inside the body of the fluid using a method that had been reported in previous studies.[16,32] These observations have been reported in Table 4.2.

TABLE 4.2 CO_2 Bubble Retention in Fluids at 8 bar and 30°C

Fluid	Duration (h)
Water	0
1000 ppm PAM	3
0.05 wt.% CNT	6.4
0.1 wt.% CNT	6.2

From the presented results, it is evident that not only did the CNT nanofluids improve the absorption of CO_2 in the body of the fluid but they also improved its retention by a significant duration. However, the CO_2 absorbed in the water escaped soon after removing the confining pressure indicating its unviability for use as CO_2 carrier fluids. The retention of CO_2 in 1000 ppm PAM increased to 3 h while in 0.05 and 0.1 wt.% CNT nanofluids, it was more than 6 h, indicating a greater likelihood of flow application. Given the increased presence of NPs in the 0.1 wt.% CNT nanofluid, the CO_2 retention was a bit longer due to more NPs per unit area stabilizing the bubbles. Greater retention of CO_2 in the fluid translates into a larger areal coverage of the reservoir which would ultimately yield increased storage of carbon and much higher oil recovery. Thus, the use of MWCNT nanofluids for carbonation is highly recommended.

4.5   ENHANCED OIL RECOVERY AND CARBON SEQUESTRATION USING CNTs

For the oil recovery experiments, usually, a porous media is synthesized using glass beads, sand-grains, or carbonate flakes to mimic an oil reservoir. The sand-pack flooding system generally comprises of the several fluid accumulators, syringe pumps, and sand-pack holder with temperature control (Fig. 4.7).

The experimental setup used for oil recovery experiments is initially saturated by water, followed by crude oil to conventionally establish an oil-bearing reservoir. In this reservoir, water, CO_2, nanofluids, or other chemicals are injected in a controlled manner via a syringe pump at a low flow rate (a lower flow rate ensures an adequate areal spread of injected fluid in the reservoir). The collected oil is carefully collected and compared between individual fluids to obtain the incremental oil recovery.[35–37]

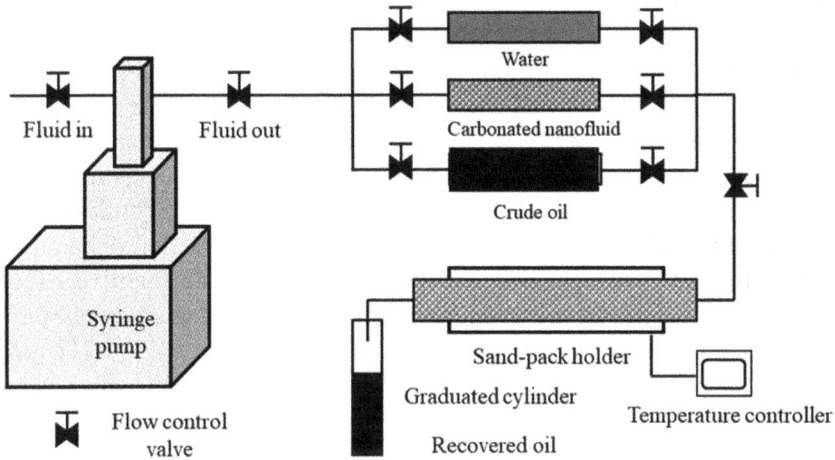

FIGURE 4.7   Schematic of the representative experimental set-up used for oil recovery experiments.

Given the high reduction in costs of synthesizing CNTs in the recent years, several studies in the past have explored the use of the modern, relatively cheaper MWCNTs for enhanced oil recovery (EOR). Also, given their ease of fabrication, MWCNTs are much cheaper than SWCNTs, leading them to be preferred for wider oilfield applications where a large quantity of nanofluids is required.[38] In previous studies, it has been observed that the addition of MWCNTs leads to a change in wettability to intermediate wet, which has the potential to massively increase oil recoveries.[17] The wettability alteration phenomenon, which is responsible for increased oil recovery, takes place in several other nanofluids as well.[39,40] In CNTs, the wettability alteration takes place when the CNTs adhere to the surface of the rock or sand-grains, enabling an easier and greater mobilization of oil. Another factor that significantly influences oil recovery is IFT reduction. The presence of CNTs enables the easier deformation of oil bubbles, which become elongated and pass through narrow rock pores where earlier they would have been trapped.[41,42] Previously, the effect of MWCNT nanofluids for oil recovery in conjunction with SiO_2 NPs was investigated in various carbonate and sandstone rock samples who found that in both reservoirs, the introduction of CNTs was found to reduce the IFT and altered the wettability of the rocks from oil-wet to water-wet.[43] Such an alteration (i.e., from oil-wet to water-wet), makes the mobilization of oil far easier as water-wet rocks repel oils, which are then produced at the outlet. The use of hydrophobic MWCNTs for oil recovery from a glass micromodel was found to improve oil recovery by30% due to the superior role in IFT reduction due to in-situ surfactant formulation.[44] Similarly, a mixture of 30% SiO_2 and 70% MWCNTs nanofluids was investigated for stabilization of Pickering emulsions and were found to exhibit enhanced stability even in the presence of saline conditions.[45] The stable Pickering emulsion was then used for oil recovery as it could cause a reduction in IFT. Separately, Soleimani et al. investigated the use of MWCNT nanofluids for oil recovery from a glass-bead reservoir and reported an increased oil recovery for 0.30 wt.% CNT nanofluids along with a significant reduction in IFT.[17] The past studies have proven the need to carefully optimize the concentration of the MWCNTs for EOR applications. Almost all studies have stated that a higher concentration of NPs should be avoided as higher NP concentrations are highly prone to Brownian agglomeration, rock-surface adsorption, and unfavorable economics.[27,46] These observations were found to be repeated in recent studies where two-step MWCNT nanofluids were able to improve oil recovery in a porous media by 24.5%.[47] This indicates the superior viability of nanofluids for oilfield applications.

Even for carbon sequestration, it is expected that MWCNTs will play a superior role. Conventionally, the injection of CO_2 in gas form is plagued by several challenges like fingering, mobility contrast, and inadequate areal coverage due to its properties as a gas. This greatly reduces the efficacy of CO_2 sequestration and researchers are forced to look for alternative methods to reduce the inherent mobility contrast (Fig. 4.8).

However, to the best of our knowledge, adequate studies extolling the principles of CNT nanofluids for improved CO_2 sequestration in porous media have yet to be performed. This may be attributed to the availability of cheaper foam stabilization materials like silica, the unfavorable economics of CO_2- storage, and lack of research direction. Hence, researchers must investigate the use of CNT nanofluids for foam stabilization and improved CO_2 utilization in upcoming studies.

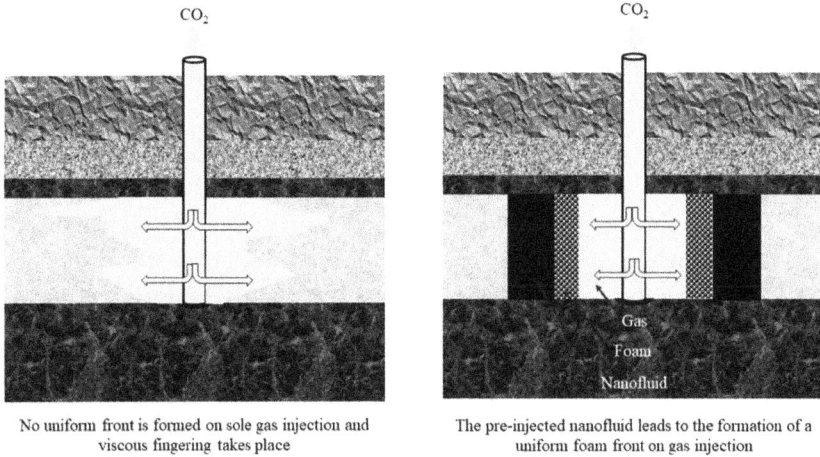

No uniform front is formed on sole gas injection and viscous fingering takes place

The pre-injected nanofluid leads to the formation of a uniform foam front on gas injection

FIGURE 4.8 Achieving mobility control using CNT nanofluids (proposed).

4.6 CURRENT LACUNAS IN EXISTING STUDIES AND FUTURE RESEARCH DIRECTIONS

While CNT NPs have established themselves as superior surface-active agents for EOR applications, a wider industrial scale adoption is yet to take. However, this problem is not unique to only CNT nanofluids and is shared by several other methods of EOR, which while improving oil recovery significantly, have unfavorable economic performance, and companies involved in oil production are hesitant to adopt them. The scenario may change in the upcoming days due to increased stress on existing oil reservoirs and the falling oil overhead. However, currently, the following lacunas exist in the research:

1. Formulation of stable and homogenous CNT nanofluids remains a challenge. Conventional methods may stabilize nanofluids upto 30 days but for wider industrial applications, a highly stable (>120 day) nanofluid is needed.
2. Rheological properties of CNT nanofluids have not been understood fully in any study.
3. The role of CNT nanofluids in EOR has not been understood fully. It is essential for field implementation that the various interactions within the reservoir be fully understood and the concentrations of constituents are properly optimized for the various reservoir rocks.

4. Other chemicals, apart from surfactants, may also be explored for CNT nanofluids stabilization as surfactants are prone to rock surface adsorption and this reduces the viability of surfactant-stabilization of nanofluids in porous media.
5. The role of CNT nanofluids for mobility control is yet to be explored and understood fully. Researchers are thus recommended to investigate the role of nanofluids and CO_2 in conjunction with improving carbon storage underground. Such studies may also involve the formulation of foams, approach oil miscibility conditions, or involve the injection of gas in immiscible conditions.

4.7 CONCLUSION

The prime objective of this chapter was to make the researchers familiar with the role of CNT nanofluids in various oilfield applications and the underlying mechanisms to facilitate a broader consensus on the inclusion of CNTs for industrial applications in EOR and carbon sequestration. While conventional methods to formulate nanofluids in base fluid water fail due to inherent NP agglomeration and sedimentation, nanofluids formulated in 1000 ppm PAM and SDS exhibit superior dispersion stability and are, hence, highly viable for flow applications. The synthesized CNT nanofluids exhibited a significant reduction in ST values, which makes them much more efficient in mobilizing crude oil from stranded reserves. The unique property of CNT NPs makes them highly unique agents for enhanced CO_2 absorption in fluids due to their superior absorption capacity and longer retention. The long retention is much desired as it will allow the greater sequestration of CO_2 in the reservoir and greatly improve the operating efficacy of the storage process. The CNT nanofluids also perform excellently in their role as EOR agents due to their property for IFT reduction, wettability alteration, and in-formation surfactant formation. Thus, CNT nanofluids emerge as viable candidates for oilfield applications in conjunction with other polymers, surfactants, and other nanofluids.

ACKNOWLEDGMENT

We would like to sincerely acknowledge the Director, Rajiv Gandhi Institute of Petroleum Technology, Jais, for his support and guidance. We would also extend our thanks to Dr. KG Biswas, Multi-phase Fluid Flow

Laboratory, Rajiv Gandhi Institute of Petroleum Technology, Jais for his help with the IFT imaging and insightful discussions.

KEYWORDS

- carbon sequestration
- enhanced oil recovery
- foams
- interfacial tension
- nanofluids
- surfactants

REFERENCES

1. Ferguson, R. C.; Nichols, C.; Leeuwen, T. Van; Kuuskraa, V. A. Storing CO_2 with Enhanced Oil Recovery. *Energy Procedia* **2009,** *1*, 1989–1996. https://doi.org/10.1016/j.egypro.2009.01.259.

2. van Bergen, F.; Gale, J.; Damen, K. J.; Wildenborg, A. F. B. Worldwide Selection of Early Opportunities for CO_2-Enhanced Oil Recovery and CO_2-Enhanced Coal Bed Methane Production. *Energy* **2004,** *29* (9–10), 1611–1621. https://doi.org/10.1016/j.energy.2004.03.063.

3. Teng, Y.; Zhang, D. Long-Term Viability of Carbon Sequestration in Deep-Sea Sediments. *Sci. Adv.* **2018,** *4* (7). https://doi.org/10.1126/sciadv.aao6588.

4. Gbadamosi, A. O.; Junin, R.; Manan, M. A.; Agi, A.; Yusuff, A. S. An Overview of Chemical Enhanced Oil Recovery: Recent Advances and Prospects. *Int. Nano Lett.* **2019,** *9* (3), 171–202. https://doi.org/10.1007/s40089-019-0272-8.

5. Suleimanov, B. A.; Ismailov, F. S.; Veliyev, E. F. Nanofluid for Enhanced Oil Recovery. *J. Pet. Sci. Eng.* **2011,** *78* (2), 431–437. https://doi.org/10.1016/j.petrol.2011.06.014.

6. Peng, B.; Zhang, L.; Luo, J.; Wang, P.; Ding, B.; Zeng, M.; Cheng, Z. A Review of Nanomaterials for Nanofluid Enhanced Oil Recovery. *RSC Adv.. R. Soc. Chem.* June 21, **2017,** 32246–32254. https://doi.org/10.1039/c7ra05592g.

7. Zhang, H.; Ramakrishnan, T. S.; Nikolov, A.; Wasan, D. Enhanced Oil Recovery Driven by Nanofilm Structural Disjoining Pressure: Flooding Experiments and Microvisualization. *Energy Fuels* **2016,** *30* (4), 2771–2779. https://doi.org/10.1021/acs.energyfuels.6b00035.

8. Emrani, A. S.; Nasr-El-Din, H. A. An Experimental Study of Nanoparticle-Polymer-Stabilized CO_2 Foam. *Colloids Surf. A Physicochem. Eng. Asp.* **2017,** *524*, 17–27. https://doi.org/10.1016/j.colsurfa.2017.04.023.

9. Kumar, R. S.; Chaturvedi, K. R.; Iglauer, S.; Trivedi, J.; Sharma, T. Impact of Anionic Surfactant on Stability, Viscoelastic Moduli, and Oil Recovery of Silica Nanofluid in Saline Environment. *J. Pet. Sci. Eng.* **2020,** *195*, 107634. https://doi.org/10.1016/j.petrol.2020.107634.

10. Li, J.; Meyyappan, M. (12) United States Patent, July 12, 2006.

11. Fazelabdolabadi, B.; Khodadadi, A. A.; Sedaghatzadeh, M. Thermal and Rheological Properties Improvement of Drilling Fluids Using Functionalized Carbon Nanotubes. *Appl. Nanosci.* **2015,** *5* (6), 651–659. https://doi.org/10.1007/s13204-014-0359-5.

12. Heidarian, J. Acid Surface Modified Carbon Nanotube-Filled Fluoroelastomers Aging Test in Oil-Based Drilling Fluids. *J. Elastomers Plast.***2017,** *49* (8), 706–737. https://doi.org/10.1177/0095244317695362.

13. Ouni, L.; Ramazani, A.; Taghavi Fardood, S. An Overview of Carbon Nanotubes Role in Heavy Metals Removal from Wastewater. *Front. Chem. Sci. Eng.* June 1, **2019,** 274–295. https://doi.org/10.1007/s11705-018-1765-0.

14. Chaturvedi, K. R.; Trivedi, J.; Sharma, T. Single-Step Silica Nanofluid for Improved Carbon Dioxide Flow and Reduced Formation Damage in Porous Media for Carbon Utilization. *Energy* **2020,** *197*, 117276. https://doi.org/10.1016/j.energy.2020.117276.

15. Devendiran, D. K.; Amirtham, V. A. A Review on Preparation, Characterization, Properties and Applications of Nanofluids. *Renew. Sustain. Energy Rev.* **2016,** 21–40. https://doi.org/10.1016/j.rser.2016.01.055.

16. Chaturvedi, K. R.; Sharma, T. Carbonated Polymeric Nanofluids for Enhanced Oil Recovery from Sandstone Reservoir. *J. Pet. Sci. Eng.* **2020,** *194*, 107499. https://doi.org/10.1016/j.petrol.2020.107499.

17. Soleimani, H.; Baig, M. K.; Yahya, N.; Khodapanah, L.; Sabet, M.; Demiral, B. M. R.; Burda, M. Impact of Carbon Nanotubes Based Nanofluid on Oil Recovery Efficiency Using Core Flooding. *Results Phys.* **2018,** *9*, 39–48. https://doi.org/10.1016/j.rinp.2018.01.072.

18. Kumar, M.; Ando, Y. Chemical Vapor Deposition of Carbon Nanotubes: A Review on Growth Mechanism and Mass Production. *J. Nanosci. Nanotech.* **2010,** 3739–3758. https://doi.org/10.1166/jnn.2010.2939.

19. Setia, H.; Gupta, R.; Wanchoo, R. K. Stability of Nanofluids. *Mater. Sci. Forum* **2013,** *757*, 139–149. https://doi.org/10.4028/www.scientific.net/MSF.757.139.

20. Al-Anssari, S.; Arif, M.; Wang, S.; Barifcani, A.; Iglauer, S. Stabilising Nanofluids in Saline Environments. *J. Colloid Interf. Sci.* **2017,** *508*, 222–229. https://doi.org/10.1016/j.jcis.2017.08.043.

21. Tkalya, E. E.; Ghislandi, M.; de With, G.; Koning, C. E. The Use of Surfactants for Dispersing Carbon Nanotubes and Graphene to Make Conductive Nanocomposites. *Curr. Opin. Colloid Interf. Sci.* **2012,** 225–232. https://doi.org/10.1016/j.cocis.2012.03.001.

22. Jiang, L.; Gao, L.; Sun, J. Production of Aqueous Colloidal Dispersions of Carbon Nanotubes. *J. Colloid Interf. Sci.* **2003,** *260* (1), 89–94. https://doi.org/10.1016/S0021-9797(02)00176-5.

23. O'Connell, M. J.; Bachilo, S. H.; Huffman, C. B.; Moore, V. C.; Strano, M. S.; Haroz, E. H.; Rialon, K. L.; Boul, P. J.; Noon, W. H.; Kittrell, C. et al. Band Gap Fluorescence from Individual Single-Walled Carbon Nanotubes. *Science* **2002,** *297* (5581), 593–596. https://doi.org/10.1126/science.1072631.

24. Geng, Y.; Liu, M. Y.; Li, J.; Shi, X. M.; Kim, J. K. Effects of Surfactant Treatment on Mechanical and Electrical Properties of CNT/Epoxy Nanocomposites. *Compos. Part A Appl. Sci. Manuf.* **2008,** *39* (12), 1876–1883. https://doi.org/10.1016/j.compositesa. 2008.09.009.

25. Matarredona, O.; Rhoads, H.; Li, Z.; Harwell, J. H.; Balzano, L.; Resasco, D. E. Dispersion of Single-Walled Carbon Nanotubes in Aqueous Solutions of the Anionic Surfactant NaDDBS. *J. Phys. Chem. B* **2003,** *107* (48), 13357–13367. https://doi.org/ 10.1021/jp0365099.

26. Duan, W. H.; Wang, Q.; Collins, F. Dispersion of Carbon Nanotubes with SDS Surfactants: A Study from a Binding Energy Perspective. *Chem. Sci.* **2011,** *2* (7), 1407–1413. https://doi.org/10.1039/c0sc00616e.

27. Gosens, I.; Post, J. A.; de la Fonteyne, L. J.; Jansen, E. H.; Geus, J. W.; Cassee, F. R.; de Jong, W. H. Impact of Agglomeration State of Nano- and Submicron Sized Gold Particles on Pulmonary Inflammation. *Part. Fibre Toxicol.* **2010,** *7* (1), 37. https://doi. org/10.1186/1743-8977-7-37.

28. Teklu, T. W.; Alameri, W.; Graves, R. M.; Kazemi, H.; AlSumaiti, A. M. Low-Salinity Water-Alternating-CO_2 EOR. *J. Pet. Sci. Eng.* **2016,** *142,* 101–118. https://doi. org/10.1016/j.petrol.2016.01.031.

29. Sharma, T.; Iglauer, S.; Sangwai, J. S. Silica Nanofluids in an Oilfield Polymer Polyacrylamide: Interfacial Properties, Wettability Alteration, and Applications for Chemical Enhanced Oil Recovery. *Ind. Eng. Chem. Res.* **2016,** *55* (48), 12387–12397. https://doi.org/10.1021/acs.iecr.6b03299.

30. Olayiwola, S. O.; Dejam, M. A Comprehensive Review on Interaction of Nanoparticles with Low Salinity Water and Surfactant for Enhanced Oil Recovery in Sandstone and Carbonate Reservoirs. *Fuel* **2019,** 1045–1057. https://doi.org/10.1016/j.fuel.2018.12. 122.

31. Chaturvedi, K. R.; Narukulla, R.; Sharma, T. CO_2 Capturing Evaluation of Single-Step Silica Nanofluid through Rheological Investigation for Nanofluid Use in Carbon Utilization Applications. *J. Mol. Liq.* **2020,** *304,* 112765. https://doi.org/10.1016/j. molliq.2020.112765.

32. Chaturvedi, K. R.; Kumar, R.; Trivedi, J.; Sheng, J. J.; Sharma, T. Stable Silica Nanofluids of an Oilfield Polymer for Enhanced CO_2 Absorption for Oilfield Applications. *Energy Fuels* **2018,** *32* (12), 12730–12741. https://doi.org/10.1021/acs.energyfuels.8b02969.

33. Chaturvedi, K. R.; Trivedi, J.; Sharma, T. Evaluation of Polymer-Assisted Carbonated Water Injection in Sandstone Reservoir: Absorption Kinetics, Rheology, and Oil Recovery Results. *Energy Fuels* **2019,** *33* (6), 5438–5451. https://doi.org/10.1021/ acs.energyfuels.9b00894.

34. Rezakazemi, M.; Darabi, M.; Soroush, E.; Mesbah, M. CO_2 Absorption Enhancement by Water-Based Nanofluids of CNT and SiO_2 Using Hollow-Fiber Membrane Contactor. *Sep. Purif. Technol.* **2019,** *210,* 920–926. https://doi.org/10.1016/j.seppur.2018.09.005.

35. Chaturvedi, K. R.; Singh, A. K.; Sharma, T. Impact of Shale on Properties and Oil Recovery Potential of Sandstone Formation for Low-Salinity Waterflooding Applications. *Asia-Pacific J. Chem. Eng.* **2019,** *14* (5). https://doi.org/10.1002/apj. 2352.

36. Goswami, R.; Chaturvedi, K. R.; Kumar, R. S.; Chon, B. H.; Sharma, T. Effect of Ionic Strength on Crude Emulsification and EOR Potential of Micellar Flood for

Oil Recovery Applications in High Saline Environment. *J. Pet. Sci. Eng.* **2018,** *170,* 49–61. https://doi.org/10.1016/j.petrol.2018.06.040.

37. Sharma, T.; Iglauer, S.; Sangwai, J. S., Silica Nanofluids in an Oilfield Polymer Polyacrylamide: Interfacial Properties, Wettability Alteration, and Applications for Chemical Enhanced Oil Recovery. *Ind. Eng. Chem. Res.* **2016,** *55* (48), 12387–12397. https://doi.org/10.1021/acs.iecr.6b03299.

38. Sherman, L. M. Carbon Nanotubes Lots of Potential–If the Price Is Right: Plastics Technology. *Plast Technol* **2007,** *53,* 68–73.

39. Dehaghani, A. H. S.; Daneshfar, R. How Much Would Silica Nanoparticles Enhance the Performance of Low-Salinity Water Flooding? *Pet. Sci.* **2019.** https://doi.org/10.1007/s12182-019-0304-z.

40. Dong, X.; Xu, J.; Cao, C.; Sun, D.; Jiang, X. Aqueous Foam Stabilized by Hydrophobically Modified Silica Particles and Liquid Paraffin Droplets. *Colloids Surfaces A Physicochem. Eng. Asp.* **2010,** *353* (2–3), 181–188. https://doi.org/10.1016/j.colsurfa.2009.11.010.

41. Mohajeri, M.; Hemmati, M.; Shekarabi, A. S. An Experimental Study on Using a Nanosurfactant in an EOR Process of Heavy Oil in a Fractured Micromodel. *J. Pet. Sci. Eng.* **2015,** *126,* 162–173. https://doi.org/10.1016/j.petrol.2014.11.012.

42. Emadi, S.; Shadizadeh, S. R.; Manshad, A. K.; Rahimi, A. M.; Mohammadi, A. H. Effect of Nano Silica Particles on Interfacial Tension (IFT) and Mobility Control of Natural Surfactant (Cedr Extraction) Solution in Enhanced Oil Recovery Process with Nano—Surfactant Flooding. *J. Mol. Liq.* **2017,** *248,* 163–167. https://doi.org/10.1016/j.molliq.2017.10.031.

43. Ershadi, M.; Alaei, M.; Rashidi, A.; Ramazani, A.; Khosravani, S. Carbonate and Sandstone Reservoirs Wettability Improvement without Using Surfactants for Chemical Enhanced Oil Recovery (C-EOR). *Fuel* **2015,** *153,* 408–415. https://doi.org/10.1016/j.fuel.2015.02.060.

44. Alnarabiji, M. S.; Yahya, N.; Shafie, A.; Solemani, H.; Chandran, K.; Hamid, S. B. A.; Azizi, K. The Influence of Hydrophobic Multiwall Carbon Nanotubes Concentration on Enhanced Oil Recovery. *Procedia Eng.* **2016,** *148,* 1137–1140. https://doi.org/10.1016/j.proeng.2016.06.564.

45. AfzaliTabar, M.; Alaei, M.; Ranjineh Khojasteh, R.; Motiee, F.; Rashidi, A. M. Preference of Multi-Walled Carbon Nanotube (MWCNT) to Single-Walled Carbon Nanotube (SWCNT) and Activated Carbon for Preparing Silica Nanohybrid Pickering Emulsion for Chemical Enhanced Oil Recovery (C-EOR). *J. Solid State Chem.* **2017,** *245,* 164–173. https://doi.org/10.1016/j.jssc.2016.10.017.

46. Kumar, R. S.; Sharma, T. Stability and Rheological Properties of Nanofluids Stabilized by SiO_2 Nanoparticles and SiO_2-TiO_2 Nanocomposites for Oilfield Applications. *Colloids Surf. A Physicochem. Eng. Asp.* **2018,** *539,* 171–183. https://doi.org/10.1016/j.colsurfa.2017.12.028.

47. Belhaj, A. F.; Elraies, K. A.; Janjuhah, H. T.; Tasfy, S. F. H.; Yahya, N.; Abdullah, B.; Umar, A. A.; Ghanem, O. Ben; Alnarabiji, M. S. Electromagnetic Waves-Induced Hydrophobic Multiwalled Carbon Nanotubes for Enhanced Oil Recovery. *J. Pet. Explor. Prod. Technol.* **2019,** *9* (4), 2667–2670. https://doi.org/10.1007/s13202-019-0653-6.

CHAPTER 5

CARBON NANOMATERIAL EMBEDDED MEMBRANES FOR HEAVY METAL SEPARATION

PALLAVI MAHAJAN-TATPATE[1], SUPRIYA DHUME[2], and YOGESH CHENDAKE[2*]

[1]School of Petroleum, Polymer and Chemical Engineering, Polymer Engineering Department, MIT, World Peace University, Pune, India

[2]Department of Chemical Engineering, Bharati Vidyapeeth (Deemed to be) University, College of Engineering, Pune, India

*Corresponding author. E-mail: yjchendake@bvucoep.edu.in

ABSTRACT

The scarcity of pure water is a major issue in today's world. The quality of food and water, which are the basic needs of humans are still a challenge in this twenty-first century. Human activities and rapid industrialization have released many harmful wastes in the environment along with natural calamities. Among these, heavy metal–contaminated water is life-threatening. The solid waste and effluents from the refinery, mining, power plant, metal processing, fertilizer, etc., are major sources of heavy metal contamination in the environment. These metallic components are nonbiodegradable and have the tendency for bioaccumulation. It can disturb and damage the soil biota and affect nutritional value. Their incorporation in the food chain can prove to be toxic to living organisms—humans, animals, and plants. It can affect photosynthesis, protein synthesis, normal growth, and the yield of the plant. In aquatic life, it stimulates reactive oxygen species, which damage fish and other organisms in water by getting accumulated in organs/tissues. Through this food chain, it will be transferred to humans, where it can result in a disorder of liver, lungs, kidney, brain functioning, and even cause cancer.

On the other hand, these heavy metals have various uses as raw materials in the production of explosive, fertilizer, pigment, fuel, etc. It is also essential for the growth and development of humans and animals. Hence, their separation and recovery are necessary. Many methods viz. conventional—precipitation, flotation, adsorption, and ion exchange and advanced—membrane-based methods are used to remove it. Conventional method has issues of chemical consumption and toxic sludge production—its disposal, recovery, and high cost considering treatment of secondary pollutant. Membrane methods have high separation efficiency, less space requirement, linear upscaling, need low energy as direct recovery as possible and is sustainable, economical, and environment friendly—no secondary pollution. The main limitation of the membrane method is its fouling and the tradeoff between selectivity and transport rate. Though various materials are reported for the preparation of membranes viz. cellulose acetate, polycarbonate, polyacrylonitrile, polyetherimide, polytetrafluoroethylene, polyethersulfone, polyethylene, polypropylene, polyvinyl chloride, polyamide, polysulfone and polyvinylidene fluoride, and polyetherketone; the limitations are still persistent. These can be overcome by the use of nanomaterial while forming membrane. These nanomaterials will modify the morphology of membranes, thus enhancing the transport property while modifying the selective properties.

Large research is going on all over the world. Various nanoparticles like silver, gold, zinc oxide, salicylate-alumoxane, and carbon are reported for the purpose. Among these, carbon nanomaterial in different forms are found to be more applicable and highly popular. The heavy metal removal efficiency of carbon nanomaterial embedded membranes is excellent, with more than 90% removal efficiency using single membrane composition. This makes them special compared to conventional systems for the formation of membranes to the removal of each component. Additionally, these nanomaterials are also known to provide extraordinary mechanical, chemical, electrical, and thermal stability to the membrane. Carbon nanomaterial incorporated membranes will exhibit high flux, selectivity, conductivity, and resistant to fouling. Thus, they improve the applicability of membranes at actual process conditions. This makes the carbon nanoparticle embedded membranes highly important for actual industrial or domestic applications for heavy metal removal. The current chapter reviews heavy metal removal using carbon nanomaterials incorporated membranes made up of different materials.

5.1   INTRODUCTION

Nowadays, the quality of water, soil, and air is continuously degrading. An increase in population leads to urbanization and growth in industrialization. A large amount of industrialization and population results in the need for a huge quantity of water for the human usage and production processing. Though the huge quantity of water is available on earth crust out of which only two to three percent of water is available for human usage. This usage would result in a large amount of waste generation. Such waste generation ultimately results in pollution of water bodies available for human and other usages. Pollution is nothing but the discharge of unwanted, undesirable material/chemicals in the environment. Such release of effluent would affect water quality. This reduction in quality would enhance water scarcity due to pollution/contamination by bacteria, virus, organic materials, chemicals, and heavy metals. The presence of heavy metals is one of the greatest threats to humans, flora, animals, and environment. It causes an adverse impact on living organisms and the environment. Additionally, the impact would multiply many times due to the bioaccumulation of heavy metals. Separation of these pollutants is highly essential for effective and better utilization of water resources for the human, animal, agricultural, and industrial applications. Identification of the source of pollution and separation and treatment of pollutant is necessary to protect living creatures and environment.

Heavy metals are natural materials with a density of more than 5 g/cm^3 Koller et al.[1] or specific gravity higher than five[2] or atomic mass between 63.5 and 200.6 Dalton.[3] These ions possessing ionic values between 2 and 6 and placed in group 2 or 6 of the periodic table are termed as heavy metals.[4] These heavy metals are essential materials in many industrial and human applications. At the same time, their presence in excess concentration would cause many health hazards. It would also affect soil biota and plant growth. Their composition, concentration, and activity in the water resources and environment would be dependent upon the source of effluent and waste contaminations. Human activities, industrial waste are the major sources of heavy metals in the environment apart from natural calamities like a volcanic eruption. This generates the need for effective separation of these components before further applications. Various conventional and advanced methods are used for its separation. Each method has its own benefits and limitations. Membrane methods are proved to be better and green technology as compared to other methods. These separation methodologies work on physical separation, without the addition of any chemicals. This results in the separation and recovery of components without any further chemical contaminations. The recovered components can be utilized in further applications. There is a need for optimization of membranes for efficient recovery of components and heavy metals. The problem/issues associated with those membranes are improved by the use of different additives either in pre-synthesis or post-synthesis of the membrane. Carbon nanomaterial embedded membranes are proven to be the best, sustainable, eco- and environment-friendly option for the separation of heavy metal components and their further utilization.

5.2   SOURCES AND OBJECTIVES FOR HEAVY METAL SEPARATION

Natural calamities and human activities generate heavy metal pollutions in the environment. These heavy metals have the characteristic of bioaccumulation. This results in an increase in its amount in the food chain and enhances its effect on surrounding conditions, humans, and atmospheres. Table 5.1 shows the permissible limit and effect of a few heavy metals on the living organisms.

5.3   SOURCES OF HEAVY METALS

Naturally, heavy metals are found in the earth crust. Effluent and solid waste from different industries viz. mining, smelting, refinery, textile, paper, paint, pigment, fertilizer, pesticides, fuel, power plant, pharmaceutical plant, metallurgical industrial waste causes release of heavy metals in the environment.[2,3] Also, acid rain causes the breaking of soil and rocks, which releases a large amount of heavy metals in the water body and groundwater.[4] Fuel and power plant, metal processing, agriculture, waste disposal industry introduces respectively 2.4, 0.39, 1.4, 0.72 million tons/ year heavy metals in the environment.[5]

TABLE 5.1 Heavy Metals and Their Effect on Living Organism.

Heavy metals	Permissible concentration limit (PPM)	Effect on living organism	References
Lead	0.01	Anemia, headache, weakness, damaged liver, kidney, central nervous system, and brain functioning	[8,9,29]
Chromium	0.05	Skin irritation, lung cancer, headache, diarrhea, and vomiting	[9,29]
Cadmium	0.003	Cancer, renal disorder, and kidney damage	[9,29]
Arsenic	0.01	Cancer, skin irritation, and vascular disease	[8,9,29]
Mercury	0.006	Damage central nervous system, kidney function, and rheumatoid arthritis	[8,9,29]
Copper	2	Required for normal growth and metabolism Cramps, vomiting, convulsion, chronic asthma, liver damage, and insomnia	[9,29]
Zinc	3	Required for physiological function Skin irritation, vomiting, nausea, stomach cramps, and depression	[8,9,29]
Nickel	0.07	Required for normal growth Required for better yield, and photosynthesis of plant Lung, kidney, skin infection and cancer, chronic asthma, and coughing	[6–9,29]

5.4 OBJECTIVE OF HEAVY METAL SEPARATION

The separation of heavy metals from water resources and environments is highly essential. Separation objectives can be defined into two distinct requirements:

• Diseases caused by its contamination
• Use of it for human growth and industrial/medical application

There are many members in the periodic table which lie in the category of heavy metals. A few examples of heavy metals are aluminum (Al), antimony (Sb), arsenic (As), bismuth (Bi), cadmium (Cd), chromium (Cr), copper (Cu), iron (Fe), lead (Pb), manganese (Mn), mercury (Hg), nickel (Ni), platinum (Pt), silver (Ag), selenium (Se), tin (Sn), vanadium (V), zinc (Zn), etc. Among these heavy metals, few are essential like copper for animal metabolism and normal growth,[2,8] zinc for physiological function and normal growth of living tissue,[2] Co, Fe, Mn, Mo, Ni for normal growth, better yield and metabolism of plants unless its concentration is not above optimum value.[6,7] These heavy metals are essential for protein synthesis, photosynthesis, and cell membrane functionality.[7] These heavy metals are beneficial untilits level is below permissible value. They act adversely on human health, soil, plants, etc., instead of their positive impact when their concentration rises beyond a certain value. Heavy metals are nonbiodegradable and have bioaccumulation property due to which their concentration in the environment will go on increase above the upper limit specified by World Health Organization (WHO). Permissible limit by WHO for lead, chromium, cadmium, arsenic, mercury, copper, zinc and nickel are 0.01, 0.05, 0.003, 0.01, 0.006, 2.3 and 0.07 PPM (parts per million) respectively.[9] These heavy metals are soluble in the aquatic environment and easily adsorbed by vegetables and aquatic animals.[8] The very low concentration of Pb in soil affects the photosynthesis process whereas its high concentration in soil affects productivity. Also, if it is carried by plants through soil, it will be passed to humans through the food chain. Lead can cause anemia, headache, weakness, and damage liver, kidney, central nervous system, and brain functioning.[8] Heavy metal–contaminated food causes many health issues such as nutrient deficiency, affects immunity and cause disability, lung/liver damage, cancer, etc.[6] Excessive amount of zinc cause skin irritation, vomiting, nausea, stomach cramps, etc.[8] Similarly, if the copper amount exceeds, it results in cramps, vomiting, convulsion or death. Chromium (Cr^{3+} and Cr^{6+}) exposure can cause skin irritation, lung cancer. Nickel contamination causes lung, kidney, and skin infection and cancer. Mercury affects kidney function, damage central nervous system, etc. Whereas arsenic is carcinogenic.[8]

These heavy metals affect adversely to humans, animals, plants, and soil. It is necessary to separate them from groundwater and effluent to avoid contamination and accumulation/deposition. Apart from this, they are industrially important. Chromium is industrially used in the textile industry, leather tanning, metal finishing industry, and electroplating industry. Whereas to prepare nickel-cadmium batteries, Cd-Te solar cells and pigments, and in steel and plastic industries cadmium is used.[9] Arsenic trioxide is useful in the treatment of leukemia, the mercurous oxide is antibacterial, antiseptic, and used in skin ointment, gold used as tonic and is anti-infective, detoxicant, and has antiaging properties.[10]

Thus, as seen above a high concentration of heavy metals, above the permissible limit has an adverse effects and are toxic/harmful to humans and the environment, it is necessary to separate/remove them. Apart from their disadvantage, they have various uses—ranging from biological functioning, industrial, and medical use. Importantly, it is a naturally available material. So, its separation and recovery are important.

5.5 APPROACHES TO REMOVE HEAVY METALS

WHO has defined stringent standards and regulations for the admissible levels of heavy metals before the discharge of industrial effluent. Hence, effective separation techniques are required for the reduction of heavy metal concentration.[2] Those techniques should be green technology, it should reduce the concentration of heavy metals and enables the recovery of metals (Aroua et al., 2007).

Different methods are reported for removal of heavy metals.

- Conventional methods
- Advanced/Membrane separation methods

5.6 CONVENTIONAL METHODS

Conventional methods namely chemical precipitation, ion exchange, adsorption, coagulation, flocculation, flotation, electrochemical treatments are used for heavy metal separation.[2] Figure 5.1 shows a generalized flow diagram for heavy metal separation by the conventional method.

In the chemical precipitation method, initially suspended particles are removed by passing through a filtration unit. Then sulfide, hydroxide, and carbonate. precipitants are added to the continuous stirred tank reactor and the pH of the solution is maintained. The precipitant converts metallic ions into the insoluble complex. Which further separated by coagulation / sedimentation and filtration (Lawrence et al., 2005). This is a simple and inexpensive process with good separation efficiency.[2,8] In the ion exchange process, resin-cation /anion/amphoteric exchangers are used to adsorb heavy metals. Based on chemical structure or charge on ions, they bind to resins. It is a fast process and shows good separation efficiency.[5,9] In the adsorption process, different carbon base or bio-based materials are used as adsorbents. Heavy metals physically or chemically get collected on the surface of the adsorbent. It is a low-cost operation and carried out over a wide pH range.[9,11] In coagulation, flocculation method coagulants—aluminum, ferrous sulphate magnesium chloride ($MgCl_2$), etc., (Pang et al., 2009)/flocculants—polyferricsulfate, polyacrylamide, polyampholyte chitosan, etc., (Kurniawan et al., 2006) are added. It forms easy settling sludge. In the floatation method, the air is bubbled through the liquid and metal ions attach to these bubbles, forming froth, which is separated. Concentrated sludge is produced, it has dewatering characteristics which cause easy separation.[2,9] Whereas the electrochemical method is a simple and fast process. Here anode and cathode are inserted in the water bath and current is supplied. The heavy metals get collected at the electrode and are separated (Azimi et al., 2017). Table 5.2 shows principles, benefits and limitations of conventional processes.

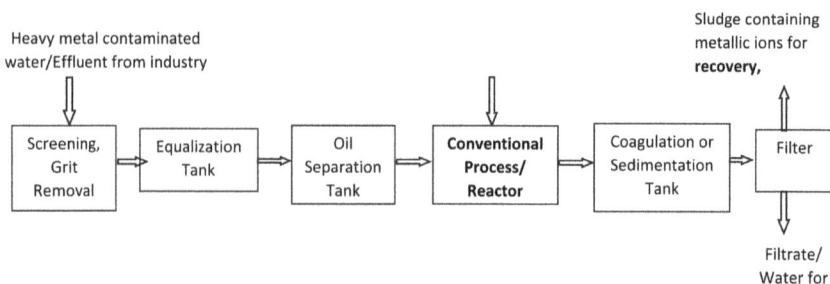

FIGURE 5.1   Process flow diagram for heavy metal separation by the conventional method.

5.7   ADVANCED/MEMBRANE SEPARATION METHODS

In the membrane separation process feed to be treated is passed over the membrane which selectively separates material. The material that remained on the membrane is called retentate and which passed through the membrane is called permeate. Depending on feed to be treated, either retentate or permeate or both are products. In the case of water containing heavy metals as feed, retentate is heavy metals and permeate is purified water. Both heavy metals and purified water are useful products that are used further (Mulder, 1998). Different membrane processes are reverse osmosis (RO), nanofiltration (NF), ultrafiltration (UF), and electrodialysis (ED). They are classified depending on the operating condition-applied

TABLE 5.2 Conventional Methods Principles, Advantages, and Limitations.

Conventional process	Principles	Advantages	Limitations	References
Chemical precipitation	Precipitant converts metallic ions into insoluble complex	Simple, inexpensive process, good separation efficiency	Requires chemical addition, large volume of sludge generation, sludge disposal/metal ions recovery issue	Lawrence et al. (2005) [11]
Ion exchange	Resin exchangers are used to adsorb heavy metals	Fast process, good separation efficiency	Requires expensive resin, resin regeneration, metal ions disposal/recovery issue	[5,9]
Adsorption	Metal ions adsorbed on the surface of the adsorbent	Low cost process, operation over wide pH range, good separation efficiency	Energy for adsorbent regeneration/desorption, metal ions disposal/recovery issue	[9,11]
Coagulation, Flocculation	Coagulant, flocculants forms metallic ions sludge	Sludge easily settles	A large volume of sludge production, metal ions disposal/recovery issue	Kurniawan et al. (2006)
Floatation	Metal ions attach to air bubbles, froth is separated	Easy separation, low-cost operation, good separation efficiency	High capital/maintenance cost, metal ions disposal/recovery issue	[9]
Electrochemical	Metal ions collected on electrode	No additional chemicals required, simple and fast process	Performance depends on current density, pH, concentration and temperature of the solution, the expensive current requirement	Azimi A. et al. (2017)

pressure /transmembrane pressure and pore size. The operating pressure varies in the range of 20–100, 5–20, 1–5 bar for RO, NF, and UF, respectively (Mulder, 1998). All these membranes work on the same principle of sieving mechanism under transmembrane pressure difference. In ED, metal ions are collected on the charged membrane using electric current as a driving force. It is a simple process for the recovery of metal ions.

5.8   SIGNIFICANCE OF MEMBRANE METHODS OVER OTHER METHODS

The chemical precipitation process requires the addition of chemicals for separation. Also, this process requires pH of solution to maintain and pH of the solution for the separation of one metal ion may not suitable for other (Taylor et al., 2014; Bhattacharyya et al., 1979). Even, it requires extra cost for sludge disposal and retreatment for recovery of metal ions and it is less suitable for the solution with low concentration.[2,11] In the ion exchange process, to regenerate resins, regenerant-acid or caustic solution is used, causing secondary pollution. Metal ion disposal and recovery is again an issue.[9] In the adsorption process, regeneration/desorption of adsorbents and disposal/recovery issues of heavy metals are observed.[9,11] Coagulation, flocculation, and floatation process requires further processing to treat sludge produced and has regeneration and recovery issues.[2,9] Whereas electrochemical method requires expensive electric supply for separation and electrodes needs to be cleaned/replaced periodically.[11]

In short, these conventional processes have their own separation conditions. They have limitations like chemical consumption, low efficiency, incomplete or partial separation, high-energy requirements, production of hazardous sludge, disposal issues, and need for recovery. It affects their applicability during real-life issues.

Whereasmembrane separation in green technology and is the physical separation process. With help of membrane process reduction in byproduct/waste due to less usage of chemicals during separation. The separation is carried out in a single stage without phase inversion hence generation of hazardous waste is eliminated. This process even can be combined with the synthesis processes for continuous separation which gives the benefit of recovery and reuse of waste/byproduct. This reduces the generation of waste, and effective recycling conserves energy along with conservation of natural resources.

Thus, membrane processes do not require any addition of chemicals/ change in the composition of effluent to be treated. It requires less space; linear upscaling is possible. It consumes less energy/less cost comparing the secondary treatment and quantity of water treated. It has high efficiency. Membrane properties can easily be varied/improved and adjustable. Finally, it does not have any recovery issue, retentate is a pure heavy metal that can be further utilized for different applications. Table 5.3 shows principles, benefits, and limitations of membrane processes.

5.9   SYNTHETIC/POLYMER MEMBRANES

Different polymers like cellulose acetate (CA), polycarbonate (PC), poly-acrylonitrile (PAN), polyetherimide (PEI), polytetrafluoroethylene (PTFE), polyethersulfone (PES), polyethylene (PE), polypropylene (PP), polyvinyl chloride (PVC), polyamide (PA), polysulfone (PS) and polyvinylidene fluoride (PVDF), polyetherketone (PEK), polyetheretherketone (PEEK), etc., have been reported for commercial preparation of membrane (Mulder, 1997)[12,13].

Those materials are used to prepare microfiltration to RO membranes. It requires less space, low energy, low cost, and high efficiency as compared to inorganic membranes. These are flexible and properties can be easily optimized. Their disadvantages as compared to inorganic material are less lifespan/fouling properties. It can be overcome by surface modification by physical or chemical blending with additives or hydrophilic groups.[12]

PP, PTFE, PVDF, PEI,[15] PS, and PES membranes are hydrophobic and hence have excellent chemical and thermal stability (Mulder, 1998). Hydrophobic membranes possess nonspecific adsorption capacity, resulting in fouling, rapid deterioration.[14] This causes blockage of membrane pores and ultimately decreases permeability and membrane lifetime.[13] Hydrophobicity of those membranes is reduced by modification of membrane surface by use of nanoparticles/hydrophilic groups in membrane matrix or by use of polymer blends.[13] PS, PES, and PVDF membranes can be used over a wide pH range. Glass transition temperature (Tg) for PS and PES are 195 and 230°C and of PVDF is −40°C. PVDF membranes are flexible and stable at a working temperature of −50 to 140°C. PEI has much better strength than PVDF having Tg 200°C.[14] Membranes obtained byCA and its derivative are hydrophilic in nature, but they have poor thermal and chemical stability and biologically easily degrade, to

TABLE 5.3 Membrane/Advanced Methods, Principles, Advantages, and Limitations.

Membrane processes	Principles	Advantages	Limitations	References
Reverse osmosis	20–100 bar pressure	Sustain high temperature, good separation efficiency	High pressure-high energy, not suitable for concentrated solution, separates healthy minerals, membrane fouling	[9] Mohammadi et al. (2004), Mulder (1996)
Nanofiltratrion	5–20 bar pressure	Less energy as compared to RO, separate viruses, good separation efficiency	Not possible to separate soluble components, separates healthy minerals, membrane fouling	[9], Mulder (1996)
Ultrafiltration	1–5 bar pressure	Less energy, separates viruses, dissolved colloidal components	Membrane fouling	[9], Mulder (1996)

avoid biodegradation- working pH should be maintained acidic (Mulder, 1998). Whereas PEK and PEEK have good chemical and thermal stability, but they are difficult in processing because of their chemical resistance. Also, PEEK at room temperature only soluble in concentrated sulfuric/ chlorosulfuric acid (Mulder, 1998). PA, PI, PEI, and PAN also possess excellent thermal, chemical, and mechanical stability and are hydrophilic in nature (Mulder, 1996).

5.10 DIFFERENT NANOMATERIALS USED IN MEMBRANE PREPARATION

As mentioned above, polymers are the most popular materials used for membrane preparation. Though these materials have many advantages like flexibility, easy processing, selective separation chemical, mechanical, thermal stability. It has some limitations of high cost, hydrophobic nature, fouling, in some cases low permeability-transport properties (Mulder, 1998).[13,14]

Membrane performance is optimized by the use of different additives, surface treatment, chemical modification, etc. This additive can be water, low molecular weight organics, surfactant, inorganic salt, polymers, mineral fillers, or their mixture (Kickelbick, 2003). During membrane formation porogen agent is added and selectively leached out, it forms additional pores and results in improved flux (Kusworo, 2014). The porogen additive affects membrane formation and structure, it enhances pore formation, prevents macro void formation, improves pore interconnectivity, introduces hydrophilicity, and maintains selectivity (Zhang, 2014).

Similarly, challenges/limitations of the membrane can be overcome by nanomaterial incorporation in membrane matrix. Nanomaterials such as carbon nanotubes, titanium nanotubes, graphene, silica, zeolites, metal oxides, etc., can be used to improve membrane properties/performance.[16,17] It modifies the membrane morphology-pore size has reduced whereas porosity and interconnectivity between pores have increased. The addition of nanomaterials while preparing membranes improves properties like hydrophilicity, reusability, better selectivity-rejection, better permeability-transport, less cost-low energy consumption, enhances membrane resistance to fouling, reduce the cleaning cost of the membrane.[17] Also, it improves antimicrobial properties.[16]

These nanomaterials are introduced into membrane matrix either pre- or post-formation of the membrane. The pre-formation method is by physical blending, nanomaterials are added in the casting solution. Whereas in post-formation, it is by coating, that is by deposition of nanomaterial on the membrane surface.

Amine imparted silica introduces hydrophilicity to the membrane and improves the fouling resistance. But silica nanomaterials are costly, and they reduces thermal stability. Even if the silica is inhaled, it causes cellular inflammation and increases the lung cancer rate.[30] ZnO is environment-friendly material, usually used as a catalyst in the industry. It is well reported by Javdaneh et al. (2016) that nano-sized ZnO is effectively used to remove heavy metals from water. ZnO nanoparticles improve water flux means it forms a more porous structure of the membrane, and it also helps to improve rejection. The equal distribution of these metal nanoparticles in dope solution improves membrane hydrophilicity and decreases its fouling (Javdaneh et al., 2016). Polysulfone membrane performance has improved by the addition of ZnO nanoparticles (Moradihamedani et al., 2013). Research by Leo et al. (2012) show that the hydrophobic nature of PS membrane has changed to hydrophilic by the addition of ZnO nanoparticles which results in the reduction in fouling of membrane. Also, it increases the thermal stability of the membrane (Leo et al., 2012). Oxides of metals such as silver, titanium, etc., improve the antibacterial property of the membrane.[16] Addition of ZnO, titanium oxide in membrane increases the water flux, affects the retention property and antifouling property of membrane. Using carbon nanotubes, graphene, and zeolite, the pore size of the membrane is reduced results in good rejection, keeping/maintaining the water flux. CNTs embedded membranes have good flexibility and transport.[17] The amount of nanomaterial in the membrane matrix is needed to optimize for better retention-rejection and permeability. Table 5.4 shows different nanomaterials, their application areas, and benefits.

5.11 CARBON NANOMATERIAL IMPORTANCE

Carbon nanomaterials possess good chemical, mechanical, thermal strength. They have good conductivity, optical strength and have low density, high surface area, and are easy to recycle. Carbon nanomaterial

TABLE 5.4 Nanomaterials, Their Applications, and Benefits.

Nanomaterials	Applications	Benefits	References
Silica	Semiconductors, Solar cells, Lithium-iron battery, Luminescent display device, and in drug delivery	Enhances fouling resistance	[30]
Zeolites	Catalyst in petroleum refining and chemical production, and separation/gas adsorption	Enhances porocity; good rejection	[17,30]
AgO	Medical application—anti-infection, Food industry—avoid/control food spoilage	Improves antibacterial property of the membrane	[16]
TiO$_2$	Purifying agent, Antimicrobial agent, Coating agent	Improves antibacterial property of membrane, increases water flux, and antifouling property	[16]
ZnO	To prepare sunscreen, toothpaste, and cosmetics, coating agent in food, plastic, and paint	Environment friendly, as a catalyst, increases flux, gives hydrophilicity to the membrane, enhances fouling resistance, and good separation efficiency	Javdaneh et al. (2016), Moradihamedani et al. (2013)
Carbon nanotubes	Biomedical field, microelectronics, and composite reinforcement	Enhances porocity, good rejection and flexibility of membrane, increases antifouling, and thermal and mechanical properties of membrane	[16,17]
Graphene	Building block for GO, CNFs, and fullerene	Enhances porosity and good rejection	[17]
Graphene oxide	Anticorrosive coatings, Energy storage, Conductive films, flexible electronics, Chemical sensors	Improves processability to form the membrane, improves antibacterial, and antifouling property of the membrane	[18,27]
Fullerene	Semiconductors, solar cells, biomedical science, surface coatings, and biological application—drug delivery, etc.	Modify/improves membrane surface property, and strength	[9]

incorporated membranes shows increase in hydrophilicity, water flux, and salt rejection.[18] Apart from this they do not have any negative impact on the environment and they are biocompatible with a living organism and the environment.[26]

5.11.1 CLASSIFICATION OF CARBON NANOMATERIAL

CNTs, graphene, graphene oxide (GO), carbon nanofibers, MXene, carbides of carbon, and fullerene are the most common types of carbon nanomaterial.[18] Broadly, they are classified into two groups: graphene and carbon nanotubes based.[16] CNTs are vertically and randomly aligned.[16] CNTs are single-walled (SWCNT) and multi-walled (MWCNT).[18]

5.11.2 CARBON NANOMATERIAL STRUCTURE, PROPERTIES, AND APPLICATIONS

The change in structure results in the change in properties and ultimately the area of application. Even for the same amount of CNTs, graphene, or GO in membrane varies membrane properties.

CNTs: The sheet of carbon atoms rolled into a hollow smooth cylindrical tube is termed as CNTs.[18] They are allotropes of carbon. Single-walled and multiwalled CNTs are single layer and multilayer of graphene sheet, respectively.[20] Multiwall carbon nanotubes are concentric cylinders of carbon sheets. The diameter of CNTs is 1–100 nm.[19] MWCNTs have comparatively high purity than SWCNTs.[20] SWCNT- and MWCNT-modified membranes show the increase in water flux and decrease in membrane fouling. CNTs improve the membrane selectivity and hence rejection, keeping the better water flux. Also, it increases the antifouling, thermal, and mechanical properties of the membrane. But CNTs are not easily soluble in any organic solvent. It has a poor dispersion property. Also, the interaction between polymer and CNTs is poor. Such membranes degrade fast and result in a decrease in their performance.[16] Because of the high strength and conducting properties of CNTs, they are used in the biomedical field, microelectronics, and as composite reinforcement.[18] Whereas release of CNTs in the environment adversely affects human, aquatic life, and plants. It can cause respiratory irritation and lung cellular proliferation in rats.[20]

Graphene: The packed form of carbon atoms layer which is joined together in a two-dimensional hexagonal lattice is termed as graphene. It is light in weight, 1 m^2 of graphene weighs 0.77 mg. It has good thermal and electrical conductivity, good tensile strength/mechanical property, and has good corrosion and bacteria resistance. Also, it has a high surface area.[18,25]. It is used as a building block for GO, CNFs, and fullerene.[18] When graphene is functionalized with nitrogen shows high metal ion rejection than with functionalize group fluorine or hydroxide.[31]

Graphene Oxide: Graphene modified version in which carbon atom has bonded with oxygen and hydrogen is termed as GO. It has a hexagonal carbon structure similar to graphene. Also, it contains a hydroxyl (-OH), carboxyl (C=O), carboxylic acid (-COOH), alkoxy (C-O-C), and other oxygen base functional groups. When graphene is oxidized with strong acid and oxidizer, it produces hydroxyl, epoxy, and carboxyl as functional groups. These functional groups improve the dispersion of nanomaterial in solvent and processability to form membrane. Those oxygen base groups increase solubility and give surface charge and selectivity properties in the membrane. GO-modified membrane prevents bacterial growth and reduces membrane fouling.[18]

Carbon nanofibers (CNF): Hollow core nanofibers with single or double graphite layer stack either parallel or with an angle from fiber axis are termed as carbon nanofibers. The diameter of CNFs is 50–200 nm.[19]

Fullerene: Allotrope of carbon having hollow structure, closed caged either pentagonal or hexagonal ring and spherical or ellipsoid in shape and look like a ball are termed as fullerene. It has carbon atoms (60, 70, 78, or more) on its surface, and when it has 60 atoms on its surface, it has a stable and spherical shape. They possess good electron affinity, hydrophilic, good strength, and surface-to-volume ratio is high. Because of these properties, fullerene is used in semiconductors, solar cells, biomedical science, and surface coatings.

5.12 UNIQUE METHODS TO PREPARE CARBON NANOMATERIAL INCORPORATED MEMBRANES

Various techniques are used to prepare nanomaterial-embedded membranes. Phase inversion and interfacial polymerization are most common.

5.13   PHASE INVERSION TECHNIQUE

Among different techniques, phase inversion is the most versatile technique which allows the preparation of all kinds of membranes and is the base for the synthesis of most of the commercial membranes. By this technique, only polymeric membranes are prepared. The phase inversion process was first time introduced in the 1960s by Loeb and Sourirajan. Here solute (polymer) is dissolved in the solvent. For the formation of membrane, the solution passes from fluid state to solid state under controlled operating conditions. The process involves a controlled transformation of a thermodynamically stable polymer solution from a liquid to a porous solid state. Membrane morphology is determined by the rate of removal of solvent and velocity of phase separation.[22] Phase inversion membranes are prepared by precipitation by solvent evaporation, precipitation from the vapor phase, precipitation by controlled evaporation, thermal precipitation, and immersion precipitation (Mulder, 1996).

In precipitation by solvent evaporation, the polymer is dissolved in suitable solvent, and solution is cast on a suitable support. The solvent is allowed to evaporate in an inert atmosphere and forms a dense membrane. The increase in the incompatibility of non-solvent with polymer results in increase in membrane porosity. In precipitation from the vapor phase, cast film which contains polymer and solvent is placed in a non-solvent vapor atmosphere. Penetration of the non-solvent vapor phase leads to precipitation. In thermal precipitation, evaporation of solvent from membrane forms skinned membrane. In immersion precipitation, the film is immersed in a non-solvent bath, here solvent-nonsolvent exchange occurs, and polymer precipitates (Mulder, 1996).[21]

The nanomaterial is incorporated in the membrane either by physical blending, while preparing casting solution or by post-synthesis of the membrane by dip/spray coating (Mulder, 1996).[20] Choi et al.[24] prepared polysulfone-MWCNT membrane using N-methyl-2-pyrrolidinone and water as solvent and coagulant respectively by physical blending phase inversion method.

5.14   INTERFACIAL POLYMERIZATION TECHNIQUE

This technique produces a composite membrane. Here two different polymers are used to form two different layers. The bottom layer is porous support, and the top layer is formed on the bottom support layer. The bottom support layer is generally prepared by phase inversion. The top layer is a thin dense film of a selective polymer. The bottom layer is immersed in a solution, where another monomer is dissolved in water-immiscible solvent. Heat is provided so that reaction/polymerization (interfacial) between monomer and support layer takes place and forms a composite membrane. Transport rate through the membrane depends on a thin dense layer and mechanical strength depends on the bottom porous support. The nanomaterial is generally used along with monomers to form a top thin selective layer (Mulder, 1996).[20,22] Kim et al.[23] prepared polyamide-CNT membrane using interfacial polymerization.

Those are called composite membranes, which are prepared by different procedures like dip coating, spray coating, spin coating, and plasma polymerization.

5.15 CARBON NANOMATERIAL EMBEDDED DIFFERENT MEMBRANES

The effect of carbon nanomaterials on membrane-RO/NF/UF properties, morphology-pore size, porosity, etc., and performance-salt rejection, fouling, etc., is discussed below.

5.15.1 CARBON NANOMATERIAL EMBEDDED RO MEMBRANES

RO membranes are tighter than NF and UF membranes. They have very small pores ranging from 0.0001 to 0.001 μm. Such membranes can reject multivalent, divalent as well monovalent ions, in addition, they can remove viruses, microorganisms, and a wide range of dissolved species from feed solution-water to be treated. To operate such membranes, it requires high pressure, 20–100 bar.

Polyamide-CNT membrane was prepared by interfacial polymerization. To add functionality to CNT, acid treatment is given. Sulfuric and nitric acid in a 3:1 ratio (volume) is taken and CNTs were treated with this mixture. CNTs used in this process were treated by varying amounts of acid mixture, reaction temperature/time, etc. In one experiment, 2/10th g of CNT and 60 mL of acid mixture mixed together and was continuously stirred in a sonicator for 30 min. The mixture is then placed in an oil bath at 65°C with stirring for 4 h. The solution is then cooled at room temperature with 1.5 L water. This dilute solution is then filtered through the anodic aluminum oxide and dried in a vacuum oven at 35°C. Those CNTs were used to prepare the membrane. PS membrane treated with IPA for 10 min to enlarge pores and then washed with water for several times and kept in water for 3 h to stabilize pores. An aqueous solution of 2 wt.% MPD (m-Phenylenediamine) and 0.002 wt.% of CNTs were prepared and polysulfone membrane is placed in this aqueous solution for 3 h. Then it is taken out and it is wiped with a rubber roller to remove air bubbles and aqueous solution. A TMC (trimesoyl chloride) solution was prepared. A 1500 g of TMC was added to a flask containing argon gas and using a syringe 149.85 g of n-hexane is added and it is stirred continuously at room temperature. This TMC solution was poured on the membrane for 1 min, then it is taken out and the excess organic solution is removed, and the membrane is placed in an oven at 100°C for 5 min for polymerization. It is observed that rejection is good at CNT treatment time in acid is 4 h at 65°C, lesser the time CNTs dispersion in polyamide active layer is less and if time/temperature increased, it results in pieces of CNTs of smaller size and its aggregation. It is observed that sodium salt rejection is almost 90% for PA-CNT membrane from zero to 50 h, whereas it tends to decrease up to 50% for PA-pure membrane without CNT.[23]

5.15.2   CARBON NANOMATERIAL EMBEDDED NF MEMBRANES

NF membranes are tighter than UF membranes but not than RO membranes. The pore size is ranging from 0.001 to 0.01 um. These membranes can reject multivalent, divalent as well as some monovalent ions and viruses. They can be operated at pressure more than UF and less than RO, between 5 and 20 bar.

Torlon-GO membrane was prepared by dip coating, this composite increases the heavy metal removal percent. GO is deposited on the membrane by layer by layer formation using dip coating. Here torlon powder is kept in a vacuum oven at 120°C. Solvent N-methyl-2-pyrrolidone and non-solvent polyethylene glycol were used to fabricate the hollow fiber membrane supports. To remove excess/residual solvent, this fiber is kept in tap water. Then it is soaked in glycerol (40%)-water (60%) mixture for 2 days to avoid collapsing of pores. To improve adhesion between GO and fiber, fiber is cross-linked with 1% HPEI (hyperbranched polyethyleneimine) at 70°C for 30 min. GO solution is prepared in 50 ppm distilled water with sonication for 1 h for proper dispersion of GO. Hollow fiber is then dipped in GO solution, after 5 s it is taken out, blotted with tissue paper to remove excess GO. Then, the fiber is dipped in 1% EDA (ethylenediamine) solution for 5 s. The coating of GO is repeated to form layer by layer deposition of GO. In this way NF composite membrane is formed by dip coating. For a single layer of GO, $MgCl_2$ rejection increases from 55.5% to 84.3%, whereas water flux decreases from 50.2 to 8 LMH. With the increase in time for dipping fiber in GO solution from 5 s to 10 min, rejection has improved to 97% and water flux obtained 4 LMH. The sequence of rejection is $MgCl_2 > MgSO_4 > NaCl > Na_2SO_4$.[27]

5.15.3   CARBON NANOMATERIAL EMBEDDED UF MEMBRANES

UF membranes have pore size more than that of NF and RO, ranging from 0.1 to 0.01 μm and operating pressure is lesser than RO, NF range from 1 to 5 bar.

The phase inversion method is used to prepare the PS-MWCNT membrane. PS was dried in a vacuum oven for 24 h at 80. Dimethylformamide (DMF) is used as solvent and coagulation bath as water with 1% isopropanol. 18% PSF, 82% DMF and CNTs (oxidized, amide and azide) in different weight ratio 0, 0.1, 0.2, 0.5 and 1% were added. The solution is stirred at 60°C for 72 h. To remove gas bubbles solution is placed in an ultrasonic bath. The solution is cast on a PET base using a glass rod or steel knife by maintaining the precise gap. The membrane is immediately placed in coagulation bath. Then, it is dried at 60°C and preserved in deionized water.

MWCNT taken for membrane formation have more than 90% purity. They were cleaned to remove amorphous carbon and traces of metal ions. 2 g of MWCNT were added to a 500 mL solution of 1% Brij 98 surfactant. It is kept in a sonicator for 2 h and then allowed to stand for 6 h. To obtain clean MWCNTs, it is centrifuged at 3000 rpm for 15 min for twice or thrice. CNTs were removed and washed with water and brine to remove surfactant. To add functional group, amide, azide or oxidation treatment is given to CNTs. For oxidized CNTs, CNTs treated with a 3:1 mixture of 40 mL concentrated sulfuric and nitric acid for 24 h at 33°C. Then those CNTs washed with double distilled water Then it is dried in a vacuum oven at 80°C for 4 h. To introduce acyl group, oxidized CNTs treated with $SOCl_2$. In 20 mL thionyl chloride and 2 mL of DMF, 200 mg CNTs are added. It is then refluxed at 60°C for 36 h. It is then washed with toluene and dried in a vacuum oven at 80°C for 4 h. Acrylated CNTs then sonicated in 10 mL EDA for 16 h at 33°C. It is then diluted with 200 mL methanol and then it is dried in a vacuum oven for 4 h at 80°C. To add azide functional group, acrylated CNTs treated with 5 mg NaN3 in DMF for 24 h and then precipitated in toluene and then dried in an oven for 4 h at 80°C. It is observed that without MWCNT membrane degradation temperature is 312°C, the addition of 1% MWCNT cause an increase in degradation temperature to 369 and 374 for amide and oxidized MWCNT. This is due to good compatibility between PSF and nanotube which results in an increase in thermal stability of the membrane. Similarly, for PSF membrane pore size is 0.1079 μ which has reduced to 0.0257 μ for 1% addition of MWCNT. Metal ion rejection without MWCNT is below 11% for Cr, Cu, Pb, Cd, and As. It is increased to more than 90% for Cr, Cu, Pb and approximately to 80% for Cd and As using 1% amide and azide MWCNT at 2.6 pH and at 0.49 MPa and for 1% oxidized MWCNTs it is in the range of 70–80%, except Pb is 40%. It is observed that as the percent of CNTs is increased, it results in a decrease in pore size and an increase in rejection. MWCNT also imparts hydrophilicity to the membrane, thus it results in a decrease in its fouling.[28]

5.16   FUTURE SCOPE AND BENEFITS

Carbon nanomaterial embedded membranes showed excellent rejection and separation properties for heavy metals. Most important these materials are separated in their physical form, without the addition of any chemicals. Due to such separation, these materials are separated in pure form and they can be used in further processes or applications. Though there are some challenges to synthesizing those membranes. The dispersion and alignment of nanomaterial when production is on large scale is an issue. Improper dispersion and alignment result in a decrease in membrane performance— salt rejection and transport properties. These dispersion and formation issues improve the cost of nanomaterial. It is a necessity to critical observation on nanomaterial impact on living organisms and the environment.

Apart from this, a standard procedure to analyze membrane performance is necessary. There is a need to define models which can work with the number of variables. The polymer-support material and its concentration, a solvent used, coagulation bath used, nanomaterial used and its concentration, operating condition-temperature, pH, evaporation time, precipitation time, the method of preparation-physical blending, coating, interfacial polymerization, etc., by which membrane are formed. Apart from obtaining good rejection maintenance of transport property is a challenge and is necessary for its applicability in real-life processes. More efforts on reduction in membrane cost, keeping this green technology eco and environment friendly.

To preserve the environment from harmful waste generated by human activities and industrial waste, stringent action/strict laws should be implemented. Though this green advanced technology is the solution to overcome the problem, today the society is facing, the industry is still reluctant to accept it instead of traditional, conventional methods. There is a need to develop a research program to facilitate the technology transfer.

ACKNOWLEDGMENT

This work is financially supported by DST- Nano Mission (Sanction No. SR/MN/NT/-1029/2015). We also thank Bharati Vidyapeeth (Deemed to be University) College of Engineering, Pune and MIT-WPU, Pune for providing us the useful resources.

KEYWORDS

- heavy metals
- health/environmental issues
- membrane separation methods
- recovery
- carbon nanomaterial
- eco-friendly
- economic
- sustainable
- removal efficiency

BIBLIOGRAPHY

1. Koller, M.; Saleh, H. M.; Saleh, M.; Koller, M. Introductory Chapter : Introducing Heavy Metals, 2018; pp 3–12. https://doi.org/10.5772/intechopen.74783
2. Fu, F.; Wang, Q. Removal of Heavy Metal Ions from Wastewaters: A Review. *J. Environ. Manage.* **2011,** *92* (3), 407–418. https://doi.org/10.1016/j.jenvman.2010.11.011
3. Bessbousse, H.; Rhlalou, T.; Verch, J. F.; Lebrun, L. Removal of Heavy Metal Ions from Aqueous Solutions by Filtration with a Novel Complexing Membrane Containing Poly(Ethyleneimine) in a Poly(Vinyl Alcohol) Matrix. *J. Membr. Sci.* **2008,** *307* (2), 249–259. https://doi.org/10.1016/j.memsci.2007.09.027
4. Alluri, H. K.; Ronda, S. R.; Settalluri, V. S.; Singh, J.; Suryanarayana, V.; Venkateshwar, P. Biosorption : An Eco-Friendly Alternative for Heavy Metal Removal. *J. Biotechnol.* **2007,** *6* (25), 2924–2931. https://doi.org/10.4314/ajb.v6i25.58244
5. Hubicki, Z.; Kolodynska, D. Selective Removal of Heavy Metal Ions from Waters and Waste Waters Using Ion Exchange Methods. *Ion Exchange Technol.* 2012. https://doi.org/10.5772/51040
6. Jiwan, S.; Kalamdhad, S. A. Chemistry and Environment Effects of Heavy Metals on Soil, Plants, Human Health and Aquatic Life. *Int. J. Res.* **2011,** *1* (2), 15–21.
7. Arif, N.; Yadav, V.; Singh, S.; Singh, S.; Ahmad, P. Influence of High and Low Levels of Plant-Beneficial Heavy Metal Ions on Plant Growth and Development, 2016. 4 (November). https://doi.org/10.3389/fenvs.2016.00069
8. Bazrafshan, E.; Mohammadi, L.; Ansari-Moghaddam, A.; Mahvi, A. H. Heavy Metals Removal from Aqueous Environments by Electrocoagulation Process—a Systematic Review. *J. Environ. Health Sci. Eng.* **2015,** *13* (1). https://doi.org/10.1186/s40201-015-0233-8
9. Ihsanullah Abbas, A.; Al-Amer, A. M.; Laoui, T.; Al-Marri, M. J.; Nasser, M. S.; Atieh, M. A. Heavy Metal Removal from Aqueous Solution by Advanced Carbon Nanotubes: Critical Review of Adsorption Applications. *Sep. Purif. Technol.* **2016,** *157*, 141–161. https://doi.org/10.1016/j.seppur.2015.11.039
10. Singh, R.; Gautam, N.; Mishra, A.; Gupta, R. Heavy Metals and Living Systems: An Overview. **2011,** *43* (3). https://doi.org/10.4103/0253-7613.81505
11. Barakat, M. A. New Trends in Removing Heavy Metals from Industrial Wastewater. *Arab. J. Chem.* **2011,** *4* (4), 361–377. https://doi.org/10.1016/j.arabjc.2010.07.019
12. Padaki, M.; Surya Murali, R.; Abdullah, M. S.; Misdan, N.; Moslehyani, A.; Kassim, M. A. Ismail, A. F. Membrane Technology Enhancement in Oil-Water Separation. A Review. *Desalination* **2015,** *357*, 197–207. http://doi.org/10.1016/j.desal.2014.11.023
13. Garcia-Ivars, J.; Alcaina-Miranda, M. I.; Iborra-Clar, M. I.; Mendoza-Roca, J. A.; Pastor-Alcañiz, L. Enhancement in Hydrophilicity of Different Polymer Phase-Inversion Ultrafiltration Membranes by Introducing PEG/Al2O3 Nanoparticles. *Sep. Purif. Technol.* **2014,** *128*, 45–57. http://doi.org/10.1016/j.seppur.2014.03.012
14. Nunes, S. P.; Peinemann, K. V. *Membrane Technology in Chemical Industry*; WILEY-VCH: Weinheim; New York; Chichester; Brisbane; Singapore; Toronto, 2001; pp 17–18.

15. Kanagaraj, P.; Nagendran, A.; Rana, D.; Matsuura, T.; Neelakandan, S.; Karthik-kumar, T.; Muthumeenal, A. Influence of N-Phthaloyl Chitosan on Poly (Ether Imide) Ultrafiltration Membranes and Its Application in Biomolecules and Toxic Heavy Metal Ion Separation and Their Antifouling Properties. *Appl. Surf. Sci.* **2015,** *329,* 165–173. https://doi.org/10.1016/j.apsusc.2014.12.082

16. Amin, M.; Shahmirzadi, A.; Kargari, A. 9 - Nanocomposite Membranes. In *Emerging Technologies for Sustainable Desalination Handbook.* 2018. https://doi.org/10.1016/B978-0-12-815818-0.00009-6

17. Subramaniam, M. N.; Goh, P. S.; Lau, W. J.; Ismail, A. F. The Roles of Nanomaterials in Conventional and Emerging Technologies for Heavy Metal Removal : A State-of-the-Art Review. 2019.

18. Manawi, Y.; Kochkodan, V.; Ali, M.; Khaleel, M. A.; Khraisheh, M.; Hilal, N. Can Carbon-Based Nanomaterials Revolutionize Membrane Fabrication for Water Treatment and Desalination? *DES.* 2016. https://doi.org/10.1016/j.desal.2016.02.015

19. Rakesh, B. M.; Bhanu, P. S.; Shailaja P. *Carbon Nanomaterials: Synthesis, Structure, Properties and Applications*; CRC Press, 2017; p 97.

20. Ullah, I. Carbon Nanotube Membranes for Water Purification : Developments, Challenges, and Prospects for the Future. *Sep. Purif. Technol.* 2018. https://doi.org/10.1016/j.seppur.2018.07.043

21. Richards, H. L.; Baker, P. G. L.; Iwuoha, E. Metal Nanoparticle Modified Polysulfone Membranes for Use in Wastewater Treatment: A Critical Review. *Journal of Surface Engineered Materials and Advanced Technology* **2012,** *02* (03), 183–193. https://doi.org/10.4236/jsemat.2012.223029

22. Tasselli, F. Membrane Preparation Techniques. **2004,** (c), 1–3. https://doi.org/10.1007/978-3-642-40872-4

23. Kim, H. J.; Choi, K.; Baek, Y.; Kim, D.; Shim, J.; Yoon, J.; Lee, J. High-Performance Reverse Osmosis CNT/Polyamide Nanocomposite Membrane by Controlled Interfacial Interactions. 2014.

24. Choi, J.; Jegal, J.; Kim, W. Fabrication and Characterization of Multi-Walled Carbon Nanotubes/Polymer Blend Membranes. *J. Membr. Sci.* **2006,** *284,* 406–415. doi:10.1016/j.memsci.2006.08.013.

25. Ji, X.; Xu, Y.; Zhang, W.; Cui, L.; Liu, J. Composites: Part A Review of Functionalization, Structure and Properties of Graphene/Polymer Composite Fibers. *Composites Part A* **2016,** *87,* 29–45. https://doi.org/10.1016/j.compositesa.2016.04.011

26. Baby, R.; Saifullah, B. Carbon Nanomaterials for the Treatment of Heavy Metal-Contaminated Water and Environmental Remediation. 2019.

27. Zhang, Y.; Zhang, S.; Gao, J.; Chung, T. Author's Accepted Manuscript Layer-by-Layer Construction of Graphene Oxide (GO) Framework Composite Membranes for Highly Efficient Heavy Metal Removal. *J. Membr. Sci.* 2016. https://doi.org/10.1016/j.memsci.2016.05.035

28. Shah, P.; Murthy, C. N. Studies on the Porosity Control of MWCNT/Polysulfone Composite Membrane and Its Effect on Metal Removal. *J. Membr. Sci.* **2013,** *437,* 90–98. https://doi.org/10.1016/j.memsci.2013.02.042

29. Gunatilake, S. K. Methods of Removing Heavy Metals from Industrial Wastewater. *Multidiscipl. Eng. Sci. Stud. (JMESS)* **2015,** *1* (1), 12–18. https://doi.org/10.13140/RG.2.1.3751.1848

30. Lehman, S. E.; Larsen, S. C. *Environmental Science Nano Greener Syntheses, Environmental Applications and Biological Toxicity.* 2014; pp 200–213. https://doi.org/10.1039/c4en00031e

31. V. A.; Namsani, S.; Singh, J. K. Removal of Heavy Metal Ions Using Functionalized Graphene Membranes: A Molecular Dynamics Study. *RSC Adv.* 2016. https://doi.org/10.1039/C6RA06817K

CHAPTER 6

CARBON NANOTUBES FOR BIOMEDICAL APPLICATIONS

SHRIKAANT KULKARNI

Faculty of Science & Technology, Vishwakarma University, Pune, India

Corresponding author. E-mail: srkulkarni21@gmail.com

ABSTRACT

The carbon-based nanomaterials (CBNs) have been studied for their applications in the field of biomedicine and have drawn attention because of their distinct physicochemical properties such as thermal, mechanical, electrical, optical, and structural ones. Given such unique intrinsic properties, CBNs, covering carbon nanotubes (CNTs), graphene oxide (GO), and graphene quantum dots (GQDs), have been explored extensively for their biomedical applications. This chapter takes a review of the latest studies in preparing of CBNs for numerous biomedical applications ranging from bio-sensing, drug delivery to cancer therapy.

The unique properties of CBNs namely, fullerenes, graphene, CNTs, and their derivatives made them versatile in their applications in a host of fields including biomedicine. The state-of-the-art applications of the CBNs in the area of biomedicine cover cancer therapy, targeted drug delivery, bio-sensing, bio-imaging as well as personalized and regenerative medicine. On functionalization, CBNs toxicity vary and become soluble in a range of solvents including water, and biomedical applications ask for it.

In this chapter, an overview of the numerous biomedical applications of carbon-based nano-materials (CBNMs) is taken. In particular, CBNs as the carriers for targeted drug delivery are described. An emphasis is given to the description of various CBNMs that are the most important for biomedical applications. Second, the critical survey on research efforts made by different researchers in the development of biomedical applications has been focused. The inclusion are drug/gene delivery, bio-imaging, biosensors, and therapeutics. For each of the case, hypothetical general information and a brief explanation have been given. Finally, important constraints associated with carbon-based nonmaterial are touched upon and possible open issues are debated for further research.

6.1 INTRODUCTION

Nanotechnology refers to structures in the 1–100 nm size in at least one dimension. With the help of nanotechnology, it is very much possible to attain improved targeted drug delivery of weakly drugs soluble in water, and of drugs to a cell or tissue, co-delivery of multiple drugs, site visualization of drug delivery by making use of therapeutic agents in tandem with imaging tools, etc.[1] The development of nano-carriers for gene and drug delivery has become possible due to advances in nanotechnology.

Presently, liposomes are preferentially employed as drug carriers for clinical purposes while polymer micelles, and silica nanoparticles are potential carriers under development.[2] Indeed, there is a continuous demand for novel delivery systems that can protect, transport, and release active molecules (i.e., drugs, antigens, antibodies, and nucleic acids) to specific sites of action.[3–8] Primarily it is of use, specifically, in cancer therapy. Carbon nanomaterials can be considered as novel and innovative tools in the development of alternative methodologies for the delivery of therapeutic molecules.[9]

This chapter will describe the structural characteristics and potential of carbon nanomaterials, including carbon nanotubes (CNTs), nanohorns (CNHs), and nanodiamonds (NDs) in the field of biomedicine.

6.2 STRUCTURES, CHARACTERISTICS, AND DERIVATIZATION OF CNTs

Carbon nanotubes resemble graphene sheets rolled up to form cylinders, with different characteristics and peculiarities depending on the number of concentric cylinders and on the rolling up angle.[10] The main distinction among single-, double- and multiwalled carbon nanotubes, SWCNTs, DWCNTs, and MWCNTs, respectively, is the presence of different rolling curvatures, which confer the metallic or semiconductor characteristics to the tubes. These objects are almost unidimensional, presenting diameters in the order of nanometers and lengths up to several microns. Improvements in CNT production are still under development since to date every methodology has disadvantages, mainly represented by the presence of metallic nanoparticles and carbonaceous impurities in the production soot. The induction of selective chirality of CNTs is still far from being reached, the control of the rolling up angle has still not been obtained, and the separation of metallic or semiconducting CNTs is difficult and imprecise.

The control of length and diameter, although better, is still not achieved as well. The possibility of applications is wide, spanning various fields of science from electronics to medicine. In some cases, CNTs can be used as produced (pristine CNTs), but often derivatization is necessary, especially to improve their manipulations and solubility characteristics. From the reactivity point of view, it is possible to distinguish two different parts: the tip and the sidewall. The former resembles the fullerene hemisphere in curvature and reactivity, while the latter presents a unidirectional distortion from planarity and is much less reactive than the tip, due to lower strain energy.[11]

The chemistry underlying CNTs is evolving very fast and different methods have been realized since the early studies on their chemical change. It is possible to distinguish three main methodologies:

- the supramolecular derivatization,
- the covalent functionalization, and
- the filling of the internal cavities of the tubes themselves.[12–14]

The first two methodologies are applied on the external CNT surface and can change some characteristics such as conductivity or solubility. The latter is quite interesting for producing CNT-based reservoirs but is not effective in assisting the solubilization. Solubility is really poor in all the solvents for pristine CNTs that form tight bundles difficult to disperse. Considering that many properties are typical for individual CNTs and can be attenuated in the presence of ropes, the importance of good dispersions or solution appears to be fundamental not only for biological purposes but also in the materials chemistry.

The supramolecular approach exerts great importance when CNT use is related to the preservation of the intact aromatic structure, as in the case of metallic CNTs, where the introduction of defects on the carbon surface diminished their intrinsic capability of conducting electrons. Carbon nano-tubes interact commonly with polymers and biopolymers.

6.3   APPLICATIONS OF CNTs IN BIOMEDICINE

Phenylene vinylene was the first to show stable p–p interactions with MWNTs.[15] Nakashima et al. reported the favorable interaction to CNTs of aromatic compounds such as pyrene, properly functionalized with a positive charge to induce water solubility.[16,17] The covalent functionalization approach can be performed using two different methodologies:

- amidation and esterification of oxidized SWCNTs
- sidewall functionalization[18,19]

6.4   AMIDATION AND ESTERIFICATION OF OXIDIZED SWCNTS

The treatment with strong oxidizing acids and sonication modifies the CNT surface with the introduction of carboxylic, carbonyl, and hydroxyl functions. The tips are removed and the opening of the tubes leads to the purification from metallic nanoparticles during this process. The carboxylic acids, which are the main type of induced functionalizations, can be activated by thionyl or oxalyl chloride or in milder conditions, and then allowed to react with alcohols or amines.

6.5   SIDEWALL FUNCTIONALIZATION OF SWCNTs

Many different methods have been proposed and developed for the sidewall derivatization of CNTs. Fluorination, radical, electrophilic, and nucleophilic additions, together with cycloaddition reactions have been widely used, obtaining a good variety of sidewall functionalities.

6.6   BIOLOGICAL APPLICATIONS OF CNTs

A host of applications of CNTs in the field of medicine and related areas are explored and various research communities are involved in these investigations currently. The major driving force behind biomedical investigations is the potential of CNTs to penetrate into tissues and cells. It is essential to make a mention here that apparently adverse results can come to the fore, due to the enormous structural variations among the examined materials. For example, differences at the macroscopic levels, like the application of SWCNTs or MWCNTs, kind of surface topography, and extent of clubbing, supramolecular complexation of CNTs with the help of (bio)polymers and aromatic compounds need to be reported. Further length-related values are either not taken into account or averaged out. The biological knowledge coming to the fore today is in accordance with the field of application and not the various CNT-based materials in use.

6.7   CELL PENETRATION

It is already known that CNTs can readily translocate within a broad range of cells.[20] The underlying mechanisms of this process involve phagocytosis, endocytosis, or membrane penetration, and the variations observed in the CNT-based materials used may be attributed to the various surface coatings on the backbone of CNT. Endocytosis has been presented for DNA–CNT complexes by Kam et al.[21,22] and Heller et al.,[23] while the membrane piercing has been suggested for block copolymer-coated noncovalently functionalized MWCNTs[24] and oxidized, water-soluble CNTs.[25] Liu et al. explored the use of CNTs to pierce the cell wall and membrane of plant cells (BY-2), by making use of oxidized SWCNTs noncovalently labeled with fluorescein isothiocyanate (Ox-SWCNT/FITC) or pristine SWCNT complexed with FITC-labeled ssDNA (SWCNT/ssDNA). In both cases the Cycloaddition direct arylation outcomes were quite interesting; involves CNTs to penetrate cells without causing toxicity. Cells maintained their morphology, fluidity of cytoplasm, and spreading rate, and, further the ssDNA delivered remain to absent the BY-2 cells. The profile of distribution within the cells varies: the Ox-SWCNT/FITC accumulated in the vacuoles as the SWCNT/ssDNA penetrate into the strands of cytoplasm. The internalization was dependent upon time and temperature, due to the presence of endocytosis which in plant cells is checked by the reduction in temperature.[26]

6.8   CNTs FOR TARGETED SMART DRUG DELIVERY

As discussed earlier, the possibility of using the internal cavity of CNTs to fill allows for adopting a new strategy to develop drug-carrying vehicles. Hampel et al. investigated this frontier with the help of carboplatin as a drug to fill.[27] They made use of oxidized MWCNTs possessing varying diameters (10–30 nm) and lengths (10–30 mm). The oxidation was brought about to open up the tubes. The incorporation is found to be temperature dependent. Numerous techniques like X-ray photoelectron spectroscopy the researchers confirmed the retention of a drug by CNTs, laced with some residues of carboplatin deposited externally. The biological tests conducted using bladder cancer cells observed that on incubating with carboplatin-MWCNTs cell growth inhibition takes place, while MWCNTs in isolation are redundant.

An innovative approach for the functionalization of CNTs using two different molecules with the help of the 1,3-dipolar cycloaddition of azomethine ylides was adopted by Pastorin et al., who bonded methotrexate as well as FITC with covalent linkage to MWCNTs. The piercing of this innovative derivatized material was examined using the epifluorescence and confocal microscopic analytical tool on human Jurkat T lymphocytes, suggesting a rapid deposition in the cytoplasm.[28] McDevitt et al. synthesized derivatives of CNT loaded with tumor-specific monoclonal antibodies, radiometal-ion complexes, or electrodes coated with fluorescent materials. Compounds soluble in water, produced by dipolar cycloaddition were subjected to treatment with DOTA-NCS and LC-SMCC, followed by chelation. The conjugates developed were examined in vitro with the help of flow cytometry and then cell-based immune-reactivity assays were conducted, biodistribution was carried out in vivo in rats by using xenografted lymphoma.[29]

SWCNTs coated with modified phospholipids (PL-PEG-NH$_2$) allowed to link a platinum(IV) derivative through the amino residue to one of the Pt ligands at axial position.[30] This molecule is nontoxic almost to testicular cancer cells, but substantially cytotoxic when bound to the surface of soluble SWCNTs functionalized with an amine. This is attributed to the availability of CNTs making the prodrug cellular uptake, and the complex gets confined into endosomes, at which the pH is less than in the cell incubation medium. The acidic environment promotes the reduction of Pt from oxidation state (IV) to (II) responsible for the release of active cisplatin, and elimination of axial ligands.

The same authors recently developed a new chelate in which one Pt axial ligand is connected with folic acid, as target species.[31] The synthesis of high molecular weight chelates is instrumental in increasing the blood circulation time. They explored Folate Receptor-positive [FR(+)] human choriocarcinoma (JAR) and human nasopharyngeal carcinoma (KB) cell lines, as negative control FR-negative testicular carcinoma cells (NTera-2), showing that FR(+) cells are influenced to a greater extent. Further, the so-developed prodrug was found to be far more active than cisplatin, forming, following intracellular reduction, cisplatin 1,2-d(GpG) intrastrand cross-linked with DNA. Moreover, recent studies show that doxorubicin immobilized on MWCNT surface using of p–p interaction shows far more efficiency than the free one, due to amelioration of delivery into cells.[32]

6.9   CNTs IN GENE DELIVERY

An approach first of its kind for genetic material delivery was undertaken by Lu et al. who chelated RNA polymer poly(rU) with SWCNTs with the help of interactions nonspecific in nature.[33] The translocation across MCF7 breast cancer cells was found to take place and the genetic material was embedded across the membranes both cellular and nuclear. Derivatives of SWCNTs, synthesized using amidation of the CNTs carboxylic groups with a chain possessing an ammonium group at the terminal position, were examined as telomerase reverse transcriptase RNA (TERT-siRNA) carrier into cells.[34] The performance was evaluated considering the spreading of tumor cells (Lewis lung cancer and HeLa cells), in vitro and in vivo. the targeted gene is silenced shows an improved intracellular delivery of siRNA, which gets released at the site from CNTs and can cause the necessary effect.

Dai and co-investigators carried siRNA to human T cells and peripheral blood mononuclear cells. SWCNTs for this purpose were deposited with PL-PEG-NH$_2$. Variations in efficacy were observed depending upon hydrophilicity degree and functionalization, based on the PEG chain length in the PL-PEG molecule.[35] SWCNTs, functionalized with PL-PEG using noncovalent linkages, were employed. A cleavable bond was found to be formed, connected by a disulfide bond between the PL-PEG molecule and the siRNA delivered.[36] The siRNA functionality created in this model is more potent over lipofectamine as a transfection agent to control. Oxidized SWCNTs, bearing positive charge are amenable to interact with fluorescein tagged dsDNA (dsDNA-FAM) and were deposited with phospholipids (PL-PEG-NH$_2$) modified using folic acid, provided with the alkyl chain wraps. This system possessing polyfunctionality was selectively up-taken by tumor cells, wherein folic receptors are overexpressed, forcing the oligo-nucleotide to get into the cells, as evidenced by fluorescence imaging.[37]

SWCNTs have been utilized as vehicles for ssDNA entery into cells, showing an increase in the resistance offered by the oligo-nucleotides towards nuclease digestion. The ssDNA is provided with the protection from enzyme-driven cleavage and interference from nucleic acids linking proteins, still paying its role of targeting mRNA.[38] There is an increase in the potential of CNTs as oligo-nucleotide vectors, as described by Kateb et al. They examined the use CNTs as DNA or RNA carrier systems in brain tumors, thus proving the potential of microglia to internalize MWCNTs deposited with Pluronic F108 with better efficiency as against glioma cells.

6.10   CNTs IN ANTICANCER APPLICATIONS

CNTs hold lot much of potential as a genetic material delivery system, having applications in gene therapy, their biocompatibility in general. Nanotubes readily attach macromolecules, like nucleic acids. Li et al. observed that oxidized SWCNTs resist the association of DNA duplex.[39] Further it is shown that they selectively make human telomeric i-motif DNA creation, connecting to the 50-end groove and thereby stabilizing the base pairs. These results can be made use of in designing new materials for cancer therapy, wherein changing the human telomeric DNA plays an important role, although the biological impacts of i-motif structure introduction are yet to be ascertained. The presence of apolyadenylic acid [poly(rA)] group in eukaryotic cells and human poly(rA) polymerase (PAP) is responsible for a tumor-specific target, and compounds instrumental in interfering with such structure hold large potential in cancer therapy. Zhao et al. examined oxidized SWCNTs and observed that CNTs bring about single-stranded poly(rA) to translate into duplex although the underlying mechanism is yet to be unraveled and should be ascribed to the properties of the oxidized CNTs given the inability of amino-derivatized SWCNTs, to induce single strand to duplex.[40]

Antitumor immunotherapy is almost ineffective. Pantarotto et al.[41] have observed that viral peptides bound to CNTs hasten the antipeptide immunity. Meng et al. further study to enhance immunotherapy. Tumor lysate protein, easily synthesized from various solid tumors, is bonded to MWCNTs and subsequently used as a tumor-cell vaccine in a rat model having the H_{22} liver cancer, reflecting upon the cure rate and cellular anti-tumor immunity.[42] Phospholipid–PG chains bonded to a residue of folic acid were connected with the help of hydrophobic forces to CNTs. The folate, if present, make an uptake of the chelate selectively into cells over-expressing folate receptors, which takes place in cancer cells. The action is selective due to the uptake of CNTs in cells is responsible for increasing the temperature, leading to thermal ablation on irradiation. CNTs carrying two specific monoclonal antibodies (insulin-like growth factor 1 receptor, IGF1R and human endothelial receptor 2, HER2), synthesized using supramolecular approach on using functionalized pyrene units, are used to destroy breast cancer cells with near-infrared (NIR) irradiation.[43] The major problem of NIR application is its weak tissue penetration ability, allowing for the treatment superficially of selective cancer lesions. Radio-frequency (RF) waves, on the other hand, have better penetration capacity, and their interaction with internalized CNTs generates a rise in cells temperature, causing cell death. Gannon et al. prescribed the application of RF waves but demands the direct administration of SWCNT laced with polyphenylene–ethynylene polymeric chain molecules into the tumor.[44]

Another approach employed in cancer therapy is concerned with angiogenesis. It refers to the inhibition by making interference in the growth parameter balance prohibit the cause of diseases, including tumors. A sound model to explore angiogenesis is the chick chorioallantoic membrane (CAM). A work carried out by Murugesan et al. used carbon-based materials (CNT, fullerene, and graphite) to check their suitability to resist FGF2- or VEGF-induced vessel development in the CAM approach. These materials exhibited improved performance in the induced angiogenesis, with better efficiency as against VEGF, with no effect on the basal process when the growth factors are absent. The underlying mechanism in this process is yet to be elicited although it is a fact that this antiangiogenic effect can be used to advantage in cancer treatment.[45]

6.11   NEURON INTERACTIONS WITH CNTs

Lovat et al. reported that the neural signal transfer can be improved by bringing about the growth of neuronal circuits on CNTs as substrates, by taking the advantage of better electrical conductivity of CNTs for the cause of this behavior.[46] Further, the same researchers designed an integrated SWCNT–neuron circuit to check whether electrical activity generated using SWCNTs can cause neuronal signaling. For example, the patch-clamped hippocampal cells, grown on SWCNT substrates, have exhibited stimulation with the help of the SWCNT layered substrates. The resistive coupling between biomembranes and SWCNTs was different from a coupling between SWCNTs and the patch pipette qualitatively via the patch seal way to earth.[47] The single-cell electrophysiology, electron microscopy techniques, and modeling, the same investigators showed that CNTs exhibit an improved response to neurons, because of the creation of close interfacing between CNTs and the cell membranes. However, this can result in electrical shortcuts between the proximal and distal parts of the neuron.[48]

The CNTs deposited on glass surface slips for cells grown, and therefore another approach has been used by Keefer et al. wherein CNTs are coated electrochemically on tungsten and stainless steel wire probes. This kind of treatment brought about improvement in recording and electrical stimulation of neuronal cultures.[49] Kotov and coinvestigators produced a layered composite using SWCNTs and laminin, which is a glycoprotein found in extracellular matrix of the human being. The application of such surface for cellular culture generated neural stem cells (NSCs) distinction and the formation of functional neural network. Further, passing the current through CNTs enhanced the action potentials.[50] CNTs can also be aligned vertically to the gate insulator of an ion-sensitive field-effect transistor so as to make it work as electrical interfaces to neurons, to examine the interactions, and to exacerbate the efficiency which can be recorded in terms of improved neuronal electrical activity.[51]

Water-soluble SWCNTs graftings with PEG have been in use in the culture media for bringing about neural growth. The derivatives of CNTs showed that they have the ability to inhibit the stimulated membrane endocytosis in neurons which justifies the extension of neurite length,[52] while MWCNTs laced with Pluronic F127 (PF127) administered to mouse cerebral cortex won't induce degeneration of the neurons in the proximity of the injection site, although PF127 can induce apoptosis by itself of rat primary cortical neurons, meaning that the MWCNTs inhibit PF127-induced apoptosis.[53] All such interesting studies done on neuronal growth and stimulation have shown a ray of hope of using CNTs substrates for designing neural damaged tissues in the near future, which can be used in addressing spinal cord disease and restoration of neuro-degeneration.

6.12 ANTIOXIDANT BEHAVIOR OF CNTs

CNTs, on instilling in lungs, produced an inflammatory and fibrotic response, which was attributed to oxidative stress caused by free radicals, although there is no concrete evidence found about ROS production from CNTs. Further, Fenoglio et al. showed that MWCNTs won't produce oxygen or free radicals when oxidized using H_2O_2 or formate; however, when radicals were generated, CNTs work as scavengers of radicals.[54] In the recent study the antioxidant behavior of CNTs has been suggested by Tour and coinvestigators. SWCNTs and tiny SWCNTs on derivatization with butylated hydroxytoluene (BHT), by applying two distinct approaches: one involves covalently bonded triazene to the walls of pluronic-wound SWCNT and the other involves amidation of carboxylic groups of ultra-short SWCNT derivatives. The ROS scavenging capacity of various compounds when studied led to quite exciting outcomes, although the different compounds show varied reactivity. For example, the pluronic-coated SWCNTs exhibited better efficiency than the derivatives obtained with BHT, while in oxidized US-SWCNTs, more is the degree of loading of residues of BHT, more the antioxidant behavior. This innovative result has paved the way for SWCNT to be used as novel medical therapeutics on the antioxidant frontier.[55] However, it is essential to keep in mind that latest study showed results in contradiction indicating MWCNT to be toxic in A549 cells, which can be decreased using prior treatment with antioxidants to reduce ROS formation and expression of interleukin-8 gene.[56]

6.13 USE OF CNTs IN IMAGING

SWCNTs are characterized by fluorescent properties that can be used to advantage in vitro as well as in vivo imaging, so as to estimate the uptake of CNTs in macrophages,[57] the CNT removal in rabbits,[58] and their toxic potential in fruit fly larvae.[59] The astonishing application of CNTs for imaging is made possible by preserving selective properties of CNTs, given at least 100 nm of them are unmodified and are not in bundles, as under such conditions the quenching of fluorescence takes place. Welsher et al. made use of semiconducting SWCNTs covered with PEG as NIR fluorescent tags. The PEG was conjugated to antibodies in order to recognize selectively CD_{20} cell receptors at the surface on B cells or HER2/

neuron positive breast cancer cells. The innate NIR photoluminescence made it possible the detection of the linking to the cells, although a low auto-fluorescence is present in different cells.[60]

MRI is an all-important imaging technique employed in medicine. After the advent of the excellent applications of Gd@C60, Gd@C82, and their derivativatized products in the form of contrast agents,[61-63] the use of CNT to has been done analogously by Wilson and co-investigators.[64] The CNTs, ultrashortened using fluorination to the size 20–80 nm, loaded with GdCl$_3$ were developed. The outcomes showed that CNTs have a relaxivity (r1) higher by 40 times than any present Gd^{3+} ion-based clinical agents. The aggregation, in fullerene endohedrals, although brings about change in the relaxivity, though not so by CNTs. Compounds with more solubility have been synthesized by Ashcroft et al.[65] They showed the improvement in the stability of gadonanotubes when placed in buffer solutions, serum, on varying pH and temperature. Moreover, they found an effect of pH on r1, with a great degree of sensitivity within a pH range of 7.0–7.4. This hints at the possibility of using the CNTs for the detection of cancer cells in the earlier stage wherein the pH reduces dramatically.[66]

Innovative work on the biodistribution and the effect of SWCNTs (raw and super-purified, raw-SWCNT and SP-SWCNT) on in vivo exposure has been reported very recently. The combination of ^{3}He and ^{1}H magnetic resonance imaging (MRI) has been in use in a mouse model. Hyperpolarized gases like ^{3}He working as a contrast agent get fast into lungs, allowing the estimation of ventilated airways and alveolar spaces. Having metal impurity in the raw-SWCNTs was responsible for inducing a decline in magnetic field uniformity detected in ^{3}He MRI, although not much change was detected after SP-SWCNT exposure.[67] ^{1}H MRI was used to trace raw-SWCNTs on the administration of the intravenous injection and observed them in spleen and kidneys. The metal nanoparticles if absent in the SP-SWCNT exclude signal variations. When this technique was employed to estimate the role of CNT on pulmonary instillation, no signal modifications either in the liver, spleen, and kidney were observed, meaning the deprivation of systemic circulation of CNTs on inhalation. These findings were also confirmed by the histological study, confirming the possibility of adopting noninvasive technique to detect the availability of CNT, if it is in conjunction with iron as a metal impurity.

A distinct approach has been brought to the notice by Richard et al. The researchers, rather than binding the CNTs with the Gd and its derivative, made use of amphiphilic gadolinium(III) complex, coated on the MWCNT wall, in varying concentrations. The resulting suspension was stable and r1 was determined in numerous conditions, confirming a reliance on the Gd-complex concentration. On the other hand, the transversal water proton relaxivity (T2) was not however found to be dependent on Gd concentration and frequency.[68]

6.14   OTHER APPLICATIONS OF CNTs

CNTs are excellent sorbents for a host of materials attributed to their large surface areas (BET of about 600 m^2/g, based on the type of CNT under consideration). They have been in use for remediation of pollutants in water treatment[69,70] and dioxin.[71] Chen et al. worked on the sorption of Am and Th on CNT, their kinetics and effect of pH[72,73] but exhibit variation in efficiency ascribed to varying degrees of pretreatment of CNTs, observed by other researchers,[74] and explored the possibility of using this novel allotrope of carbon in nuclear waste management. Both SWCNTs and MWCNTs were also employed the removal of bacteria, such as *Streptococcus mutans*. CNT holds on to bacteria resulting in the formation of a turbid suspension, which demonstrates that the bacteria adhere to CNTs, depending upon the CNT morphology.[75]

6.15   TOXIC POTENTIAL OF CNTs

There has been a lot of discussion and many concerns have been raised about the toxic potential of CNT however the information obtained till now is inadequate and contradictory at times pertaining to a broad range of CNT derivatives exhibiting numerous properties. In general, it is considered that toxicity reduces commensurate with the decrease in the degree of dispersion and/or solubility, but a lot of work is yet to be done in this field.[76] A systematic work on the impact of structural defects present in the MWCNT carbon cages on lung toxicity has been undertaken by Fenoglio et al., who studied CNTs with

- structural defects generated
- Functionalization of CNTs with reduced oxygen-containing functionalities and metallic oxides

- a pure annealed sample
- a carbon framework deprived of metal impurities and with generated structural defects.[77]

The hydroxyl radical transportation behavior of modified CNTs was studied by EPR showed that the pulverized material showed a better activity towards hydroxyl radicals, which remained absent in the refined annealed sample, but was again surfaced in carbon framework without metal but with generated structural defects. In vitro experiments on mouse lung epithelial cells were carried out to evaluate the genotoxicity of various CNTs. Acute pulmonary toxicity and the genotoxicity of CNTs were decreased on heating but recovered on grinding, showing that the innate toxicity of CNTs is primarily influenced by the defects in the CNT carbon framework.[78] The scavenging activity is associated with the presence of defects and further with the genotoxic and inflammatory potential of CNT.

Tabet et al. dispersed MWCNT in dipalmitoyl lecithin, ethanol, and saline which is phosphate-buffered (PBS), to study CNTs toxicity on human epithelial cell line A549. The PBS makes the formation of agglomeration above the cells, but in general, MWCNTs check the cellular metabolism without affecting the permeability of the cell membrane or apoptosis, on the other hand asbestos fibers increase apoptosis by permeating into cells.[79] Wick et al. studied CNTs with varying degrees of agglomeration to estimate their cytotoxic potential on human MSTO-211H cells. CNTs on dispersing well in polyoxyethylene sorbitan monooleate were found to be less toxic as compared to asbestos, forming agglomerates that are like ropes furthered the cytotoxic potential as against asbestos fibers.[80]

Guo et al. studied the conversion of MWCNTs using glucosamine, decylamine, and irradiating it with c-ray into g-MWCNT and d-MWCNT, and the same investigators used these derivatives to conduct cytotoxicity study on *Tetrahymena pyriformis*. The decylamine derivative shows growth inhibition that is dependent on dose and is ascribed to the amine toxicity, carried into cells by the CNTs. The distinction between pure MWCNTs and g-MWCNTs showed an impetus to growth by the latter. The hydrophilic material forms complex with peptone, available in the medium, and further to carry it into cells. This implies that interactions that are nonspecific between CNTs and compounds in the culture medium have to be taken into account in assessing the cytotoxic potential of CNTs.[81] Given that very little is known about the effect of CNT on naturally occurring ecosystems, the Tetrahymena thermophila which is a ciliated protozoan has been examined due to its role in maintaining microbial populations by way of ingestion and digestion of bacteria. It is a very vital organism in the treatment of wastewater which is an indicator of the quality of sewage. SWCNTs have been observed in such microorganisms, making protozoa aggregate thereby hampering their ability to interact with bacteria.[82] *Salmonella typhimurium* and *Escherichia coli* strains were employed to conduct mutagenicity tests by using MWCNTs. The mutagenicity was missing, even on using the metabolic activating agents.[83]

Numerous tests carried out to measure the genotoxicity of MWCNTs have been conducted both in vitro and in vivo in mice on lung epithelial cells or human epithelial cells. Three days after injecting CNTs intratracheally, type II pneumocytes increase based on the dose. The MWCNTs induce clastogenic and aneugenic activities (involves rise in the rate of mutation because of DNA breakdown and loss or gain of chromosomes.[84] The clastogenicity at times is dependent upon ROS formation and the availability of metal nanoparticles such as Fe or Co; however, the MWCNT antioxidant activity demonstrates various routes and suggests that these systems need to be explored in better detail.

6.16 STRUCTURE, CHARACTERISTICS, AND FUNCTIONALIZATION OF SWCNHs

Single-walled carbon nanohorns (SWCNHs) were first observed by Iijima in 1999, during CO_2 laser ablation of graphite at room temperature without a metal catalyst.[85] SWCNHs are spherical aggregates (diameter 80–100 nm) of graphitic tubes. These tubular structures mainly consist of tubes that have closed ends with cone-shaped caps (horns) with an average cone angle of 120 nm. Each individual tube has a diameter of 2–3 nm and an average length in the range of 30–50 nm.[85] Due to their peculiar geometry, reminiscent of a sponge, carbon nanohorns can be exploited for their capacity to adsorb most types of molecules.[86] SWCNH's are soluble abysmally low in organic solvents and insoluble in water. A prerequisite for biomedical application is the increase in the solubility of such material. So, it has to be stressed that functionalization of carbon nanohorns is obligatory for rendering this material biocompatible. Indeed, carbon nanohorns behave like carbon nanotubes and can be dispersed or solubilized into physiological or water solutions, provided they are modified at their surface with suitable functional groups. Contrary to the carbon nanotube, the reaction with SWCNHs proceeds well due to better dispersion of nanohorns in the examined solvents because the presence of rough surface structures of nanohorn aggregates prohibits the increase in the aggregate–aggregate contact area, leading to their weak interactions via van der Waals forces.

SWCNHs were fluorinated in the temperature range of 303–473° K without destroying the nanohorns.[87] The nature of C–F bonds in F-SWCNHs changed from ionic to covalent with an increased temperature during fluorination. After fluorination at 473 K, several nanoscale windows were produced on the walls of SWCNHs. Tagmatarchis et al. reported organic covalent functionalization on the sidewalls of SWCNHs, via the 1,3-dipolar cycloaddition of azomethine ylides generated in situ upon thermal condensation of different aldehydes and a-aminoacids. The functionalized nanohorns (1–2) were soluble in several organic solvents and were characterized by NMR, UV/Vis-NIR, IR, TEM, and EDX spectroscopy.[88] Prato and coworkers also reported similar 1,3-dipolar cycloaddition of azomethine ylides on SWCNH. Functionalized nanohorns (1 and 3) were soluble in DMF, chloroform, and dichloromethane but insoluble in THF, diethyl ether, and methanol.[89] Tagmatarchis and coworkers also reported the functionalization procedure of SWCNH that is based on the opening of their conical and highly strained end on treatment with molecular O_2 at 550°C. The as-generated carboxylic groups terminated nanohorns, converted to the corresponding acyl chlorides, and reacted with a variety of amines, alcohols, and thiols, possesses short and long hydrophobic alkyl chains, polar oligoethylene units, aromatic chromophores such as pyrene or anthracene groups to give corresponding SWCNH-based amide (4), ester (5), and thioester (6) material.[90]

Recently, they also reported covalent functionalization of SWCNHs utilizing in situ generated aryl diazonium compounds. Due to the grafting of aryl moieties on the skeleton of SWCNH, the solubility is significantly enhanced. Water-soluble cationic aryl ammonium functionalized SWCNHs are expected to be useful in biotechnological applications such as drugs and gene delivery systems.[91] Recently, we also reported the organic functionalization of CNHs by direct addition of diamine moiety to pristine CNHs that can be further modified with a fluorescent probe (8). The resulting functionalized SWCNHs were biocompatible and were evaluated on the basis of their capacity to be integrated into a cellular system.[92] In a noncovalent modification approach, amphiphilic agents (i.e., molecules having both hydrophobic and hydrophilic moieties) have been used as surfactants to improve the solubility of SWCNHs.

6.17 BIOMEDICAL APPLICATIONS OF CNHs

SWCNHs, with their unique flower-like structure, are promising substrates for supporting materials for drug delivery systems and radiographic contrast agents. Although SWCNHs are one of the most attractive nanomaterials, their biomedical applications have not been satisfactorily achieved.[93]

6.18 CNH AS A LASER-INDUCED THERAPEUTIC AGENT

Dai et al. have described a new cancer therapy, which utilizes carbon nanomaterial, Fe/Co graphitic cell nanocrystal, and near-infrared laser irradiation. Biological systems are transparent to the NIR light region (700–1100 nm). Carbon nanomaterial, when exposed to NIR radiation shows NIR-laser driven exothermy because optically stimulated electronic excitation of SWCNHs is transferred rapidly to molecular vibration energies causing heating in a similar way to photo-induced CNTs. For selective elimination of microorganisms, molecular recognition element (MRE) [wheat germ agglutinin (WGA) for S. Cerevisiae as well as anti-lipo polysaccharide core antibody (the core of LPS) for E. coli], noncovalently linked to SWCNHs, has recently been synthesized and studied by Miyako et al.[94] They used SWCNH-COOH functionalized noncovalently with polyethyleneglycol carbamyl diasteroyl phosphatidiylethanol amine (PEG-PL) complex to attach noncovalently, respective MRE units.[95]

The MRE-PEG-PL-CNH complex binds specifically to the microbial surface and after irradiation of CNH–microbe complex using NIR laser, microbial cells were destroyed due to NIR laser-triggered exothermy of the CNH complexes. Similarly, for the removal of the virus, they reported a complex containing T7 tag antibody noncovalently linked with SWCNH-PEG-PL. This complex binds with the model T7 bacteriophage virus and on interaction with NIR light, the T7 phage capsid protein becomes thermally denatured due to the photo-exthermocity of the NIR laser-driven CNH complex albumine (ZnPC-oxSWCNH-BSA) hybrid molecule-centric cancer phototherapy which makes use of a single laser. In this system, ZnPc acts as photodynamic therapy (PDT) agent and SWCNHox as the photohyperthermia (PHT) agent. When ZnPC-oxSWCNH-BSA complex was administered into tumors which were transplanted subcutaneously into rats, followed by irradiation with 670 nm laser brought about the disappearance of tumor, while on injecting either ZnPc or SWCNHox– BSA, the tumor continued to grow.[96]

6.19   CNHs FOR DRUG DELIVERY

SWCNHs with a diameter of 80–100 nm are particularly suitable as drug carriers to the tumor tissue because their size fulfills the condition for achieving the enhanced permeability and retention (EPR) effect. They can permeate through the damaged vessels in tumor tissue and remain there because of little lymphatic drainage. In addition to an extensive surface area, carbon nanohorns have a high number of interstices, which allow the adsorption of a large amount of guest molecules. As single-walled carbon nanohorns have tubes at the end of spherical aggregates with a diameter 2–5 nm, holes can be generated at the tips of the tubes and can be exploited to insert different therapeutic agents into their empty space. Finally, SWCNHs do not exhibit cytotoxicity, making them potentially applicable in drug delivery systems.

Functionalized CNHs have been suggested for the controlled targeted drug delivery of anti-inflammatory and anticancer agents including dexamethasone (DEX), doxorubicin, and cisplatin. Shiba and coworkers reported the binding and release of the anti-inflammatory glucocorticoid dexamethasone by oxSWCNHs.[97] DEX–ox SWCNH complex brings about release of DEX into PBS (pH 7.4) in a sustained manner at 37°C. Carbon-Based Nanomaterial Applications in Biomedicine more rapid release in the culture medium. DEX–oxSWCNHs activated GR-mediated transcription in mouse bone marrow stromal ST2 cells and inducted alkaline phosphatase in mouse osteoblastic MC3T3-E1 cells. The GR-mediated transcription activation obtained with 0.1 mg/mL DEX–oxSWCNHs was comparable with that obtained with 0.01–0.05 mM free DEX.

The same group also reported noncovalent modification of ox SWCNHs using amide-linked polyethyleneglycol–doxorubicin (PEG–DXR) conjugates that results in a good dispersion.[98] Although the preparation of the conjugates required the use of organic solvents, which are not compatible with biological moieties (such as cell culture) but can eventually be eliminated using chromatographic separation, equilibrated with water. It has been demonstrated that carbon nanohorns adsorb PEG–doxorubicin via the doxo-rubicin moiety. The complexes, which have a diameter of 160 nm, contain more than 250 mg of PEG–doxorubicin per gram of nanohorns. Preliminary in vitro tests have shown that complexes exhibit DXR-dependent apoptotic activity against cancer cells. However, the incidence of apoptosis obtained with 0.2 mg/mL PEG–DXR–oxSWCNHs was lower than that obtained with 10 ng/mL DXR. It is probable that PEG–doxorubicin is retained on the surface of the nanohorns, thus reducing its therapeutic effect. The authors could not exclude the possibility that some amount of free doxorubicin remained in their preparation and was responsible for the apoptotic activity.

Noncovalent/covalent functionalization and adsorption of different therapeutic agents on/in carbon nanohorns for biomedical applications. The right end shows the release of C_{60} molecule (as a model drug) on exposure to NIR laser; adsorption of Fe and Gd nanoparticles for magnetic resonance imaging; adsorption of drugs (CDDP and DXR) molecules, whereas the lower left end showing the solubilization of SWCNHs with PEG through noncovalent functionalization. Different therapeutic agents like drugs, antibodies, and MRE (molecular recognition element for microbes and yeast) can be covalently linked with PEG before being mixed with SWCNHs and the upper left end showing the covalent attachment of fluorescent moieties for fluorescence imaging in the cells.

In another approach by the same group, cisplatin, cis-$(NH_3)_2PtCl_2$ (CDDP), an anticancer drug, was incorporated inside SWCNHs. Carbon nanohorns do not alter the structure of the anticancer agent, which was slowly released in an aqueous solution.[99] Following the liberation of the drug, the cell viability of human lung cancer cells was monitored for 48 h. The anticancer behavior of the CNH's having cisplatin was resembling with the drug alone, while nanohorns used as controls presented no cyto-toxic effects. Although carbon nanohorns can be easily dispersed in water, they have been shown to aggregate by their tendency to form clusters in the highly ionic and protein-rich cell culture media. The presence of aggregates in the micrometer scale formed by both oxidized and cisplatin-containing nanohorns is a major concern with this approach that will need to be overcome to achieve in vivo applications. Further, authors also investigated the effect of the hole-edge structure of the carbon nanohorns on cisplatin incorporation and release. They found that 70% of the cisplatin was released from SWCNHs having holes with hydrogen-terminated edges (NHh) and only 15% was released from SWCNHs having holes with oxygen-containing functional groups at the hole edge (NHox), on immersing in phosphate-buffered saline.[100] Although SWCNHs having holes with hydrogen-terminated edges show high release, unfortunately, the hydrophobic property of an NH hinders the dispersion in PBS and thus cannot be used for the biological applications, whereas the release of CDDP from inside NHox in PBS was considerably suppressed because sodium ions in PBS replaced hydrogen of the oxygen-containing functional group at the hole edge, which results in the plugging of the holes, hindering the CDDP release. For CDDP incorporating NHox to be used as a CDDP-releasing carrier in vivo, the hole-size enlargement and control of the number of the functional groups at the hole edges of NHox might be effective in producing the necessary CDDP releasing quantities at a slower rate. So, optimization of hole opening of SWCNHs, by heat treatment from room temperature to 500°C–550°C, in flowing dry air, with an increased rate of 1°C/ min, results in an 80% release of CDDP in PBS that greatly improved from the previous value of 15%.[101]

Recently, the same authors have devised an alternative methodology to maintain well-dispersed nanohorns in physiological conditions or cell culture media.[102] They used peptide aptamer that specifically binds to the surface of the nanohorns as the SWCNH-binding block (NHBP-1) and synthesized a PEG-peptide aptamer conjugate for the dispersal of the SWCNHs. It was observed that modifications of the surfaces of the oxSWCNHs by PEG-NHBP do not interfere with the ability to load and release the cisplatin drug. In addition, this complex cisplatin@ox SWCNH-PEGNHBP shows good dispersion in both PBS and culture medium and exerted a potent cytotoxic effect against cancer cells. This is probably the approach to be followed to avoid the incapacitating aggregation phenomena described previously. Carbon nanohorns modified with Gum Arabic can also be used for intracellular delivery because several biological cargoes could be linked with Gum Arabic.[103] Zhang et al. covalently attached rhodamine B to the polypeptide chain of Gum Arabic and through in vitro studies confirmed that modified SWCNHs can readily enter the cell, probably through the endocytotic pathway and were nontoxic.

Alternatively, carbon nanohorns have been modified with magnetite and administered in vivo for MRI applications. Miyawaki et al. reported a simple method for attaching Fe_3O_4 nanoparticles (6 nm in diameter) to hole opened SWCNHs,[104] by depositing first $Fe(OAc)_2$ on oxSWCNHs that thermally decompose at 400°C to yield superparamagnetic magnetite Fe_3O_4 strongly attached to oxSWCNHs. The attached Fe_3O_4 (magnetite) nanoparticles caused significant darkening of the MR images that made it possible in vivo visualization of the cumulative activity of SWCNHs in the spleen and kidneys. Similarly, the same group reported that Gd-acetate clusters inside SWCNHs can be transformed into ultrafine Gd_2O_3 nanoparticles of 2–3 nm in average diameter and they retain their particle size even after treatment at 700°C. Gd_2O_3 nanoparticles thus obtained can be useful to magnetic resonance imaging.[105]

6.20 TOXICITY OF CNHs

Miyawaki et al. carried out an extensive in vivo and in vitro toxicological assessment of the as-grown SWCNHs for various exposure pathways, showing that as-grown SWCNHs have low toxicities. SWCNHs were found to be nonirritant and non-dermal sensitizers through skin primary and conjunctival irritation test and skin sensitization test. They are not carcinogenic. The acute peroral toxicity of SWCNHs was found to be quite low. Intratracheal instillation test revealed that SWCNHs rarely damaged rat lung tissue for a 90-day test period.[106] Isobe et al. reported the aminative de-agglomerization of SWCNH agglomerates resulted in the formation of a homogenous aqueous solution of primary particles of uniform size distribution. The amino group-containing SWCNHs is taken up by mammalian cells but does not show significant cytotoxicity.

In fact, the cytotoxicity is much lower than that of the quartz micropar-ticles.[107] Alternatively, the organic functionalization of SWCNHs permits to render such material soluble in physiological conditions. Pristine SWCNHs modified with a diamine moiety further functionalized with a fluorescent probe resulted biocompatible. Such functionalized SWCNHs were phagocytozed with the help of primary murine macrophages without influencing cell sustenance. However, some signs of activation, which could lead to an inflammatory status in vivo, were visible, such as a clear oxidative burst and IL6 production. The uptake efficiency of these kinds of carbon nanoparticles can be used to advantage if an approach that target macrophages is adopted. The delivery of biologically active agents using functionalized SWCNHs, which internalize into cells and contemporarily promote the activation of the immune system, is a very promising concept in vaccine development.

6.21 CARBON NANODIAMONDS

Nanodiamonds (NDs) have been already described in the 1960s,[108] but in recent years, they have been widely investigated for their interesting properties. Diamonds are commonly known as stable and inert materials.[109] They are very difficult to manipulate being practically insoluble in any solvent. As a consequence, it was very difficult to imagine their use in nanomedicine. However recent findings have shown that if the dimensions of diamonds are reduced to the level of nanometers or microns, they can be treated as molecules that can be eventually functionalized at the surface.[110–112] This way solubility can be increased and maneuvering can be facilitated significantly. Shock wave transformation of graphite into sintered nanodiamond,[113] or detonations of certain explosives in a closed container[114] are synthetic methods for the production of nanodiamond besides commonly used CVD method.[115] Most of the applications are based on polycrystalline diamond films or diamond particles with a relatively large grain size produced by CVD, whereas, nanodiamond, powder-based particles with a much smaller size (<10 nm in diameter) can be readily generated by the detonation technique.

Nanodiamond can be functionalized with different chemical addends[116] like diazonium salts, azo-perfluoro alkyl group, fluorination, chlorination/amination, silylation, and radical reaction with acyloxy, -aryl, -alkyl peroxides to name a few. Oxidized nanodiamond films (containing hydroxyl and carboxyl groups) can be also used for the covalent attachment of bioactive moieties, whereas, organic and biological molecules, such as apo-obelin and luciferase,[117] cytochrome C,[118] lysozyme[119] can also be noncovalently immobilized on the oxidized nanodiamond surface because oxygen-containing surface groups participate in hydrogen bonding and other interaction with the adsorbed species.

Nanodiamonds can be useful in a variety of biological applications such as carriers of drugs, genes or proteins, novel imaging probes, coatings for implantable materials, biosensors, and biomedical nanorobots. Moreover, the high adsorption capacity of nanodiamonds makes them good candidates for medical application.

They can be used in protein chemistry as new adsorbents for the effective extraction and purification of proteins. Due to the presence of hydrophobic and hydrophilic interaction, nanodiamonds possess a high affinity for the proteins and, thus, can be used to capture the protein in the dilute solution that can be analyzed by MALDI-time of flight (TOF) mass spectrometry. The promise of this method for clinical proteomics research is demonstrated with an application to human blood serum.[120] It was demonstrated that the immobilization of the protein on the nanodiamonds did not alter its stability and confirmation. Huang et al.[121] reported the immobilization of anti-Salmonella and anti-Staphylococcus aureus antibodies on hydrophobic and hydrophilic nanodiamond.

6.22   CARBON NANODIAMOND AS DELIVERY VEHICLE

Huang et al.[122] described the adsorption of doxorubicin hydrochloride (DXR) on nanodiamond (2–8 nm diameter) that was highly functionalized with hydroxyl and carboxylic groups. Noncovalent complexes were formed with the addition of NaCl, whereas reversible release of the drug from the nanodiamonds was achieved by reducing the concentration of chloride ions (salt effect). The nanoparticles were able to enter into cells alone or complexed to DXR. To prove the capacity of these nanoobjects to pass the cell membrane, nanodiamonds were coated with a fluorescent polylysine derivative and then localized inside the cytoplasm. The same nanodiamonds were highly biocompatible, as demonstrated by the fact that cell viability was not reduced. The complexes with DXR were uptaken and apoptosis was assessed as a consequence of the liberation of the drug from the complex. The effects of doxorubicin-induced cell death were tested in comparison to the drug alone. Nanodiamonds sequestered DXR for a longer time, decreasing the efficacy compared to the drug alone, but they were proposed as an alternative technology for a delayed and time-controlled drug release, prolonging efficacy during the treatment.

Kossovsky et al.[123] reported surface-modified diamond nanoparticles that can be used as effective antigen delivery vehicles. The particles consist of the diamond substrate, coated with a carbohydrate film and the mussel adhesive protein (MAP). It is believed that diamond nanoparticles provide conformational stabilization as well as a high degree of surface exposure to protein antigens resulting in strong and specific antibody response in rabbits. Liu et al.[124] proposed a model of the visual system by carboxylated nanodiamond (cND)-conjugated alpha bungarotoxin (a-BTX), a neurotoxin that preserves the physiological activity of blocking the function of a7-nAChR (alpha7-nicotinic acetylcholine receptor) while targeting cells. It has been shown that a7-nAChRs located on the cellular membrane-mediated positively charged ions influx. When cND–a-BTX complexes were used, they bound to a7-nAChR, located on the cell membrane of oocytes and human lung A549 cancer cells, and blocked choline evoked a7-nAChR-mediated downstream signaling pathway.

Puzyr et al. studied the use of detonation nanodiamond particles along with light emitting-obelin and luciferase protein, for fabricating a luminescence biochip. The former bioluminescent system can be used for determining Ca^{2+} in the biological liquids, whereas the later system exhibits catalytic activity both in the bioenzyme reaction and with photoreduced flavin mononucleotide.

6.23   CARBON NANODIAMOND AS BIOMARKER FOR CELLULAR IMAGING

NDs can also be doped with other atoms or can be modified by inducing defects and holes into their structure to render them fluorescent and are therefore extremely useful as cellular biomarkers for imaging purposes. Indeed, we can exploit the fluorescence properties of the nanodiamonds functionalized at their surface with specific ligands to target cancer cells with exceptionally high sensitivity in the detection, which is fundamental for early tumor diagnosis. These NDs can exhibit two types of fluorescence:

- bright red fluorescence
- natural green fluorescence

Nanodiamonds can become fluorescent either by noncovalent/covalent functionalization and adsorption of different therapeutic agents on/in carbon nanodiamonds and can be used for biomedical applications. The right part shows the adsorption of various proteins and antibodies, whereas, drugs or antigens can also be noncovalently attached to nanodiamonds. The left end shows that doping with other atoms or the presence of nitrogen impurities renders the nanodiamond fluorescent and emits at 700 nm on irradiation with light between 510 and 560 nm, and can be used as a biomarker for cellular imaging. A luminescent biochip can also be made by attaching proteins molecule, noncovalently to nanodiamond.

Attachment of fluorescent tags or noncovalent adsorption of fluorescently labeled bioactive molecules. The most common defect is the presence of a negatively charged nitrogen-vacancy center in the nanodiamond structure that can be created by irradiation of NDs having nitrogen impurities with beams of the proton (3 MeV) or electron (2 MeV), followed by thermal annealing at 800°C. This defect center strongly adsorbs at 560 nm and emits fluorescence at 700 nm. Since the nitrogen atom is confined to an inert matrix, photobleaching is greatly reduced if not completely eliminated, and no fluorescence blinking takes place, thus rendering such nanodiamonds that are very useful as biomarkers for imaging.[125,126] The fluorescence quantum efficiency is 1, with a lifetime of 11.6 ns at room temperature. In a 10 nm size nanodiamond roughly 100 defects can be deposited without a very high energy transfer among them. Chang et al.[127] reported that fluorescent nanodiamond can also be produced by irradiation of type Ib diamond powders using 40 keV He ions. The advantage of He ions instead of proton includes first chemical inertness of He and second, 40 keV of He can produce 40 vacancies in comparison to only 13 vacancies using 3 MeV proton. The fluorescence is sufficiently bright and stable, suggesting fluoro-nanodiamonds as an ideal probe for long-term tracking and imaging. Zhao et al.[128] prepared fluorescent carbon nanocrystals by electrooxidation of graphite in an aqueous solution, which possess no photobleaching and low cytotoxicity and thus can be used in biological labeling, imaging, and disease diagnosis. Borjanovic et al.[129] described that nanodiamonds when incorporated into poly(dimethyl-siloxanes) matrix, show much stronger and more stable photoluminescence when irradiated with proton flux.

The maximum emission occurred at 520 nm on excitation at 425 nm. Another defect is the H3 defect center (N-V-N) of a type Ia diamond that produces green emission (531 nm).[126] Aharonovich et al.[130] studied the formation of Ni-related color center in CVD-grown NDs by Ni implantation into the substrate prior to the growth. The photoluminescence peak appeared at >800 nm, proved to be a single photon emitter and did not interfere with other photoluminescence peaks arising from other defects. Chang et al.[131] reported the preparation of fluorescent magnetic nano-diamond (FMND) for cellular imaging. These FMND possess a water solubility of 2.1 mg/mL.

The authors found that these water-soluble FMNDs easily entered into HeLa cells via nonreceptor-mediated endocytosis and remained in the cytoplasm without entering into the nucleus and showing low cytotoxicity toward HeLa cells. Similarly, Faklaris et al.[132] studied the internalization of NDs in HeLa cancer cells and found that they mainly localize into the extra-endosomal region thus, providing a positive indication of their suitability for application in drug delivery. Vial et al.[133] analyzed the grafting of fluorescently labeled thiolated peptides on to NDs, which were earlier deposited either by silanization or by polyelectrolyte layer. These modified NDs were nontoxic and readily enter the mammalian cells. Besides fluorescence, nanodiamond unique spectroscopic Raman signal from sp^3 carbon structure could also be used as a biomarker for studying the molecular interaction at the single-cell level via confocal Raman mapping. In recent reports, growth hormone receptor molecules in A549 human lung epithelial cells were observed by nanodiamond labeling. When cND alone interacts with cells, the cND Raman signals overlap with the cells suggesting that cND resides near the nucleus of the cell. However, when nanodiamonds were attached to growth hormone, the whole complex resides only on the surface of the cell, suggesting the presence of growth hormone receptors at the cell surface.[134] Chao et al.[135] described protein targeted cell interaction by first attaching the lysozyme to a carboxylated nanodiamond via physical adsorption using electrostatic interaction and then interaction of cND–lysozyme complex with E. coli. The intrinsic Raman signal from the cND–lysozyme complex was detected and made use of as a marker to identify the position of the protein that interacts.

6.24 BIOCOMPATIBILITY AND TOXICITY

Schrand et al.[136] found that NDs were more biocompatible than carbon fiber, and pristine, carbon nanotubes. They studied neuroblastoma and alveolar macrophage cell lines, at concentrations ranging from 25 to 100 mg/mL for 24 h, and found that NDs do not disturb the mitochondrial membrane and do not generate ROS, while carbon nanotubes can cause membrane leakage and generate ROS. Further, the same authors reported that nanodiamond (diameter 2–10 nm), with or without surface modifications by acid or base, are biocompatible with a variety of cells including neuroblastoma, macrophages, keratinocytes, and PC-12 cells at concentrations ranging between 5 and 100 mg/mL at pH between 7.2 and 7.6. Other studies confirmed that cells can grow on the nanodiamond-coated substrate without undergoing morphological changes.[137] Liu et al.[138] independently studied the biocompatibility of the nanodiamonds and carbon nanotubes on human lung A549 epithelial cells and HFL-1 normal fibroblasts. The ND particles, accumulated in A549 cells, were observed by AFM and confocal spectroscopy. Treatment of CND particles of size 5 or 100 nm in diameter with a concentration range of 0.1–100 mg/mL did not reduce cell viability or alter the protein expression profile in these cells whereas carboxylated carbon nanotubes induced cytotoxicity to these cells. Yu et al. investigated the biocompatibility of synthetic abrasive diamond powders of large diameter (100 nm) in cell cultures and reported very low cytotoxicity in kidney cells. Puzyr et al.[139] also showed that the complete replacement of water in the experimental animal diet with modified nanodiamonds (MNDs) hydrosols for 6 months did not cause the death of the mice, nor affect the growth and weight dynamics of their organs. There were no indications of the destruction of their blood cells after the intravenous administration of MNDs.

6.25   CONCLUSION

This chapter describes the biomedical applications of three different forms of carbon-based nanomaterials, which are currently under investigation. Carbon nanotubes, nanohorns, and nanodiamonds can be functionalized with different types of therapeutic molecules following alternative chemical approaches. These novel nanomaterials can be modified to facilitate their manipulation and render them biocompatible. Such soluble/dispersible carbon-based nanoobjects in physiological conditions are then able to penetrate into the cells or can be administered in vivo to deliver their cargo molecules, which eventually display the desired activity.

KEYWORDS

- **carbon nanotubes**
- **biomedical applications**
- **physicochemical properties**
- **bio-sensing**
- **drug delivery**
- **cancer therapy**

REFERENCES

1. Farokhzad, O. C.; Langer, R. Impact of Nanotechnology on Drug Delivery. *ACS Nano* **2009,** *3,* 16–20.
2. Hollinger, R. V. *Drug Delivery Systems,* 2nd ed.; CRC: Boca Raton, FL, 2004.
3. Allen, T. M.; Cullis, P. R. Drug Delivery Systems: Entering the Mainstream. *Science* **2004,** *303,* 1818–1822.
4. Lavan, D. A.; McGuire, T.; Langer, R. Small-Scale Systems for in Vivo Drug Delivery. *Nat. Biotechnol.* **2003,** *21,* 1184–1191.
5. Langer, R. Drug delivery and Targeting. *Nature* **1998,** *392* (Suppl.), 5–10.
6. Duncan, R. The Dawning Era of Polymer Therapeutics. *Nat. Rev. Drug Discov.* **2003,** *2,* 347–360.
7. Varde, N. K.; Pack, D. W. Microspheres for Controlled Release Drug Delivery. *Exp. Opin. Biol. Ther.* **2004,** *4,* 35–51.
8. Boas, U.; Heegaard, P. M. Dendrimers in Drug Research. *Chem. Soc. Rev.* **2004,** *33,* 43–63.
9. Martin, C. R.; Kohli, P. The Emerging Field of Nanotube Biotechnology. *Nat. Rev. Drug Discov.* **2003,** *2,* 29–37.
10. Iijima, S. Helical Microtubules of Graphitic Carbon. *Nature* **1991,** *354,* 56–58.
11. Niyogi, S.; Hamon, M. A.; Hu, H.; Zhao, B.; Bhowmik, P.; Sen, R.; Itkis, M. E.; Haddon, R. C. Chemistry of Single-Walled Carbon Nanotubes. *Acc. Chem. Res.* **2002,** *35,* 1105–1113.
12. Hirsch, A. Functionalization of Single-Walled Carbon Nanotubes. *Angew. Chem. Int. Ed.* **2002,** *41,* 1853–1859.
13. Pastorin, G.; Kostarelos, K.; Prato, M.; Bianco, A. Functionalized Carbon Nanotubes: Towards the Delivery of Therapeutic Molecules. *J. Biomed. Nanotechnol.* **2005,** *1,* 133–142.
14. Tasis, D.; Tagmatarchis, N.; Bianco, A.; Prato, M. Chemistry of Carbon Nanotubes. *Chem. Rev.* **2006,** *106,* 1105–1136.
15. Curran, S. A.; Ajayan, P. M.; Blau, W. J.; Carroll, D. L.; Coleman, J. N.; Dalton, A. B.; Davey, A. P.; Drury, A.; McCarthy, B.; Maier, S.; Strevens, A. A Composite from Poly(Mphenylenevinylene-co-2,5-Dioctoxy-Pphenylenevinylene) and Carbon Nanotubes: A Novel Material for Molecular Optoelectronics. *Adv. Mater.* **1998,** *10,* 1091–1093.
16. Nakashima, N.; Tomonari, Y.; Murakami, H. Water-Soluble Single-Walled Carbon Nanotubes via Noncovalent Sidewall-Functionalization with a Pyrene-Carrying Ammonium Ion. *Chem. Lett.* **2002,** *31,* 638–639.
17. Wu, P.; Chen, X.; Hu, N.; Tam, U. C.; Blixt, O.; Zettl, A.; Bertozzi, C. R. Biocompatible Carbon Nanotubes Generated by Functionalization with Glycodendrimers. *Angew. Chem. Int. Ed.* **2008,** *47,* 1–5.
18. Hirsch, A.; Vostrowsky, O. Functionalization of Carbon Nanotubes, in Functional Molecular Nanostructures. *Top. Curr. Chem.* **2005,** *245,* 193–237.
19. Peng, X.; Wong, S. S. Functional Covalent Chemistry of Carbon Nanotube *surfaces. Adv. Mater.* **2009,** *21,* 625–642.
20. Kostarelos, K.; Lacerda, L.; Pastorin, G.; Wu, W.; Wieckowski, S.; Luangsivilay, J.; Godefroy, S.; Pantarotto, D.; Briand, J. P.; Muller, S.; Prato, M.; Bianco, A. Cellular

uptake of Functionalized Carbon Nanotubes is Independent of Functional Group and Cell Type. *Nat. Nanotechnol.* **2007,** *2,* 108–113.

21. Kam, N. W. S.; O'Connell, M.; Wisdom, J. A.; Dai, H. J. Carbon Nanotubes as Multifunctional Biological Transporters and Near-Infrared Agents for Selective Cancer Cell Destruction. *Proc. Natl. Acad. Sci. USA* **2005,** *102,* 11600–11605.

22. Kam, N. W. S.; Liu, Z. A.; Dai, H. J. Carbon Nanotubes as Intracellular Transporters for Proteins and DNA: An Investigation of the Uptake Mechanism and Pathway. *Angew. Chem. Int. Ed.* **2006,** *45,* 577–581.

23. Heller, D. A.; Jeng, E. S.; Yeung, T. K.; Martinez, B. M.; Moll, A. E.; Gastala, J. B.; Strano, M. S. Optical Detection of DNA Conformational Polymorphism on Single-Walled Carbon Nanotubes. *Science* **2006,** *311,* 508–511.

24. Kateb, B.; Van Handel, M.; Zhang, L. Y.; Bronikowski, M. J.; Manohara, H.; Badie, B. Internalization of MWCNTs by Microglia: Possible Application in Immunotherapy of Brain Tumors. *NeuroImage* **2007,** *37,* S9–S17.

25. Rojas-Chapana, J.; Troszczynska, J.; Firkowska, I.; Morsczeck, C.; Giersig, M. Multi-Walled Carbon Nanotubes for Plasmid Delivery into *Escherichia coli* Cells. *Lab Chip* **2005,** *5,* 536–539.

26. Liu, Q.; Chen, B.; Wang, Q.; Shi, X.; Xiao, Z.; Lin, J.; Fang, X. Carbon Nanotubes as Molecular Transporters for Walled Plant Cells. *Nano Lett.* **2009,** *9,* 1007–1010.

27. Hampel, S.; Kunze, D.; Haase, D.; Kr€amer, K.; Rauschenbach, M.; Ritschel, M.; Leonhardt, A.; Thomas, J.; Oswald, S.; Hoffmann, V.; B€uchner, B. Carbon Nanotubes Filled with a Chemotherapeutic Agent: A Nanocarrier Mediates Inhibition of Tumor Cell Growth. *Nanomedicine* **2008,** *3,* 175–182.

28. Pastorin, G.; Wu, W.; Wieckowski, S.; Briand, J. P.; Kostarelos, K.; Prato, M.; Bianco, A. Double Functionalisation of Carbon Nanotubes for Multimodal Drug Delivery. *Chem. Commun.* **2006,** 1182–1184.

29. McDevitt, M. R.; Chattopadhyay, D.; Kappel, B. J.; Jaggi, J. S.; Schiffman, S. R.; Antczak, C.; Njardarson, J. T.; Brentjens, R.; Scheinberg, D. A. Tumor Targeting with Antibody-Functionalized, Radiolabeled Carbon Nanotubes. *J. Nucl. Med.* **2007,** *48,* 1180–1189.

30. Feazell, R. P.; Nakayama-Ratchford, N.; Dai, H.; Lippard, S. J. Soluble Single-Walled Carbon Nanotubes as Longboat Delivery Systems for Platinum (IV) Anticancer Drug Design. *J. Am. Chem. Soc.* **2007,** *129,* 8438–8439.

31. Dhar, S.; Liu, Z.; Thomale, J.; Dai, H.; Lippard, S. J. Targeted Single-Wall Carbon Nanotube-Mediated Pt(IV) Prodrug Delivery Using Folate as a Homing Device. *J. Am. Chem. Soc.* **2008,** *130,* 11467–11476.

32. Ali-Boucetta, H.; Al-Jamal, K. H.; McCarthy, D.; Prato, M.; Bianco, A.; Kostarelos, K. Multiwalled Carbon Nanotube–Doxorubicin Supramolecular Complexes for Cancer Therapeutics. *Chem. Commun.* **2008,** 459–461.

33. Lu, Q.; Moore, J. M.; Huang, G.; Mount, A. S.; Rao, A. M.; Larcom, L. L.; Chun Ke, P. RNA Polymer Translocation with Single-Walled Carbon Nanotubes. *Nano Lett.* **2004,** *4,* 2473–2477.

34. Zhang, Z.; Yang, X.; Zhang, Y.; Zeng, B.; Wang, S.; Zhu, T.; Roden, R. B. S.; Chen, Y.; Yang, R. Delivery of Telomerase Reverse Transcriptase Small Interfering RNA in Complex with Positively Charged Single-Walled Carbon Nanotubes Suppresses Tumor Growth. *Clin. Cancer Res.* **2006,** *12,* 4933–4939.

35. Liu, Z.; Winters, M.; Holodniy, M.; Dai, H. siRNA Delivery into Human T Cells and Primary Cells with Carbon-Nanotube Transporters. *Angew. Chem. Int. Ed.* **2007**, *46*, 2023–2027.

36. Shi Kam, N. W.; Liu, Z.; Dai, H. Functionalization of Carbon Nanotubes via Cleavable Disulfide Bonds for Efficient Intracellular Delivery of siRNA and Potent Gene Silencing. *J. Am. Chem. Soc.* **2005**, *127*, 12492–12493.

37. Yang, X.; Zhang, Z.; Liu, Z.; Ma, Y.; Yang, R.; Chen, Y. Multifunctionalized Single-Walled Carbon Nanotubes as Tumor Cell Targeting Biological Transporters. *J. Nanopart. Res.* **2008**, *10*, 815–822.

38. Wu, Y.; Phillips, J. A.; Liu, H.; Yang, R.; Tan, W. Carbon Nanotubes Protect DNA Strands during Cellular Delivery. *ACS Nano* **2008**, *2*, 2023–2028.

39. Li, X.; Peng, Y.; Ren, J.; Qu, X. Carboxyl-Modified Single-Walled Carbon Nanotubes Selectively Induce Human Telomeric I-Motif Formation. *Proc. Nat. Acad. Sci. USA* **2006**, *103*, 19658–19663.

40. Zhao, C.; Peng, Y.; Song, Y.; Ren, J.; Qu, X. Self-Assembly of Single Stranded RNA on Carbon Nanotube: Polyadenylic Acid to Form a Duplex Structure. *Small* **2008**, *4*, 656–661.

41. Pantarotto, D.; Partidos, C. D.; Hoebeke, J.; Brown, F.; Kramer, E.; Briand, J. P.; Muller, S.; Prato, M.; Bianco, A. Immunization with Peptide Functionalized Carbon Nanotubes Enhances Virus-Specific Neutralizing Antibody Responses. *Chem. Biol.* **2003**, *10*, 961.

42. Meng, J.; Meng, J.; Duan, J.; Kong, H.; Li, L.; Wang, C.; Xie, S.; Chen, S.; Gu, N.; Xu, H.; Yang, X.-D. Carbon Nanotubes Conjugated to Tumor Lysate Protein Enhance the Efficacy of an Antitumor Immunotherapy. *Small* **2008**, *4*, 1364–1370.

43. Shao, N.; Lu, S.; Wickstrom, E.; Panchapakesan, B. Integrated Molecular Targeting of IGF1R and HER2 Surface Receptors and Destruction of Breast Cancer Cells Using Single Wall Carbon Nanotubes. *Nanotechnology* **2007**, *18*, 315101–315110.

44. Gannon, C. J.; Cherukuri, P.; Yakobson, B. I.; Cognet, L.; Kanzius, J. S.; Kittrell, C.; Weisman, R. B.; Pasquali, M.; Schmidt, H. K.; Smalley, R. E.; Curley, S. A. Carbon Nanotube-Enhanced Thermal Destruction of Cancer Cells in a Noninvasive Radiofrequency Field. *Cancer* **2007**, *111*, 2654–2665.

45. Murugesan, S.; Shaker, A.; Mousa, S. A.; O'Connor, L. J.; Lincoln, D. W.; II; Linhardt, R. J. Carbon Inhibits Vascular Endothelial Growth Factor- and Fibroblast Growth Factor-Promoted Angiogenesis. *FEBS Lett.* **2007**, *581*, 1157–1160.

46. Lovat, V.; Pantarotto, D.; Lagostena, L.; Cacciari, B.; Grandolfo, M.; Righi, M.; Spalluto, G.; Prato, M.; Ballerini, L. Carbon Nanotube Substrates Boost Neuronal Electrical Signaling. *Nano Lett.* **2005**, *5*, 1107–1110.

47. Mazzatenta, A.; Giugliano, M.; Campidelli, S.; Gambazzi, L.; Businaro, L.; Markram, H.; Prato, M.; Ballerini, L. Interfacing Neurons with Carbon Nanotubes: Electrical Signal Transfer and Synaptic Stimulation in Cultured Brain Circuits. *J. Neurosci.* **2007**, *27*, 6931–6936.

48. Cellot, G.; Cilia, E.; Cipollone, S.; Rancic, V.; Sucapane, A.; Giordani, S.; Gambazzi, L.; Markram, H.; Grandolfo, M.; Scaini, D.; Gelain, F.; Casalis, L.; Prato, M.; Giugliano, M.; Ballerini, L. Carbon Nanotubes Might Improve Neuronal Performance by Favouring Electrical Shortcuts. *Nat. Nanotechnol.* **2009**, *4*, 126–133.

49. Keefer, E. W.; Botterman, B. R.; Romero, M. I.; Rossi, A. F.; Gross, G. W. Carbon Nanotube Coating Improves Neuronal Recordings. *Nat. Nanotechnol.* **2008**, *3*, 434–439.
50. Kam, N. W. S.; Jan, E.; Kotov, N. A. Electrical Stimulation of Neural Stem Cells Mediated by Humanized Carbon Nanotube Composite Made with Extracellular Matrix Protein. *Nano Lett.* **2009**, *9*, 273–278.
51. Massobrio, G.; Massobrio, P.; Martinoia, S. Modeling the Neuroncarbon Nanotube-ISFET Junction to Investigate the Electrophysiological Neuronalactivity. *NanoLett.* **2008**, *8*, 4433–4440.
52. Malarkey, E. B.; Reyes, R. C.; Zhao, B.; Haddon, R. C.; Parpura, V. Water Soluble Single-Walled Carbon Nanotubes Inhibit Stimulated Endocytosis in Neurone. *Nano Lett.* **2008**, *8*, 3538–3542.
53. Bardi, G.; Tognini, P.; Ciofani, G.; Raffa, V.; Costa, M.; Pizzorusso, T. Pluronic-Coated Carbon Nanotubes Do Not Induce Degeneration of Cortical Neurons in Vivo and in Vitro. *NNMB* **2009**, *5*, 96–104.
54. Fenoglio, I.; Tomatis, M.; Lison, D.; Muller, J.; Fonseca, A.; Nagy, J. B.; Fubini, B. Reactivity of Carbon Nanotubes: Free Radical Generation or Scavenging Activity? *Free Radic. Biol. Med.* **2006**, *40*, 1227–1233.
55. Lucente-Schultz, R. M.; Moore, V. C.; Leonard, A. D.; Price, B. K.; Kosynkin, D. V.; Lu, M.; Partha, R.; Conyers, J. L.; Tour, J. M. Antioxidant Single Walled Carbon Nanotubes. *J. Am. Chem. Soc.* **2009**, *131*, 3934–3941.
56. Ye, S. F.; Wu, Y. H.; Hou, Z. Q.; Zhang, Q. Q. ROS and NF-kappaB are Involved in Upregulation of IL-8 in A549 Cells Exposed to Multi-Walled Carbon Nanotubes. *Biochem. Biophys. Res. Commun.* **2009**, *379*, 643–648.
57. Cherukuri, P.; Bachino, S. M.; Litovsky, S. H.; Weisman, R. B. Near Infrared Fluorescence Microscopy of Single-Walled Carbon Nanotubes in Phagocytic Cells. *J. Am. Chem. Soc.* **2004**, *126*, 15638–15639.
58. Cherukuri, P.; Gannon, C. J.; Leeuw, T. K.; Schmidt, H. K.; Smalley, R. E.; Curley, S. A.; Weisman, R. B. Mammalian Pharmacokinetics of Carbon Nanotubes Using Intrinsic Near-Infrared Fluorescence. *Proc. Natl. Acad. Sci. USA* **2006**, *103*, 18882–18886.
59. Leeuw, T. K.; Reith, R. M.; Simonette, R. A.; Harden, M. E.; Cherukuri, P.; Tsyboulski, D. A.; Beckingham, K. M.; Weisman, R. B. Single-Walled Carbon Nanotubes in the Intact Organism: Near-IR Imaging and Biocompatibility Studies in Drosophila. *Nano Lett.* **2007**, *7*, 2650–2654.
60. Welsher, K.; Liu, Z.; Daranciang, D.; Dai, H. Selective Probing and Imaging of Cells with Single-Walled Carbon Nanotubes as Near-Infrared Fluorescent Molecules. *Nano Lett.* **2008**, *8*, 586–590.
61. Zhang, S.; Sun, D.; Li, X.; Pei, F.; Liu, S. Synthesis and Solvent Enhanced Relaxation Property of Water-Soluble Endohedral Metallofullerenes. *Fullerene Sci. Tech.* **1997**, *5*, 1635–1643.
62. Bolskar, R. D.; Alford, J. M. Chemical Oxidation of Endohedral Metallofullerenes: Identification and Separation Of Distinct Classes. *Chem. Commun.* **2003**, *11*, 1292–1293.
63. Bolskar, R. D.; Benedetto, A. F.; Husebo, L. O.; Price, R. E.; Jackson, E. F.; Wallace, S.; Wilson, L. J.; Alford, J. M. First Soluble M@C60 Derivatives Provide Enhanced

Access to Metallofullerenes and Permit in Vivo Evaluation of Gd@C60[C(COOH)2]10 as a MRI Contrast Agent. *J. Am. Chem. Soc.* **2003**, *125*, 5471–5478.

64. Sitharaman, B.; Kissell, K. R.; Hartman, K. B.; Tran, L. A.; Baikalov, A.; Rusakova, I.; Sun, Y.; Khant, H. A.; Ludtke, S. J.; Chiu, W.; Laus, S.; Tóth, E.; Helm, L.; Merbach, A. E.; Wilson, L. J. Superparamagnetic Gadonanotubes are High-Performance MRI Contrast Agents. *Chem. Commun.* **2005**, *31*, 3915–3917.

65. Ashcroft, J. M.; Hartman, K. B.; Kissell, K. R.; Mackeyev, Y.; Pheasant, S.; Young, S.; Van der Heide, P. A. W.; Mikos, A. G.; Wilson, L. J. Single-Molecule I2@US-Tube Nanocapsules: A New X-Ray Contrastagent Design. *Adv. Mater.* **2007**, *19*, 573–576.

66. Hartman, K. B.; Laus, S.; Bolskar, R. D.; Muthupillai, R.; Helm, L.; Tóth, E.; Merbach, A. E.; Wilson, L. J. Gadonanotubes as Ultrasensitive pH Smart Probes for Magnetic Resonance Imaging. *Nano Lett.* **2008**, *8*, 415–419.

67. Faraj, A. A.; Cieslar, K.; Lacroix, G.; Gaillard, S.; Canet-Soulas, E.; Cremillieux, Y. In Vivo Imaging of Carbon Nanotube Biodistribution Using Magnetic Resonance Imaging. *Nano Lett.* **2009**, *9*, 1023–1027.

68. Richard, C.; Doan, B.-T.; Beloeil, J.-C.; Bessodes, M.; Tóth, E.; Scherman, D. Noncovalent Functionalization of Carbon Nanotubes with Amphiphilic Gd3þ Chelates: Toward Powerful T1 and T2 MRI Contrast Agents. *Nano Lett.* **2008**, *8*, 232–236.

69. Stafiej, A.; Pyrzynska, K. Adsorption of Heavy Metal Ions with Carbon Nanotubes. *Sep. Purif. Technol.* **2007**, *58*, 49–52.

70. Li, Y. H.; Wang, S. G.; Wei, J. Q.; Zhang, X. F.; Xu, C. L.; Luan, Z. K.; Wu, D. H.; Wei, B. Q. Lead Adsorption on Carbon Nanotubes. *Chem. Phys. Lett.* **2002**, *357*, 263–266.

71. Long, R. Q.; Yang, R. T. Carbon Nanotubes as Superior Sorbent for Dioxin Removal. *J. Am. Chem. Soc.* **2001**, *123*, 2058–2059.

72. Wang, X.; Chen, C.; Hu, W.; Ding, A.; Xu, D.; Zhou, X. Sorption of Am(III) to Multiwall Carbon Nanotubes. *Environ. Sci. Technol.* **2005**, *39*, 2856–2860.

73. Chen, C.; Li, X.; Zhao, D.; Tan, X.; Wang, X. Adsorption Kinetic, Thermodynamic and Desorption Studies of Th(IV) on Oxidized Multi-Wall Carbon Nanotubes. *Colloids Surf. A* **2007**, *302*, 449–454.

74. Belloni, F.; K€utahyali, C.; Rondinella, V. V.; Carbol, P.; Wiss, T.; Mangione, A. Can Carbon Nanotubes Play a Role in the Field of Nuclear Waste Management? *Environ. Sci. Technol.* **2009**, *43*, 1250–1255.

75. Akasaka, T.; Watari, F. Capture of Bacteria by Flexible Carbon Nanotubes. *Acta Biomater.* **2009**, *5*, 607–612.

76. Wick, P.; Manser, P.; Limbach, L. K.; Dettlaff-Weglikowskab, U.; Krumeich, F.; Roth, S.; Stark, W. J.; Bruinink, A. The Degree and Kind of Agglomeration Affect Carbon Nanotube Cytotoxicity. *Toxicol. Lett.* **2007**, *168*, 121–131.

77. Fenoglio, I.; Greco, G.; Tomatis, M.; Muller, J.; Raymundo-Piñero, E.; Beguin, F.; Fonseca, A.; Nagy, J. B.; Lison, D.; Fubini, B. Structural Defects Play a Major Role in the Acute Lung Toxicity of Multiwall Carbon Nanotubes: Physicochemical Aspects. *Chem. Res. Toxicol.* **2008**, *21*, 1690–1697.

78. Muller, J.; Huaux, F.; Fonseca, A.; Nagy, J. B.; Moreau, N.; Delos, M.; Raymundo Piñero, E.; B_eguin, F.; Kirsch-Volders, M.; Fenoglio, I.; Fubini, B.; Lison, D. Structural Defects Play a Major Role in the Acute Lung Toxicity of Multiwall Carbon Nanotubes: Toxicological Aspects. *Chem. Res. Toxicol.* **2008**, *21*, 1698–1705.

79. Tabet, L.; Bussy, C.; Amara, N.; Setyan, A.; Grodet, A.; Rossi, M. J.; Pairon, J. C.; Boczkowski, J.; Lanone, S. Adverse Effects of Industrial Multiwalled Carbon Nanotubes on Human Pulmonary Cells. *J. Toxicol. Environ. Health A* **2009**, *72*, 60–73.

80. Wick, P.; Manser, P.; Limbach, L. K.; Dettlaff-Weglikowskab, U.; Krumeich, F.; Roth, S.; Stark, W. J.; Bruinink, A. The Degree and Kind of Agglomeration Affect Carbon Nanotube Cytotoxicity. *Toxicol. Lett.* **2007**, *168*, 121–131.

81. Guo, J.; Zhang, X.; Zhang, S.; Zhu, Y.; Li, W. The Different Bio-Effects of Functionalized Multi-Walled Carbon Nanotubes on Tetrahymena Pyriformis. *Curr. Nanosci.* **2008**, *4*, 240–245.

82. Ghafari, P.; St-Denis, C. H.; Power, M. E.; Jin, X.; Tsou, V.; Mandal, H. S.; Bols, N. C.; and Tang, X. S. Impact of Carbon Nanotubes on the Ingestion and Digestion of Bacteria by Ciliated Protozoa. *Nat. Nanotechnol.* **2008**, *3*, 347–351.

83. Di Sotto, A.; Chiaretti, M.; Carru, G. A.; Bellucci, S.; Mazzanti, G. Multi-Walled Carbon Nanotubes: Lack of Mutagenic Activity in the Bacterial Reverse Mutation Assay. *Toxicol. Lett.* **2009**, *184*, 192–197.

84. Muller, J.; Decordier, I.; Hoet, P. H.; Lombaert, N.; Thomassen, L.; Huaux, F.; Lison, D.; Kirsch-Volders, M. Clastogenic and Aneugenic Effects of Multi-Wall Carbon Nanotubes in Epithelial Cells. *Carcinogenesis* **2008**, *29*, 427–433.

85. Iijima, S.; Yudasaka, M.; Yamada, R.; Bandow, S.; Suenaga, K.; Kokai, F.; Takahashi, K. Nano-Aggregates of Single-Walled Graphitic Carbon Nanohorns. *Chem. Phys. Lett.* **1999**, *309*, 165–170.

86. Yudasaka, M.; Fan, J.; Miyawaki, J.; Iijima, S. Studies on the Adsorption of Organic Materials Inside Thick Carbon Nanotubes. *J. Phys. Chem. B* **2005**, *109*, 8909–8913.

87. Hattori, Y.; Kanoh, H.; Okino, F.; Touhara, H.; Kasuya, D.; Yudasaka, M.; Iijima, S.; Kaneko, K. Direct Thermal Fluorination of Single Wall Carbon Nanohorns. *J. Phys. Chem. B* **2004**, *108*, 9614–9618.

88. Tagmatarchis, N.; Maigne, A.; Yudasaka, M.; Iijima, S. Functionalization of Carbon Nanohorns with Azomethine Ylides: Towards Solubility Enhancement and Electron Transfer Processes. *Small* **2006**, *2*, 490–494.

89. Cioffi, C.; Campidelli, S.; Brunetti, F. G.; Meneghetti, M.; Prato, M. Functionalisation of Carbon Nanohorns. *Chem. Commun.* **2006**, 2129–2131.

90. Pagona, G.; Tagmatarchis, N.; Fan, J.; Yudasaka, M.; Iijima, S. Cone-End Functionalization of Carbon Nanohorns. *Chem. Mater.* **2006**, *18*, 3918–3920.

91. Pagona, G.; Karousis, N.; Tagmatarchis, N. Aryl Diazonium Functionalization of Carbon Nanohorns. *Carbon* **2008**, *46*, 604–610.

92. Lacotte, S.; Garcia, A.; Decossas, M.; Al-Jamal, W. T.; Li, S.; Kostarelos, K.; Muller, S.; Prato, M.; Dumortier, H.; Bianco, A. Interfacing Functionalized Carbon Nanohorns with Primary Phagocytic Cells. *Adv. Mater.* **2008**, *20*, 2421–2426.

93. Bianco, A.; Kostarelos, K.; Prato, M. Opportunities and Challenges of Carbon-Based Nanomaterials for Cancer Therapy. *Exp. Opin. Drug Deliv.* **2008**, *5*, 331–342.

94. Miyako, E.; Nagata, H.; Hirano, K.; Makita, Y.; Nakayama, K.-I.; Hirotsu, T. Near-Infrared Laser-Triggered Carbon Nanohorns for Selective Elimination of Microbes. *Nanotechnology* **2007**, *18*, 475103–475109.

95. Miyako, E.; Nagata, H.; Hirano, K.; Sakamoto, K.; Makita, Y.; Nakayama, K.-I.; Hirotsu, T. Photoinduced Antiviral Carbon Nanohorns. *Nanotechnology* **2008**, *19*, 075106–075111.

96. Zhang, M.; Murakami, T.; Ajima, K.; Tsuchida, K.; Sandanayaka, A. S. D.; Ito, O.; Iijima, S.; Yudasaka, M. Fabrication of ZnPc/Protein Nanohorns for Double Photodynamic and Hyperthermic Cancer Phototherapy. *Proc. Natl. Acad. Sci. USA* **2008**, *105*, 14773–14778.

97. Murakami, T.; Ajima, K.; Miyawaki, J.; Yudasaka, M.; Iijima, S.; Shiba, K. Drug-Loaded Carbon Nanohorns: Adsorption and Release of Dexamethasone in Vitro. *Mol. Pharm.* **2004**, *1*, 399–405.

98. Murakami, T.; Fan, J.; Yudasaka, M.; Iijima, S.; Shiba, K. Solubilization of Single-Wall Carbon Nanohorns Using a PEG-Doxorubicin Conjugate. *Mol. Pharm.* **2006**, *3*, 407–414.

99. Ajima, K.; Yudasaka, M.; Murakami, T.; Maigne, A.; Shiba, K.; Iijima, S. Carbon Nanohorns as Anticancer Drug Carriers. *Mol. Pharm.* **2005**, *2*, 475–480.

100. Ajima, K.; Yudasaka, M.; Maigne, A.; Miyawaki, J.; Iijima, S. Effect of Functional Groups at Hole Edges on Cisplatin Release from Inside Single-Wall Carbon Nanohorns. *J. Phys. Chem. B* **2006**, *110*, 5773–5778.

101. Ajima, K.; Maigne, A.; Yudasaka, M.; Iijima, S. Optimum Hole-Opening Condition for Cisplatin Incorporation in Single-Wall Carbon Nanohorns and Its Release. *J. Phys. Chem. B* **2006**, *110*, 19097–19099.

102. Matsumara, S.; Ajima, K.; Yudasaka, M.; Iijima, S.; Shiba, A. Dispersion of Cisplatin-Loaded Carbon Nanohorns with a Conjugate Comprised of an Artificial Peptide Aptamer and Polyethylene Glycol. *Mol. Pharm.* **2007**, *4*, 723–729.

103. Fan, X.; Tan, J.; Zhang, G.; Zhang, F. Isolation of Carbon Nanohorn Assemblies and Their Potential for Intracellular Delivery. *Nanotechnology* **2007**, *18*, 195103–195108.

104. Miyawaki, J.; Yudasaka, M.; Imai, H.; Yorimitsu, H.; Isobe, H.; Nakamura, E.; Iijima, S. In Vivo Magnetic Resonance Imaging of Single-Walled Carbon Nanohorns by Labeling with Magnetite Nanoparticles. *Adv. Mater.* **2006**, *18*, 1010–1014.

105. Miyawaki, J.; Yudasaka, M.; Imai, H.; Yorimitsu, H.; Isobe, H.; Nakamura, E.; Iijima, S. Synthesis of Ultrafine Gd2O3 Nanoparticles Inside Single-Wall Carbon Nanohorns. *J. Phys. Chem. B* **2006**, *110*, 5179–5181.

106. Miyawaki, J.; Yudasaka, M.; Azami, T.; Kubo, Y.; Iijima, S. Toxicity of Single-Walled Carbon Nanohorns. *ACS Nano* **2008**, *2*, 213–226.

107. Isobe, H.; Tanaka, T.; Maeda, R.; Noiri, E.; Solin, N.; Yudasaka, M.; Iijima, S.; Nakamura, E. Preparation, Purification, Characterization, and Cytotoxicity Assessment of Water-Soluble, Transition-Metal-Free Carbon Nanotube Aggregates. *Angew. Chem. Int. Ed.* **2006**, *45*, 6676–6680.

108. Shenderova, O. A.; Gruen, D. M., Eds. *Ultracrystalline Diamond*; William-Andrew Publishing: Norwich, UK, 2006.

109. Ferro, S. Synthesis of Diamond. *J. Mater. Chem.* **2002**, *12*, 2843–2855.

110. Schreiner, P. R.; Fokina, N. A.; Tkachenko, B. A.; Hausmann, H.; Serafin, M.; Dahl, J. E. P.; Liu, S.; Carlson, R. M. K.; Fokin, A. A. Functionalised Nanodiamonds: Triamantane and Tetramantane. *J. Org. Chem.* **2006**, *71*, 6709–6720.

111. Fokin, A. A.; Tkachenko, B. A.; Gunchenko, P. A.; Gusev, D. V.; Schreiner, P. R. Functionalised Nanodiamonds Part 1. An Experimental Assessment of Diamantane and Computational Predictions for Higher Diamondoids. *Chem. Eur. J.* **2005**, *11*, 7091–7101.

112. Liu, Y.; Gu, Z.; Margrave, J. L.; Khabashesku, V. N. Functionalization of Nanoscale Diamond Powder: Fluore-, Alkyl-, Amino-, and Amino Acid–Nanodiamond Derivatives. *Mater. Chem.* **2004**, *16*, 3924–3930.
113. Novikov, N. V. New Trends in High Pressure Synthesis of Diamond. *Diamond Relat. Mater.* **1999**, *8*, 1427–1432.
114. Shenderova, O. A.; Zhirnov, V. V.; Brenner, D. W. Carbon Nanostructures. *Crit. Rev. Solid State Mater. Sci.* **2002**, *27*, 227–356.
115. Frenklach, M.; Kematick, R.; Huang, D.; Howard, W.; Spear, K. E.; Phelps, A. W.; Koba, R. Homogeneous Nucleation of Diamond Powder in the Gas Phase. *J. Appl. Phys.* **1989**, *66*, 395–399.
116. Krueger, A. New Carbon Materials: Biological Applications of Functionalized Nanodiamond Materials. *Chem. Eur. J.* **2008**, *14*, 1382–1390.
117. Puzyr, A. P.; Pozdnyakova, I. O.; Bondar, V. S. Design of Luminescent Biochip with Nanodiamonds and Bacterial Luciferase. *Phys. Solid State* **2004**, *46*, 761–763.
118. Huang, L.-C. L.; Chang, H.-C. Adsorption and Immobilization of Cytochrome C on Nanodiamonds. *Langmuir* **2004**, *20*, 5879–5884.
119. Chung, P.-H.; Perevedentseva, E.; Tu, J. S.; Chang, C. C.; Cheng, C.-L. Spectroscopic Study of Bio-Functionalized Nanodiamonds. *Diamond Relat. Mater.* **2006**, *15*, 622–625.
120. Kong, X. L.; Huang, L. C. L.; Hsu, C.-M.; Chen, W.-H.; Han, C.-C.; Chang, H. C. High-Affinity Capture of Proteins by Diamond Nanoparticles for Mass Spectrometric Analysis. *Anal. Chem.* **2005**, *77*, 259–265.
121. Huang, T. S.; Tzeng, Y.; Liu, Y. K.; Chen, Y. C.; Walker, K. R.; Guntupalli, R.; Liu, C. Immobilization of Antibodies and Bacterial Binding on Nanodiamond and Carbon Nanotubes for Biosensor Applications. *Diamond Relat. Mater.* **2004**, *13*, 1098–1102.
122. Huang, H.; Pierstorff, E.; Osawa, E.; Ho, D. Active Nanodiamond Hydrogels for Chemotherapeutic Delivery. *Nano Lett.* **2007**, *7*, 3305–3314.
123. Kossovsky, N.; Gelman, A.; Hnatyszyn, H. J.; Rajguru, S.; Garell, R. L.; Torbati, S.; Freitas, S. S. F.; Chow, G.-M. Surface-Modified Diamond Nanoparticles as Antigen Delivery Vehicles. *Bioconjugate Chem.* **1995**, *6*, 507–511.
124. Liu, K.-K.; Chen, M.-F.; Chen, P.-Y.; Lee, T. J. F.; Cheng, C.-L.; Chang, C.-C.; Ho, Y. P.; Chao, J.-I. Alphabungarotoxin Binding to Target Cell in a Developing Visual System by Carboxylated Nanodiamond. *Nanotechnology* **2008**, *19*, 205102.
125. Fu, C.-C.; Lee, H.-Y.; Chen, K.; Lim, T. S.; Wu, H.-Y.; Lin, P.-K.; Wei, P.-K.; Tsao, P.-H.; Chang, H.-C.; Fann, W. Characterization and Application of Single Fluorescent Nanodiamonds as Cellular Biomarkers. *Proc. Natl. Acad. Sci. USA* **2007**, *104*, 727–732.
126. Yu, S.-J.; Kang, M.-W.; Chang, H.-C.; Chen, K.-M.; Yu, Y.-C. Bright Fluorescent Nanodiamonds: No Photobleaching and Low Cytotoxicity. *J. Am. Chem. Soc.* **2005**, *127*, 17604–17605.
127. Chang, Y.-R.; Lee, H.-Y.; Chen, K.; Chang, C.-C.; Tsai, D.-S.; Fu, C.-C.; Lim, T.-S.; Tzeng, Y.-K.; Fang, C.-Y.; Han, C.-C.; Chang, H.-C.; Fann, W. Mass Production and Dynamic Imaging of Fluorescent Nanodiamonds. *Nat. Nanotechnol.* **2008**, *3*, 284–288.
128. Zhao, Q.-L.; Zhang, Z.-L.; Huang, B.-H.; Peng, J.; Zhang, M.; Pang, D. W. Facile Preparation of Low Cytotoxicity Fluorescent Carbon Nanocrystals by Electrooxidation of Graphite. *Chem. Commun.* **2008**, 5116–5118.

129. Borjanovic, V.; Lawrence, W. G.; Hens, S.; Jaksic, M.; Zamboni, I.; Edson, C.; Vlasov, I.; Shenderova, O.; McGuire, G. E. Effect of Proton Irradiation on Photoluminescent Properties of PDMS Nanodiamond Composites. *Nanotechnology* **2008**, *19*, 455701.

130. Aharonovich, I.; Zhou, C.; Stacey, A.; Treussart, F.; Roch, J.-F.; Prawer, S. Formation of Color Centers in Nanodiamonds by Plasma Assisted Diffusion of Impurities from the Growth Substrate. *Appl. Phys. Lett.* **2008**, *93*, 243112.

131. Chang, I. P.; Hwang, K. C.; Chiang, C.-S. Preparation of Fluorescent Magnetic Nanodiamonds and Cellular Imaging. *J. Am. Chem. Soc.* **2008**, *130*, 15476–15481.

132. Faklaris, O.; Garrot, D.; Joshi, V.; Druon, F.; Boudou, J.-P.; Sauvage, T.; Georges, P.; Curmi, P. A.; Treussart, F. Detection of Single Photoluminescent Diamond Nanoparticles in Cells and Study of the Internalization Pathway. *Small* **2008**, *4*, 2236–2239.

133. Vial, S.; Mansuy, C.; Sagan, S.; Irinopoulou, T.; Burlina, F.; Boudou, J.-P.; Chassaing, G.; Lavielle, S. Peptide-Grafted Nanodiamonds: Preparation, Cytotoxicity and Uptake in Cells. *ChemBioChem* **2008**, *9*, 2113–2119.

134. Cheng, C.-Y.; Perevedentseva, E.; Tu, J.-S.; Chung, P.-H.; Cheng, C.-L.; Liu, K.-K.; Chao, J.-I.; Chen, P.-H.; Chang, C.-C. Direct and in Vitro Observation of Growth Hormone Receptor Molecules in A549 Human Lung Epithelial Cells by Nanodiamond Labeling. *Appl. Phys. Lett.* **2007**, *90*, 163903.

135. Chao, J.-I.; Perevedentseva, E.; Chung, P. H.; Liu, K.-K.; Cheng, C.-Y.; Chang, C.-C.; and Cheng, C.-L. Nanometer-Sized Diamond Particle as a Probe for Biolabeling. *Biophys. J.* **2007**, *93*, 2199–2208.

136. Schrand, A. M.; Dai, L.; Schlager, J. J.; Hussain, S. M.; Osawa, E. Differential Biocompatibility of Carbon Nanotubes and Nanodiamonds. *Diamond Relat. Mater.* **2007**, *16*, 2118–2123.

137. Schrand, A. M.; Huang, H.; Carlson, C.; Schlager, J. J.; Osawa, E.; Hussain, S. M.; Dai, L. Are Diamond Nanoparticles Cytotoxic? *J. Phys. Chem. B* **2007**, *111*, 2–7.

138. Liu, K.-K.; Cheng, C.-L.; Chang, C.-C.; Chao, J.-I. Biocompatible and Detectable Carboxylated Nanodiamond on Human Cell. *Nanotechnology* **2007**, *18*, 325102.

139. Puzyr, A. P.; Baron, A. V.; Purtov, K. V.; Bortnikov, E. V.; Skobelev, N. N.; Mogilnaya, O. A.; Bondar, V. S. Nanodiamonds with Novel Properties: A Biological Study. *Diamond Relat. Mater.* **2007**, *16*, 2124–2128.

CHAPTER 7

CARBON NANOTUBES FOR GREENING THE ENVIRONMENT

SUKANCHAN PALIT

¹Department of Chemical Engineering, University of Petroleum and Energy Studies, Energy Acres, Dehradun 248007, Uttarakhand, India.

E-mail: sukanchan68@gmail.com, sukanchan92@gmail.com, sukanchanp@rediffmail.com

ABSTRACT

The world of science and technology is moving toward one scientific regeneration over another. Green engineering and green chemistry are the utmost needs of the progress of human civilization today. A deep introspection into the field of carbon nanotubes, nanomaterials, and engineered nanomaterials will surely open up new doors of scientific innovation and scientific instinct in the field of both nanotechnology and green sustainability in decades to come. Environmental degradation, loss of ecological biodiversity, and frequent environmental catastrophes are urging the scientific community to gear forward toward newer challenges and newer vision. Carbon nanotubes and carbon-based nanomaterials have been found to be highly effective in environmental remediation. The author in this treatise targets with vast scientific vision and scientific ardor the needs of nanotechnology applications in human society. This treatise delineates with vision and scientific alacrity the recent advances in carbon nanotubes applications in environmental protection. Green nanotechnology is the coin word of scientific research pursuit globally today. A deep scientific comprehension in the field of green engineering, nanotechnology, and environmental remediation will open new doors of innovation and scientific profundity in global research and development initiatives. These issues are deeply validated with vision and lucidity in this treatise.

7.1   INTRODUCTION

The vision of global science, technology, and engineering of nanotechnology and nanoengineering are surpassing vast and versatile scientific frontiers. Rapid industrialization, progress of human civilization, and burgeoning human population have veritably urged scientists and engineers to move toward more environmental engineering solutions. Environmental degradation, growing concerns for loss of ecological biodiversity, and global climate change have vociferously propelled global research initiatives to move toward green chemistry, green engineering, and green nanotechnology. Mankind and human civilization today stands in the critical juncture of vision, scientific forbearance, and engineering validation. Nanomaterials and engineered nanomaterials are the marvels of science and engineering today. Carbon nanotubes and other nanomaterials will open up new vistas of scientific research and scientific vision in years to come. Water purification science, drinking water treatment, and industrial wastewater treatment are the scientific needs of civilization today and will be the true forerunners of global research and development initiatives in environmental remediation. Scientific revelation and deep scientific provenance in United Nations Sustainable Development Goals need to be implemented in human society as science and engineering moves forward. In this treatise, the author deeply investigates the significant advances in the field of carbon nanotubes applications, green engineering, and green sustainability. Sustainable development whether it is social, economic, energy, or environmental is the utmost needs of the human society today. The scientific and academic rigors behind the environmental and green sustainability are vastly explored in this treatise. Provision of basic human needs such as food, water, shelter, education, and sustainability will all go a long way in the true realization of the science of sustainability. Today, the world of challenges in green chemistry and green engineering are vast and versatile. These areas of scientific research pursuit are deeply elucidated with scientific might and scientific determination in this treatise. A scientific world of true glory and true emancipation will veritably emerge if sustainability whether it is energy or environmental are vastly implemented in human society. These areas are focused pointedly in this treatise.

7.2 THE AIM AND OBJECTIVE OF THIS STUDY

Environmental protection science and the science of green or environmental sustainability are veritably linked with each other. Man and mankind are today in the middle of vision and scientific perseverance. The engineering applications of carbon nanotubes and other nanomaterials are today replete with scientific and academic rigor. Environmental degradation today is highly thought provoking. Global climate change, loss of ecological biodiversity, and frequent environmental catastrophes are surely urging scientists and engineers to move toward newer innovations and vast scientific ingenuity. The main aim and objective of this treatise is to target the visionary areas of nanotechnology, green engineering, and green sustainability. Environmental remediation and the vast world of environmental engineering science are the burgeoning concerns of civilization and mankind today. Thus, the need of an investigative study in the field of nanomaterials, green engineering, and green chemistry. The visionary words of Dr Gro Harlem Brundtland, former Prime Minister of Norway on the science of sustainability needs to be envisioned and re-organized with the passage of scientific history and the visionary time frame. Green engineering and green nanotechnology are revolutionizing the vast scientific firmament in the global scientific perspectives. Thus, the need of a thorough exploration in the field of green engineering and environmental remediation. A new dawn in civilization will surely emerge is concerted efforts from scientists, engineers, government and the civil society are envisioned and properly implemented.[1–4]

7.3 THE VAST SCIENTIFIC DOCTRINE OF CARBON NANOTUBES

Man and mankind are today progressing at a rapid pace surpassing one visionary boundary over another. Nanotechnology, nanomaterials, and engineered nanomaterials have diverse applications in every branch of science and technology. In the similar vein, science and technology are moving forward with a drastic pace. Carbon nanotubes are tubes made of carbon with diameters typically measured in nanometers. Today, carbon nanotubes are the marvel and vision of global research and development initiatives. Carbon nanotubes often refers to single-walled carbon nanotubes with diameters in the range of a nanometer.[25,26] Single-walled carbon nanotubes are one of the allotropes of carbon, intermediate between fullerene cages and flat graphene. Carbon nanotubes also often refer to multiwalled carbon nanotubes (MWCNTs) consisting of nested single-walled carbon nanotubes. Multiwalled carbon nanotubes are also sometimes referred to double- and triple-walled carbon nanotubes. Human scientific struggle and determination and the validation of science and engineering will all be the pallbearers toward a new era in the field of nanomaterials and engineered nanomaterials.[23,24] Carbon nanotubes can also refer to tubes with undetermined carbon wall structure and diameters less than 100 nm. Carbon nanotubes can exhibit absolutely remarkable electrical conductivity. They also have exceptional tensile strength and thermal conductivity because of their nanostructure and strength of the bonds between carbon atoms. In addition, they can be chemically modified with vast vision and validation.[1-4]

7.4   GREEN ENGINEERING, GREEN SUSTAINABILITY, AND THE VAST VISION FOR THE FUTURE

Green engineering and green sustainability are the powerful and magnificent tools of science and engineering today. The vast vision for the future needs to be revamped and re-organized with regards to environmental or green sustainability. In the similar vein, green chemistry and green nanotechnology are the path breakers of science and technology globally. Application of green engineering, nanomaterials and engineered nanomaterials in vast and varied avenues of science and technology are the vision and academic rigor of today's science. The visionary words of Dr Gro Harlem Brundtland, former Prime Minister of Norway on the science of sustainability needs to be envisioned and organized as civilization trudges forward. The world of science and technology is today in the middle of an unimaginable scientific catastrophe. Millions of citizens around the world are without pure drinking water. Public health engineering in the global scenario is in the midst of a crisis. Green engineering and green chemistry are veritably changing the scientific landscape.[1-4]

Green engineering is the design, commercialization, and use of processes and products in a manner that reduces pollution, promotes environmental and green sustainability, and minimizes risk to human health and the surrounding environment without sacrificing economic viability and efficiency of the processes and the products.[25,26]

7.5   THE SUCCESS OF SCIENCE AND ENGINEERING OF ENVIRONMENTAL SUSTAINABILITY

The success of science and engineering of environmental sustainability needs to be re-envisioned and readdressed as the people around the world are in the middle of an unimaginable environmental crisis. Science and engineering of environmental sustainability are the marvels of civilization today. The vast scientific doctrines and the scientific sagacity of nanotechnology and green engineering needs to be epitomized as man and mankind tread forward. The author throughout this chapter focuses on the success of green engineering, green nanotechnology, and carbon nanotubes in environmental remediation. This treatise will widely pronounce the scientific need of carbon nanotubes and other nanomaterials in the true realization and the true vision of environmental engineering science.[1-4]

Barrow[27] deeply discussed with scientific vision and farsightedness environmental management and sustainable development. Today, in the global scenario and in the scientific horizon, environmental management and sustainability are integrated to each other. The continual degradation of the planet's environment is something that affects every country whether it is developed or developing. Rapid industrial advancements in modern human society are veritably threatening the ecology and environment.[27] Thus, the need of a comprehensive treatise in the field of environmental management and environmental sustainability. Human scientific stance and ingenuity, the challenges which lie ahead and the global vision of environmental management will surely open new doors of innovation in environmental sustainability.[27] This book clarifies the definition, nature, and role of environmental management in development and in developing countries. The author of this book lucidly deals with theory and approaches of environmental management and developing countries, resource management issues by sector, and environmental tools and policies. Resource management issues involve water, coastal and island resources, agriculture, land degradation and food security, biodiversity resources, atmospheric issues, urban environments and industrial pollution issues, and environmental threats.[27] Environmental management and environmental sustainability are evolving rapidly and is being increasingly applied in developing and developed nations and to transboundary and global issues. Humanity's progress and scientific and academic rigor of environmental sustainability are touching visionary heights as civilization moves forward.[27] This is a watershed text in the field of environmental management and development. The process of development takes place in the global environment, using resources, generating wastes, and causing other severe impacts. This book explores the meaning of development and environmental management and how they correlate. The evolution of interest in development and environmental issues and management are deeply reviewed. Environmental management veritably helps steer the vast transition from environmental exploitation to large challenges and possibilities to stewardship of nature and global environment. Today, environmental management and environmental sustainability are the utmost needs of human civilization and human scientific progress.[27] Environmental management and sustainable development are both difficult to define. The former can be a goal or vision, a veritable attempt to steer the process, the application of a set of tools, a vastly more philosophical exercise seeking to identify and establish new scientific outlooks. This treatise is an eye-opener to the vast and difficult world of environmental sustainability and opens up new windows of innovation in both environmental management and sustainable development.[27]

The success of science and engineering of energy and environmental sustainability are in the avenues of newer scientific regeneration. The provision of clean drinking water and water and environmental remediation are the utmost needs of humanity today. In the similar vision, groundwater remediation and industrial wastewater treatment will go a long and visionary way in the true realization and the true emancipation of environmental sustainability and environmental management techniques.

7.6 TECHNOLOGICAL CHALLENGES IN THE FIELD OF GREEN ENGINEERING AND ENVIRONMENTAL PROTECTION

Technological challenges and the verve and vision of science are today unfolding the new genre in the field of green engineering and environmental protection. Shannon et al.[5] deeply discussed with vision and insight science and technology of water purification. The many problems associated worldwide with the lack of clean and fresh water are well known: 1.2 billion people lack access to safe drinking water, 2.6 billion have no sanitation, millions of people die annually, some 3900 children die every day due to diseases transmitted through unsafe water or human excreta.[5] Thus, the evergrowing concerns for environmental sustainability and environmental engineering innovations. Green engineering and green chemistry are the zenith of human scientific progress and human vision today. Technological and engineering challenges in environmental engineering are immense and surpassing frontiers. Water strongly affects energy and food production, industrial output, and the quality of human environment, affecting developed as well as developing economies around the world.[5] Many freshwater aquifers are highly contaminated or suffer saltwater intrusions along coastal regions. With agriculture, livestock and energy consuming more than 80% of all water for human use, demand of fresh water source is expected to increase with population growth, thus further stressing public health engineering. Reliability engineering and environmental engineering are today linked with each other. The vast shift to biofuels for energy may add further scientific demands for irrigation and farming.[5] Alarming and with deep adjudication, within 30 years receding glaciers may cause major rivers, such as the Brahmaputra, Ganges, Yellow and Mekong rivers, which serve China, India and South East Asia, to become intermittent, resulting in an unimaginable disaster. It will affect 1.5 billion people during the dry months.[5] Fortunately, a recent flurry of research and development initiatives offers deep hopes in mitigating the impact of impaired waters around the globe. Conventional methods of water disinfection, decontamination, and desalination can veritably address many of these scientific issues with water quality and supply. The authors[5] deeply discussed the techniques of disinfection, decontamination, and reuse and reclamation. The overarching goal for providing safe and pure drinking water is affordable and robust to disinfect water from traditional and emerging pathogens, without creating more problems due to the disinfection process itself. Waterborne pathogens have a devastating effect on human health, public health engineering, and the global environmental engineering scenario in the developing countries of sub-Saharan Africa and southeast Asia.[5] Shannon et al.[5] deeply discussed the new, conventional, and nonconventional disinfection strategies. New disinfection strategies involve physicochemical removal, such as coagulation, flocculation, sedimentation, and media and membrane filtration. The situation in developing countries is absolutely grave and thought provoking. Technological challenges and the vast scientific profundity in conventional and nonconventional environmental engineering techniques are deeply dealt with in this treatise.[5]

The application of nanotechnology in water purification science is a marvel of science and technology today. Scientific revelation and deep scientific sagacity are the pillars of science and mankind. This well-researched treatise unfolds the scientific needs of sustainability science and nanotechnology in the furtherance of environmental engineering science.[25,26]

7.7 SIGNIFICANT SCIENTIFIC ADVANCES IN THE FIELD OF CARBON NANOTUBES AND ENVIRONMENTAL PROTECTION

Environmental engineering and environmental remediation are today integrated with the science of nanotechnology. Advancements in the science of nanotechnology are in the process of newer regeneration. Ren et al.[6] reviewed with scientific forbearance and insight carbon nanotubes as adsorbents in environmental pollution management. Carbon nanotubes have vastly aroused wide attention as a new type of adsorbent due to their outstanding ability for the removal of various organic and inorganic pollutants and radionuclides from industrial wastewater.[6] This review summarizes the properties of carbon nanotubes and their properties related to the adsorption of various pollutants from large volumes of aqueous solutions.[6] According to the authors, this review is not comprehensive as the main focus was on exploiting the exceptional adsorption properties of carbon nanotubes.[6] This research pursuit detailed gas adsorption, sites for gas adsorption on carbon nanotubes, adsorption mechanisms of gases on carbon nanotubes, liquid adsorption, heavy metal ion adsorption, radio-nuclide adsorption, organic pollutant adsorption, and factors affecting adsorption.[6] Today, adsorption is a scientific need in gas separation. This scientific ingenuity is deeply investigated with vision and insight in this paper.[6]

Attar et al.[28] discussed with immense scientific and engineering vision about carbon nanotubes and their environmental applications. Recent developments in nanotechnologies have veritably helped to benchmark carbon nanotubes as one of the best smart materials and ecomaterials of future. In this short review, the vast contributions of carbon nanotubes is deeply addressed in terms of sustainable environment and green engineering and green technological perspective, such as water and wastewater treatment, air pollution monitoring, biotechnologies, renewable energy technologies, supercapacitors, and green nanocomposites.[28] Scientific vision and ingenuity, scientific transcendence and engineering vision will all lead a long, effective and visionary way in the true emancipation of environmental applications of carbon nanotubes. Today, carbon nanotubes have immense potential are also the most promising material for application in various environmental fields.[28] Nanotechnology is a vast field which deals with materials at their atomic level and is presently considered as the boundary of quantum mechanics. Carbon nanotubes are relatively new but are broadly studied materials in the nanotechnology field. The authors discussed in minute details about carbon nanotubes and energy storage and energy engineering. Carbon nanotubes are allotropes of carbon with a cylindrical nanostructure. Nanotubes are members of the fullerene structural family.[28] Applied quantum chemistry, specifically orbital hybridization veritably describes the chemical bonding in nanotubes. Water treatment applications involve carbon nanotubes applications for microbial fuel cells and air pollution control applications. Carbon nanotubes endeavor has today wide research scope in its synthesis with cost reduction and applications in diverse areas.[28] Production cost and health impacts are major scientific concerns. Carbon nanotubes have today covered vast areas of environmental remediation and in near future mankind and science can expect the dominance of nanoscience and nanotechnology.[28]

Wang et al.[29] described with scientific foresight and scientific determination environmental remediation applications of carbon nanotubes and graphene oxides and adsorption and catalysis. Environmental issues such as wastewater have deeply influenced each aspect of human life on earth. Coupling the vastly existing remediation solutions along with exploring new functional carbon nanomaterials (example carbon nanotubes, graphene oxides, graphene) by various perspectives shall open up new avenues of scientific understanding of the environmental problems and their phenomenon. This review depicts profoundly an overview of potential environmental remediation solutions and the diverse and vast challenges happening by using low-dimensional carbon nanomaterials and their adsorbents, catalysts or catalyst supports for ensuring social and environmental sustainability. Along with the growing global population, industrialization, and vast urbanization, the lack of fresh and clean water is becoming a grave problem around the world.[29] Meanwhile, the shortage of water resources envisions new technologies and newer innovations for

decontamination of wastewater and also technologies for sea water desalination. Water is precious for human civilization and needs to be preserved
in developing and developed nations around the world. Over one-half of
the world population, mainly in Asia has to face the severity of contaminated water. Mankind's immense scientific ingenuity and introspection
in the field of drinking water and groundwater remediation needs to be
re-envisioned as scientific progress in environmental protection moves
forward.[29] Low-dimensional novel carbon nanomaterials example carbon
nanotubes and graphene oxide have been stimulating immense interest in
various scientific communities ever they have been discovered. Carbon
nanotubes have excellent physical and chemical properties.[29] The application of carbon nanotubes are not limited to electrical, electronics, sensors,
and thermal devices but they are emerging materials for environmental
protection and water remediation. Carbon nanotubes are indeed a new
class of smart materials useful for environmental applications because
of their hollow cylindrical structure, large surface area, high-length to
radius, ratio and hydrophobic wall and surface that can be easily modified.
The authors discussed in details of carbon nanotubes-based composite
materials for water remediation, adsorption of organic dyes, adsorption of
organic pollutants, adsorption of heavy metal ions, adsorption of cesium
and strontium ions, carbon nanotubes in catalysis reactions for water
remediation, electrocatalysis, and other catalytic oxidation.[29] Graphene
oxide-based composite materials for water remediation are the other
cornerstones of this treatise. Surface functionalization with graphene oxide
as host material is another area of this well-researched paper. In spite of
the enormous progress already achieved in preparation, processing, and
applications of the low-dimensional carbon nanomaterials, for example,
carbon nanotubes, graphene oxide and its derivatives in environmental
protection and water remediation, there are tremendous challenges in its
application and much research work needs to be done.[29] The technologies
of novel and ecomaterials, such as carbon nanotubes and graphene oxides
will surely open new dimensions in research endeavor and vast scientific
ingenuity in nanoscience and nanotechnology in decades to come.

Rahman et al.[30] deeply discussed with scientific foresight and immense
lucidity about the recent progress in the synthesis and applications of carbon
nanotubes. Carbon nanotubes are known as nanoarchitectured allotropes of
carbon having graphene sheets that are wrapped forming a cylindrical shape.
A new era in the field of nanotechnology and environmental protection

is evolving as inroads into research pursuit in carbon nanomaterials are deeply envisioned.[30] Carbon nanotubes exhibit some unusual properties like a high degree of stiffness, a large length-to-diameter ratio, and exceptional resilience, and for these reasons, they are used in diverse areas of science and engineering. Technological and engineering vision are today at its pinnacle as nanotechnology and environmental engineering science move in the right research directions.[30] Carbon is an astonishing element not only due to the reason the element is necessary for life on earth but also it can occur in various allotropic forms. In the conventional sense, carbon materials consist of graphene blocks the category under which activated carbons, carbon black, and diamonds are present and classified. Mankind and science are today in the path of newer scientific provenance and scientific redemption. The recently developed materials of carbon include nanotextured and nanosized carbons. Nanotextured carbons cover a wide variety of carbon structures from carbon fibers, pyrolytic carbons, or glass-like carbons to diamond-like carbon materials. Scientific research pursuit in the field of nanomaterials and engineered nanomaterials are scaling new heights today.[30] The nanosized carbons comprise fullerenes, graphene, and carbon nanotubes. The authors discussed in minute details the synthesis of carbon nanotubes, arc discharge method, laser ablation method, flame synthesis method, saline solution method, spray pyrolysis method, and characterization techniques for carbon nanotubes which include Raman spectroscopy, transmission electron microscopy, atomic force microscopy, and other shapes and applications of carbon nanotubes are the other pivots of this research pursuit. Energy storage and water treatment applications of carbon nanotubes stand as major cornerstones of this paper.[30] Biomedical applications, electronic applications, and carbon nanotube-based diodes are the other research areas of this paper. The promising results achieved in the synthesis, functionalization and design of carbon nanotubes are the challenges of the wide world of nanotechnology today. Further improvements in the synthesis methods are required to obtain carbon nanotubes for desired applications. The constant miniaturization of electronic devices is the most energetic force of the microelectronic and electronic industry today. A new dawn in the field of nanotechnology and electronics will evolve and science and engineering will surely usher in a new scientific horizon.[30]

Application domain of carbon nanotubes in chemical engineering and environmental engineering science are vast and versatile. Chemical process technology, unit operations of chemical engineering, and the visionary world of nanotechnology are today integrated with each other. A new era in science and engineering of nanotechnology is ushering in as man and mankind moves forward at diverse scientific goals.[25,26] Biomedical engineering and electronics science and engineering are in the path of vast scientific regeneration. Application of carbon nanotubes, fullerenes, and graphene oxides in environmental and water remediation will surely be the future challenges.

7.8 SIGNIFICANT SCIENTIFIC ADVANCES IN THE FIELD OF APPLICATIONS OF CARBON NANOTUBES

Carbon nanotubes, nanomaterials, and engineered nanomaterials are the scientific vision, scientific ingenuity, and scientific profundity of tomorrow's world. Application areas of carbon nanotubes in diverse areas of science and engineering needs to be re-envisioned and re-organized as man and mankind moves forward. In this section, the author deeply enunciates the needs of nanotechnology and nanoengineering in the true emancipation of environmental remediation and environmental engineering science. Some significant scientific research pursuit are dealt with vision and insight in this section.[25,26]

Ajayan et al.[7] discussed and redefined with insight applications of carbon nanotubes. Carbon nanotubes have highly attracted the fancy of scientists around the world. The small dimensions, strength, and remarkable physical properties of these structures make them a very unique material with a wide range of engineering and scientific applications.[7] In this review, the authors described with vision and foresight about material science applications of carbon nanotubes. The authors deeply discussed the electronic and electrochemical applications of nanotubes, nanotubes as mechanical reinforcements in high performance composites, nanotube-based field emitters, and their vast use as nanoprobes in metrology and biological and chemical investigations and as templates for the creation of other nanostructures. Civilization and technology's immense verve, vision, and stance will surely open new doors of innovation in the field of material science and nanotechnology in years to come.[7] These challenges are enumerated in this paper. Electronic properties and device applications of nanotubes are dealt deeply in this paper.[7] The discovery of fullerenes provided exciting insight into carbon nanostructures. Quasi-one-dimensional carbon whiskers or nanotubes are perfectly straight tubules with diameters of nanometer size and have properties close to that of an ideal graphite element. The authors[7] with deep scientific wonder and glory elucidate the structure and the properties of carbon nanotubes.[7] Carbon nanotubes were discovered accidently by Sumio Ijima in 1991 while deeply investigating the surfaces of graphite electrodes used in an electric arc discharge. Today, nanoscience and nanotechnology are the scientific needs of civilization. Sumio Ijima's vision, observation, and analysis of the nanotube structure started a new visionary direction in research and development initiative in nanotechnology and also opened newer avenues in fullerene research.[7] These tiny carbon tubes with incredible strength and fascinating electronic properties appear to overtake the areas of fullerenes research. Thus, a new scientific direction emerged. The uniqueness of the nanotube arises from its structure and the inherent subtlety in the structure, which is the helicity in the arrangement of the carbon atoms in hexagonal arrays on their surface honeycomb lattices. Carbon nanotubes are today in the forefront of a scientific revolution and vast regeneration. The authors discussed in details the potential applications of carbon nanotubes in vacuum microelectronics, prototype electron emission devices based on carbon nanotubes, energy storage, electrochemical intercalation of carbon nanotubes with lithium, hydrogen storage, filled composites, nanoprobes and sensors, templates, and the vast challenges and potential of carbon nanotubes applications. Carbon nanotubes have come a long way since their discovery in 1991.[7] A scientific regeneration and a scientific vision emerged later. The structures that were first reported in 1991 were multiwalled nanotubes with a range of diameters and lengths.[7] These were essentially the distant relatives of the highly defective carbon nanofibers grown in the laboratory via catalytic chemical vapor deposition. Thus, ushered in a new era in nanoscience and nanotechnology.[7] This review described in details several possible applications of carbon nanotubes, with vast emphasis on material science-based applications. Material science and composite science are two opposite sides of the visionary coin. The overarching goal of this paper is to pronounce the unique structure, topology, and dimensions of nanotubes which resulted in a superb all-carbon material which may be considered as the most perfect fiber.[7] The remarkable physical properties of nanotubes create a host of applications possibilities some due to enhancement of novel electronic and mechanical behavior of carbon nanotubes. Nanotubes vastly and truly bridge the gap between the molecular and sub-atomic realm and the macrolevel of nanoscience and in future will be destined as a star in future science and engineering.[7]

Popov[8] discussed in details properties and applications of carbon nanotubes. Civilization, science, and engineering are in the path of new regeneration today. Carbon nanotubes are unique tubular structures of nanometer diameter and a veritably large length/diameter ratio. The nanotubes may consist of one up to tens and hundreds of concentric shells of carbons with adjacent shells separations of approximately 0.34 nm.[8] The carbon network of the shells is closely related to the honeycomb arrangement of the carbon atoms in the graphite sheets. These nanomaterials have amazing mechanical and electronic properties and this is due to quasi one-dimensional structure and graphite-like arrangement of the carbon atoms in the shells.[8] This well-researched treatise is intended to summarize some of the major advancements in the field of carbon nanotube research both experimental and theoretical in connection with the vast engineering applications of carbon nanotubes. This treatise discusses in details the synthesis of carbon nanotubes, arc discharge, laser ablation, catalytic growth, multiwalled nanotubes growth mechanisms, single-walled nanotubes, electronic band structure of single-walled nanotubes, and the vast domain of electrical transport in perfect nanotubes.[8] In this report, the major developments in both the research and the industrial applications of carbon nanotubes are deeply reviewed. Nanotubes have a wide range of applications in various technological areas, such as aerospace, energy, automobile, medicine, and the chemical industry. They can be used as adsorbents in chemical process industries, chemical sensors, nanopipes, and nanoreactors. A visionary era in the field of nanotechnology and chemical process engineering will evolve if proper emancipation and applications of nanotubes are envisioned.[8]

Seetharamappa et al.[9] discussed and elucidated with vision and scientific determination carbon nanotubes as the next generation electronic materials. The current interest in carbon nanotubes is vastly a direct consequence of the synthesis of buckminsterfullerene in 1985 and its derivatives thereafter. A new scientific regeneration and deep rejuvenation evolved after that discovery. Sumio Iijima discovered fullerene-related carbon nanotubes in 1991 using a single evaporator. Mankind's immense scientific investigations and knowledge prowess thus ushered in a new era. The word nanotube is derived from their size, because the diameter of a nanotube is on the order of a few nanometers (approximately 50,000 times smaller than the width of a human hair) and can be upto several micrometers in length. A nanotube (also known as buckytube) is a member of the fullerene structural family.[9] The authors in this paper discussed in details the preparation and purification of carbon nanotubes through various techniques, electrochemistry, and the applications of carbon nanotubes as biosensors.[9] In conclusion of this chapter, the authors deeply discussed about the vast research and development activities in the area of fullerenes and carbon nanotubes for the past one and a half decade.[9] A new scientific genre is emerging in the field of nanotechnology and thus carbon nanotubes application in diverse areas of science and technology is vastly envisioned as civilization treads forward.

Baughman et al.[10] discussed and elucidated with scientific farsightedness the route toward applications in carbon nanotubes. Many potential applications have been widely proposed for carbon nanotubes, including conductive and high strength composites, energy storage and energy conversion devices, sensors, field emission displays, hydrogen storage media, and nanometer-sized semiconductor devices, probes, and interconnects. Man and mankind are today in the middle of scientific introspection and technological validation.[10] Carbon nanotubes applications need to be validated as science and civilization move forward. There are two main types of carbon nanotubes that can have high and vast structural perfection. Single-walled nanotubes and multiwalled nanotubes are the classifications. The authors in this paper discussed with scientific girth and determination nanotube synthesis and processing, carbon nanotube composites, electrochemical devices, hydrogen storage, nanometer-sized electronic devices and a detailed investigation of sensors and probes.[10] Today, application domain of carbon nanotubes needs to be re-envisioned and re-organized with the passage of history of science and visionary time frame. Nanotechnologies of the future in many diverse areas will veritably be built on the advancements of carbon nanotubes.[10]

The challenges, the vision and the scientific grit and determination in the field of applications of carbon nanotubes are vast and beyond scientific imagination. In this entire treatise, the author deeply focuses on the immense scientific advances in the field of nanotechnology. This includes research forays in the field of carbon nanotubes. A new regeneration will surely start in the field of nanomaterials and engineered nanomaterials.

7.9   RECENT ADVANCEMENTS IN NANOMATERIALS APPLICATIONS IN WATER AND WASTEWATER TREATMENT

The global challenges in nanotechnology and water and wastewater treatment are in the process of scaling new heights in present day scientific horizon. In this section, the author deeply deals with some recent advances in nanomaterials applications.

Gopakumar et al.[31] discussed with vision, perseverance, and scientific determination of the state-of-the-art, new challenges and opportunities in the field of nanomaterials. Nanotechnology today is considered an interdisciplinary area of science and engineering integrating engineering aspects with applied science, such as physics, chemistry, and biology. Nanotechnology veritably descends from the present trend of miniaturization in devices and technology and vastly combines all other branches of science and engineering.[31] Nanotechnology is generally considered a tool of science and engineering that steps across the limit of miniaturization where the materials exhibit different behavior as compared with macroscopic level. In the past few decades, the term nanotechnology has been extensively discussed and described and has almost become synonymous for things that are highly innovative and highly promising.[31] On the other hand, the area of nanoscience is a subject of considerable debate and scientific introspection regarding the question of toxic and health hazards of nanoparticles and nanoobjects. The vast alacrity of science, the question of scientific validation in nanotechnology applications, and the challenges of environmental protection will veritably open new windows of innovation and scientific instinct in decades to come. The authors discussed in details about nanomaterials for water filtration, classification of nanomaterials for water filtration, fullerenes, carbon nanotubes, graphene, the application areas of nanocellulose, nanochitin, noble metal nanoparticles, metal-based nanoadsorbents, nanodendrimers, and the vast and wide world of nanostructured membranes. Nanofiber membranes and its applications are the other hallmarks of this treatise.[31] The authors also discussed challenges and limitations of nanotechnology in water and wastewater treatment. Hygienic and clean water are vital to human health and civilization's progress and is a critical feedstock in various industries including electronics, pharmaceuticals, and food technology.[31] The entire planet is today facing tough challenges in provision of pure drinking water due to extended droughts, burgeoning population, more stringent health-based regulations, and competing demands from a wide variety of users. Thus, human civilization's immense scientific stances are in a state of disaster. Nanotechnology for water and wastewater treatment is increasing momentum day by day. Most of the nanotechnologies that have been reported in literature are still in the laboratory research stage and some have succeeded in setting pilot plants. Among the nanomaterials, three categories show the most promising candidates in large-scale applications due to their commercial availability, cost and compatibility with existing infrastructure.[31] A new dawn in science and mankind will emerge if nanotechnology applications move forward in the right scientific direction.[31]

Werkneh et al.[32] deeply discussed with scientific and engineering vision about the applications of nanotechnology and biotechnology for sustainable water and wastewater treatment. Environmental and sustainable biotechnology is the future scientific research endeavor. Water pollution and freshwater scarcity have become a serious problem worldwide causing serious concerns to public health engineering and human health. To eradicate these challenges, various treatment technologies have been adopted.[32] Among these technologies, nanotechnology and biotechnology-based techniques are usually applied separately for water (domestic) purposes and wastewater (reuse) treatment. The conventional water treatment technologies used for the remediation of water pollutants are the activated carbon-based adsorption, membrane filtration, ion exchange, coagulation and flocculation, reverse osmosis, flotation and extraction, electrochemical treatment, advanced oxidation processes, and biosorption.[32] The authors discussed in details of the nanomaterials for the disinfection of pathogenic microbes, photocatalytic applications of nanomaterials, applications of nanomaterials as adsorbent, and their potential use in the field of environmental biotechnology. Bioenergy and environmental biotechnology are the other hallmarks of this chapter. Bioremediation techniques and biotransformation techniques and bioreactor configurations in water and wastewater treatment stand as major pillars of this treatise. Nanotechnology and biotechnology are the two most promising technologies in this century. Nanotechnologies have clearly demonstrated higher removal efficiency of the pollutants from water and wastewater but their toxic effects are questionable. This paper deeply deals with the toxicological perspectives of nanomaterials. A new visionary era in the field of nanomaterials applications in water treatment and diverse areas of science and engineering will evolve, and science and technology will veritably usher in new avenues of scientific might and determination.[32]

Environmental biotechnology and environmental nanotechnology are the most promising areas of research pursuit in the global scenario. In this entire treatise, the author deeply elucidates the present day scientific needs and the scientific comprehension in the field of nanoscience and nanotechnology.

7.10 ARSENIC AND HEAVY METAL GROUNDWATER REMEDIATION AND THE APPLICATION OF NANOTECHNOLOGY

Currrently, environmental remediation is the greatest need of the hour in the critical juncture of an unimaginable scientific crisis. Arsenic groundwater contamination and drinking water poisoning are ravaging and destroying the scientific landscape. In south Asia, mainly in Bangladesh and the Indian state of West Bengal, the situation is absolutely grave. Thus, the need arises of innovations and scientific ingenuity in the domain of nano-technology. Carbon nanotubes and other nanomaterials have tremendous applications in environmental pollution control management. This domain needs to be deeply reorganized as research foray to move forward. Water and wastewater treatment thus are in the midst of scientific introspection and vision. Application of nanotechnology, forays in integrated water resource management and wastewater management, and imperatives of science and technology are the needs currently to be considered. Ground-water remediation and nanotechnology envisioning will surely mitigate this arsenic-related crisis. Biological and biochemical treatments, conven-tional and nonconventional environmental engineering tools will surely go a long and visionary way toward true scientific realization.[11–15]

Hassan[33] described and elucidated the poisoning and risk assessment of arsenic in groundwater. Arsenic groundwater and drinking water contamination are creating havoc in many developing and developed nations around the world. It is world's largest drinking water poisoning in history. The author discussed in details the global scenario of groundwater arsenic catastrophe, spatial mapping, spatial planning, and public participation, and the environmental health concern of chronic arsenic exposure to drinking water. Epidemiological and spatial assessment of risk from groundwater is the other area of discussion in this book.[33] Policy response and arsenic mitigation in Bangladesh are also discussed. Groundwater is the main source of drinking water in many countries of the world but much of the drinking water has been found to be contaminated with high levels of arsenic. It is estimated that more than 300 million people in 70 countries worldwide are at risk of groundwater arsenic contamination. Apart from Bangladesh and neighboring state of West Bengal, India, there have been warnings and grave concerns from Argentina, Chile, Taiwan, Vietnam, China, Pakistan, Thailand, and even the southwestern part of USA.[33] This book deals with the methodological problems of spatial, quantitative, qualitative enquiries on arsenic poisoning, for example, using Geographical Information System to investigate the distribution of arsenic laced water in space-time. The world of scientific and technological challenges in the field of groundwater remediation will surely open new windows of innovation in environmental engineering science and chemical process engineering in decades to come.[33]

Application of nanoscience and nanotechnology in groundwater and drinking water remediation are the promising areas of research pursuit today. Arsenic groundwater poisoning needs to be eradicated at the utmost. Nanotechnology and water treatment are today integrated with vast interdisciplinary areas of science and engineering especially physics, chemistry, and biology. Human scientific vision will thus be a serious eye-opener to the application of nanotechnology in environmental and groundwater remediation.

7.11 THE VISION OF WATER PURIFICATION SCIENCE AND THE APPLICATION OF NANOTECHNOLOGY

Heavy metal groundwater and drinking water contamination are huge burdens of civilization and science today. Man and mankind have practically no answers to the evergrowing concerns of water and air pollution in the global scenario. Thus, the need of a visionary effort in water purification and the application of nanotechnology. Nanoscience and nanoengineering are the two opposite sides of the visionary scientific coin. The scientific needs, the scientific provenance, and the scientific ingenuity of water purification science need to be envisioned and reorganized as mankind moves forward. These futuristic thoughts are the cornerstones of this well-researched treatise.[16–24]

Hashim et al.[23] deeply discussed with vision, insight and lucidity remediation technologies for heavy metal contaminated groundwater. The contamination of groundwater by heavy metal, originating from either natural soil sources or from anthropogenic sources is a matter of immense concern to human health and public health engineering globally. Remediation of contaminated water is of highest priority in the global scientific scenario as millions of people around the world are without clean drinking water.[23] In this paper,[23] 35 approaches for groundwater treatment are reviewed and classified under three large categories which are (1) chemical, (2) biological/biochemical, and (3) physicochemical treatment techniques. A larger scientific vision and scientific ingenuity is today ushering in a new domain called water and wastewater treatment. Public health engineering in developing and developed nations around the world is in the critical juncture of scientific introspection and engineering vision.[23] Selection of a suitable technology for contaminant remediation and groundwater treatment at a particular site is one of the most challenging jobs due to complex soil chemistry and aquifer characteristics and there is no rule of thumb. Civilization's knowledge prowess and scientific stance are thus in the midst of scientific contemplation.[23] Keeping the green sustainability issues in mind, the technologies encompassing natural chemistry, bioremediation, and biological treatments are highly recommended in appropriate case.[23] The authors in this paper[23] elucidate the sources, chemical property and speciation of heavy metals in groundwater, technologies for the treatment of heavy metal contaminated groundwater, in situ treatment, reduction by iron-based technologies, in situ soil flushing, in situ chelate flushing, biological, biochemical, and biosorptive treatment technologies. Scientific vision, scientific and academic rigor in water and wastewater treatment are today in the vistas of newer regeneration. Physicochemical treatment technologies particularly permeable reactive barriers are the other hallmarks of this well-researched treatise.[23] Electrokinetic remediation of soil are the other cornerstones of this research pursuit. Heavy metals are highly toxic for living beings and a burden to public health engineering in the global scenario. It becomes impossible to find out the right causes of groundwater contamination because of the complex speciation chemistry. In this paper,[23] the newer innovations and the newer techniques in minute details.

Singh et al.[24] deeply discussed with vision, scientific fortitude, and scientific determination applications of nanoparticles in wastewater treatment. Emerging pollutants in vast wastewater streams are mostly chemical substances that are highly nonbiodegradable and persist in the environment, bioaccumulate through the food web and highly pose a risk of causing dangerous and adverse effects not only to humans but also to environment, microflora, etc.[24] Nanomaterials are materials measuring between 1 and 100 nm in at least one dimension. The properties of nanoparticles—such as their magnetic, optical, and electrical properties are dissimilar veritably from conventional materials. The authors discussed in details zerovalent metal nanoparticles, zinc nanoparticles, iron nanoparticles, and the vast domain of metal oxide nanoparticles.[24] Wastewater treatment and reuse help to largely maintain and manage the environmental water balance but also raise serious questions. Nanomaterials equipped with chemical, physical, and electronic properties can remove recalcitrant pollutants in water and wastewater. A new scientific discernment and a newer scientific redeeming will surely usher in a new era in nanomaterials and engineered nanomaterials.[24]

7.12 FUTURISTIC FLOW OF SCIENTIFIC THOUGHTS AND FUTURISTIC RECOMMENDATIONS OF THIS STUDY

Environmental engineering science and green sustainability are the immediate needs of man and mankind today. Futuristic flow of scientific thoughts and futuristic recommendations of this study should be targeted toward greater scientific emancipation in environmental sustainability and environmental remediation. Rapid advancements of human civilization and industrialization have truly led to a deep scientific and environmental engineering catastrophe. Water purification science, drinking water treatment, and industrial wastewater treatment are in the middle of a scientific abyss and a deep catastrophe. Thus, the needs arise of a vast scientific and academic rigor in the field of environmental protection science and environmental sustainability. The future of environmental engineering lies in the hands of the science of sustainability and the vast world of environmental nanotechnology. Human life and public health engineering are in the midst of a scientific conundrum as arsenic and heavy metal groundwater contamination devastates the scientific and engineering scenario globally. The futuristic recommendations of this well-researched treatise are vast, varied, and visionary. A new chapter in the field of environmental engineering and environmental nanotechnology will ensue if proper steps of the provision of basic human needs are effectively taken. Economic, social, energy, and environmental sustainability should be the forerunners toward a new era in science and technology.[25,26]

7.13 CONCLUSION, OUTLOOK, AND ENVIRONMENTAL ENGINEERING PERSPECTIVES

Environmental protection science and environmental engineering perspectives are today in the avenues of new scientific regeneration. Global warming and global climate change are urging scientists and engineers around the world to target green or environmental sustainability. The scientific outlook of green engineering and chemical process engineering needs to be revamped as the monstrous issue of global climate change destroys the scientific fabric. The needs of the human civilization are social, economic, energy, and environmental sustainability. The status of research findings in environmental protection and nanotechnology are in the vistas of vision, alacrity, and farsightedness. The future of civilization lies in the hands of scientists, engineers, and governments around the world. The vision, the challenges, and the targets of environmental protection and environmental sustainability are vast and varied. In this treatise, the author with vision and insight focuses on the needs of environmental or green sustainability in the furtherance of science and civilization. Today's world is a world of advancements in space technology and nuclear engineering. Nanotechnology and the science of sustainability are at the forefront of scientific progress of civilization. Thus, the outlook of this treatise targets the vision of the science of carbon nanotubes and the world of challenges in the field of environmental sustainability. Green engineering and green nanotechnology are the forerunners toward a newer visionary era in science and engineering. A newer dawn in the field of nanotechnology and environmental protection will surely emerge if concerted efforts from scientists and engineers toward true realization of green sustainability are emancipated. This treatise truly validates the environmental engineering issues of climate change and loss of ecological biodiversity. Thus, man and mankind will definitely witness the drastic and visionary changes in environmental engineering science.

ACKNOWLEDGMENT

The author wishes to acknowledge the contributions of his late father Shri Subimal Palit, an eminent textile engineer who taught the author rudiments of chemical engineering. The author also wishes to acknowledge the contributions of students, teachers, and the management of University of Petroleum and Energy Studies, Dehradun, India.

KEYWORDS

- **carbon**
- **nanotubes**
- **environmental**
- **protection**
- **vision**
- **sustainability**

REFERENCES

1. Hussain, C. M. Carbon Nanomaterials as Adsorbents for Environmental Analysis (Chapter-14). In *Nanomaterials for Environmental Protection*; Kharisov, B. I.; Kharissova, O. V.; Rasika Dias, H. V., Eds.; John Wiley and Sons, USA, 2014; pp 217–236.
2. Badawy, A. E.; Salih, H. H. M. Nanomaterials for the Removal of Volatile Organic Compounds from Aqueous Solutions (Chapter-5). In *Nanomaterials for Environmental Protection*; Kharisov, B. I.; Kharissova, O. V.; Rasika Dias, H. V., Eds.; John Wiley and Sons, USA, 2014; pp 85–93.
3. Kharisov, B. I.; Kharissova, O. V.; Mendez, U. O. Nanomaterials on the Basis of Chelating Agents, Metal Complexes, and Organometallics for Environmental Purposes (Chapter-7). In *Nanomaterials for Environmental Protection*; Kharisov, B. I., Kharissova, O. V., Rasika Dias, H. V.; John Wiley and Sons, USA, 2014; pp 109–124.
4. Yee, K. F.; Yeang, Q. W.; Ong, Y. T.; Vadivelu, V. M.; Tan, S. H. Water Remediation Using Nanoparticle and Nanocomposite Membranes (Chapter-17). In *Nanomaterials for Environmental Protection*. In Kharisov, B. I.; Kharissova, O. V.; Rasika Dias, H. V., Eds.; John Wiley and Sons, USA, 2014; pp 271–291.

5. Shannon, M. A.; Bohn, P. W.; Elimelech, M.; Georgiadis, J. G.; Marinas, B. J.; Mayes, A. M. *Science and Technology for Water Purification in the Coming Decades*; Nature Publishing Group: London, 2008; pp 301–310.

6. Ren, X.; Chen, C.; Nagatsu, M.; Wang, X. Carbon Nanotubes as Adsorbents in Environmental Pollution Management: A Review. *Chem. Eng. J.* **2011,** *170*, 395–410.

7. Ajayan, P. M.; Zhou, O. Z. Applications of Carbon Nanotubes (Chapter). In *Carbon Nanotubes: Topics in Applied Physics*; Dresselhaus, M. S., Dresselhaus, G., Avouris, Ph., Eds., Vol. 80; Springer-Verlag Berlin Heidelberg, Germany, 2001; pp 391–425.

8. Popov, V. N. Carbon Nanotubes: Properties and Applications. *Mater. Sci. Eng.* **2004,** *R 43*, 61–102.

9. Seetharamappa, J.; Yellappa, S.; D'Souza, F. Carbon Nanotubes: Next Generation of Electronic Materials. *Electrochem. Soc. Interf.* Summer 2006, 23–26.

10. Baughman, R. H.; Zakhidov, A. A.; de Heer, W. A. Carbon Nanotubes: The Route toward Applications. *Science* **2002,** *297*, 787–792.

11. Rickerby, D. G. Nanostructured Titanium Dioxide for Photocatalytic Water Treatment. (Chapter-10). In *Nanomaterials for Environmental Protection*; Kharisov, B. I., Kharissova, O. V., Rasika Dias, H. V., Eds.; John Wiley and Sons, USA, 2014; pp 169–182.

12. Ramanathan, R.; Shukla, R.; Bhargava, S. K.; Bansal, V. Green Synthesis of Nano-materials Using Biological Routes (Chapter-20). In *Nanomaterials for Environmental Protection*. In Kharisov, B. I., Kharissova, O. V., Rasika Dias, H. V., Eds.; John Wiley and Sons, USA, 2014; pp 329–348.

13. Verma, V. C.; Anand, S.; Gangwar, M.; Singh, S. K. Engineered Nanomaterials for Purification and Desalination of Palatable Water (Chapter- 23). In *Nanomaterials for Environmental Protection*; Kharisov, B. I., Kharissova, O. V., Rasika Dias, H. V., Eds.; John Wiley and Sons, USA, 2014; pp 389–400.

14. Hussain, C. M. *Handbook of Nanomaterials for Industrial Applications*; Elsevier: Amsterdam, The Netherlands, 2018.

15. Palit, S.; Hussain, C. M. Environmental Management and Sustainable Development: A Vision for the Future, Chapter. In *Handbook of Environmental Materials Management*; Hussain, C. M., Ed.; Springer Nature Switzerland A. G., 2018; pp 1–17.

16. Palit, S.; Hussain, C. M. Nanomembranes for Environment, Chapter. In *Handbook of Environmental Materials Management*; Hussain, C. M., Ed.; Springer Nature Switzerland A. G., 2018; pp 1–24.

17. Palit, S.; Hussain, C. M. Remediation of Industrial and Automobile Exhausts for Environmental Management, Chapter. In *Handbook of Environmental Materials Management*; Hussain, C. M., Ed.; Springer Nature Switzerland A. G., 2018; pp 1–17.

18. Palit, S. Hussain, C. M. Sustainable Biomedical Waste Management, Chapter. In *Handbook of Environmental Materials Management*; Hussain, C. M., Ed.; Springer Nature Switzerland A. G., 2018; pp 1–23.

19. Palit, S. Industrial vs Food Enzymes: Application and Future Prospects, Chapter. In *Enzymes in Food Technology: Improvements and Innovations*; Kuddus, M., Ed.; Springer Nature Singapore Pte. Ltd.: Singapore, 2018; pp 319–345.

20. Palit, S.; Hussain, C. M. Green Sustainability, Nanotechnology and Advanced Materials- a Critical Overview and a Vision for the Future, Chapter-1. In *Green and*

Sustainable Advanced Materials, Volume-2, Applications; Ahmed, S., Hussain, C. M., Eds.; Wiley Scrivener Publishing: Beverly, MA, USA, 2018; pp 1–18.

21. Palit, S. Recent Advances in Corrosion Science: A Critical Overview and a Deep Comprehension, Chapter. In *Direct Synthesis of Metal Complexes*; Kharisov, B. I., Ed.; Elsevier: Amsterdam, The Netherlands, 2018; pp 379–410.

22. Palit, S. Nanomaterials for Industrial Wastewater Treatment and Water Purification, Chapter. In *Handbook of Ecomaterials*; Springer International Publishing, AG: Switzerland, 2017; pp1–41.

23. Hashim, M. A.; Mukhopadhyay, S.; Sahu, J. N.; Sengupta, B. Remediation Technologies for Heavy Metal Contaminated Groundwater. *J. Environ. Manage.* **2011,** *92,* 2355–2388.

24. Singh, S.; Kumar, V.; Romero, R.; Sharma, K.; Singh, J. Applications of Nanoparticles in Wastewater Treatment, Chapter-17. In *Nanobiotechnology in Bioformulations, Nanotechnology in the Life Sciences*; Prasad, R. et al., Eds.; Springer Nature Switzerland AG, 2019.

25. www.wikipedia.com (accessed on Feb 1, 2020).

26. www.google.com (accessed Feb 1, 2020).

27. Barrow, C. J. *Environmental Management and Development*; Routledge, Taylor and Francis Group: Abingdon, UK, 2005.

28. Attar, S.; Ranveer, A. Carbon Nanotubes and Its Environmental Applications. *J. Environ. Sci. Comput. Sci. Eng. Technol.*, March–May **2015,** Sec C, *4* (2), 304–311.

29. Wang, Y.; Pan, C.; Chu, W.; Vipin, A. K.; Sun, L. Environmental Remediation Applications of Carbon Nanotubes and Graphene Oxide: Adsorption and Catalysis. *Nanomaterials* **2019,** *9,* 439. doi:10. 3390/nano9030439, pp-1-25.

30. Rahman, G.; Najaf, Z.; Mehmood, A.; Bilal, S.; Shah, A. H. A.; Mian, S. A.; Ali, G. An Overview of Recent Progress in the Synthesis and Applications of Carbon Nanotubes. *J. Carbon Res.* **2019,** *5* (3), 1–31. doi:10. 3390/c5010003.

31. Gopakumar, D. A.; Pai, A. R.; Pasquini, D.; Leu, S-Y. H. P. S. Khalil, T. S. Nanomaterials-State of Art, New Challenges and Opportunities, Chapter-1. In *Nanoscale Materials in Water Purification*; Thomas, S. et al., Eds.; Elsevier: The Netherlands, 2019; pp 1–23.

32. Werkneh, A. A.; Rene, E. R. Applications of Nanotechnology and Biotechnology for Sustainable Water and Wastewater Treatment, Chepter-19. In *Water and Wastewater Treatment Technologies, Energy, Environment and Sustainability*; Bui, X. T. et al., Eds.; Springer Nature Singapore Pte. Ltd., 2019; pp 405–430.

33. Hassan, M. *Arsenic in Groundwater: Poisoning and Risk Assessment*; CRC Press, Taylor and Francis Group: Boca Raton, Florida, USA, 2018.

IMPORTANT WEBSITES FOR REFERENCE

https://www.hindawi.com/journals/jchem/2013/676815/
https://academic.oup.com/toxsci/article/92/1/5/1642931
https://en.wikipedia.org/wiki/Carbon_nanotube
https://www.nanowerk.com/nanotechnology/introduction/introduction_to_nanotechnology_22.php
https://www.sciencedirect.com/topics/materials-science/carbon-nanotubes

https://www.cheaptubes.com/carbon-nanotubes-properties-and-applications/
https://www.intechopen.com/books/carbon-nanotubes-polymer-nanocomposites/
    functionalization-of-carbon-nanotubes
https://www.researchgate.net/publication/325786477_Carbon_Nanotubes_A_Review_Article
http://globalresearchonline.net/journalcontents/v13-1/022.pdf
https://pdfs.semanticscholar.org/350d/b5de1bb5c39446c60760a2c9422f297a57ca.pdf
https://iopscience.iop.org/article/10.1088/1757-899X/270/1/012027
https://core.ac.uk/download/pdf/81585491.pdf
http://pubs.sciepub.com/nnr/4/4/1/
http://www.insituarsenic.org/
http://www.insituarsenic.org/details.html
https://www.environmentalscience.org/sustainability
https://en.wikipedia.org/wiki/Sustainability
https://www.mdpi.com/journal/sustainability
https://www.epa.gov/
https://en.wikipedia.org/wiki/United_States_Environmental_Protection_Agency
https://www.epa.gov/environmental-topics

CHAPTER 8

RHEOLOGICAL BEHAVIOR OF CARBON NANOTUBES-BASED MATERIALS AND ITS ROLE IN PROCESSING INTO VARIOUS PRODUCTS

ANDREEA IRINA BARZIC

"Petru Poni" Institute of Macromolecular Chemistry, Laboratory of Physical Chemistry of Polymers, 700487, Iasi, Romania

E-mail: irina_cosutchi@yahoo.com

ABSTRACT

The processing of carbon nanotubes (CNTs)-derived composites is imposing a close knowledge on the fluid phase properties. Solution or melt compounding involves a specific response to the shear deformation, which can be investigated by means of the rheology. Basic notions on this experimental method are presented with accent on the flow phenomena occurring in the CNTs-reinforced fluids. Some case studies are described, revealing the effects induced by the degree of reinforcement, CNTs functionalization, and used matrix on the rheological parameters. In addition, the rheological response will be accounted for the selection of the proper processing conditions of the CNTs composites from solution or melt state into uniform and stable coatings, films, or fiber-like materials. Such investigations lie at the basis of the production of a wide range of commercial products.

8.1   INTRODUCTION

The appearance of nanometric fillers, like carbon nanotubes (CNTs), in the scientific community led to remarkable breakthroughs, which were reflected in a surprisingly increased number of publications and the emergence on the market of a huge amount of products categories based on nanotechnologies.[1-4] The development of various sorts of materials containing one or more nanocomponents is still continuing to the present days since there are multiple aspects that are not completely elucidated.[5] Moreover, there are still fresh aspects to be emphasized and which could potentially allow upgrading and expanding the current knowledge. This is expected to enhance the performance of the present technologies and also to provide the premises to create new modern ones.

Each commercial product must fulfill well-established criteria and hence the synthesis and processing steps must be carefully monitored to ensure the accomplishment of the quality standards. Therefore, fundamental research must be made for clarification of the parameters affecting the structure–properties relation. Processing of the materials, particularly those made of several components, requires deep investigation of the flow behavior in specific conditions. Besides the microscopy techniques, rheology is an outstanding tool for the analysis of the microstructure changes in fluid phase (solution or melt) as a result of variations in the material's composition, applied deformation, temperature, solvent type, additive properties, and many other aspects.[6] The use of reinforcement agents, such as CNTs, into a matrix imposes a fine control of their state of dispersion because this is mirrored onto the physicochemical characteristics of the composite. For example, electrical or thermal conduction abilities of CNTs-containing materials are not dependent only on the CNTs quantity, they are better when the filler is uniformly distributed within the matrix.[7] These aspects are linked to the balance between the polymer–filler and filler–filler interactions. Occurrence of the filler network in the matrix generates extraordinary modifications in the properties of the composite.[8] The percolation of anisotropic CNTs and the shift from isotropic to nematic state bound the domain of system composition over which the shear flow characteristics are governed by the mesoscale structure and dispersion,[9] which are of great relevance to the processing of CNTs-based composites. The percolation point denoting the change from isotropic to nematic transition is highly dependent on the inverse of the effective aspect ratio of the inserted CNTs and consequently limits the concentrations zones over which the reinforced materials could be deployed.[10,11] However, in some cases, even strong doping does not lead to the desired performance of the composite because there are certain limitations due to small interactions or improper interphase compatibility.[12] For such situations, chemical functionalization of nanotubes was performed through a variety of approaches to solve the aforementioned drawback.[13,14]

All the above aspects are reflected in the flow behavior since the CNTs network is sensitive to the intensity of the applied shear rate producing changes in the composite microstructure.[15,16] Depending on the material composition, shear viscosity might become very dependent on the applied shear rate and the flow curves are displaying a pronounced non-Newtonian behavior since the friction inside the composite is modified. Another rheological parameter that is sensitive to the composite composition is the storage or elastic modulus.[17] Progressive reinforcement determines a change in the rheological response through the underlying structure of the nanocomposite material and interplay of several forces. Shear oscillatory measurements are good indicative of the transition from isotropic phase to percolation one within the material. Flow behavior in solution or melt state of the CNTs-based composites is important for establishing the processing window that allows the preparation of uniform coatings, films, fibers, or other types of products.[18–21]

The present chapter covers important aspects concerning the rheological behavior of CNTs-based materials and their role on processing from fluid to solid phase. The first section of the manuscript will provide a concise overview of basic notions on rheology of reinforced materials with CNTs, emphasizing the relevance of shear flow experiments. The following section will describe various reports showing the importance of the chosen rheological test, system composition and interaction forces between CNTs and matrix, particularly for the case of functionalized CNTs. The last section will present the implications of the shear flow behavior of some CNTs composites in obtaining uniform and stable coatings, films, or fibers. The chapter concludes with the ongoing challenges and future outlook.

8.2   BASICS IN RHEOLOGY OF CNTs-REINFORCED FLUIDS

During compounding, a composite material must be brought in fluid phase and in order to shape, it is mandatory to know its response to deformation. Rheology is the most adequate tool for characterization of the processes happening when liquids are deformed and start to flow. This method establishes a connection among the force, induced deformation, and time.[22–24] For fluids, deformation is not entirely reversible, like in the case of elastic solids, since they display a viscous nature. Ideally, viscous fluids present an irreversible strain when the imposed forces are applied to them. Among the rheological properties, it is paramount to achieve data on viscosity and viscoelastic characteristics. Flow behavior of a fluid can be described by the connection among shear stress and shear rate. Besides the properties of the examined sample, there are some factors, such as temperature, shear rate, pH, and electromagnetic field intensity, which affect the viscosity.[25,26] Depending on the shear level, the viscosity might change (non-Newtonian fluids) or not (Newtonian fluids). Another essential aspect is that fluids often present a combination of viscous or elastic properties. In linear viscoelastic zone, solutions and melts are keeping a temporary network of entanglements upon shear, which in non-viscoelastic domain is destroyed.

Rheology is widely used in several applications areas that deal with fluid deformation under external forces. Depending on the pursued response to deformation, there are known at least three main sorts of devices[26]:

- Rotational or shear rheometers: act through rotary motion. The device enables the control of the magnitude of the shear stress or shear strain. Such instruments were generally projected as a strain-controlled device (control and use a predefined shear strain and subsequently records the corresponding shear stress) or a stress-controlled device (control and use a precise shear stress and registers the shear strain).
- Extensional rheometers: act through rotating drums. The device involves the use of extensional stress or extensional strain to analyze the fluid sample.
- Capillary rheometers: use the capillary action under the piston movement. The fluid sample is constrained to flow out of a cylinder via the capillary die.

For a clearer image of the rheometer devices, Figure 8.1 depicts the main categories of rheometer devices. Taking into account the sort of employed measuring system, further classification of shear rheometers can be made as follows[26]:

- Cone and plate rheometer
- Rotating cylinders rheometer
- Plate–plate rheometer

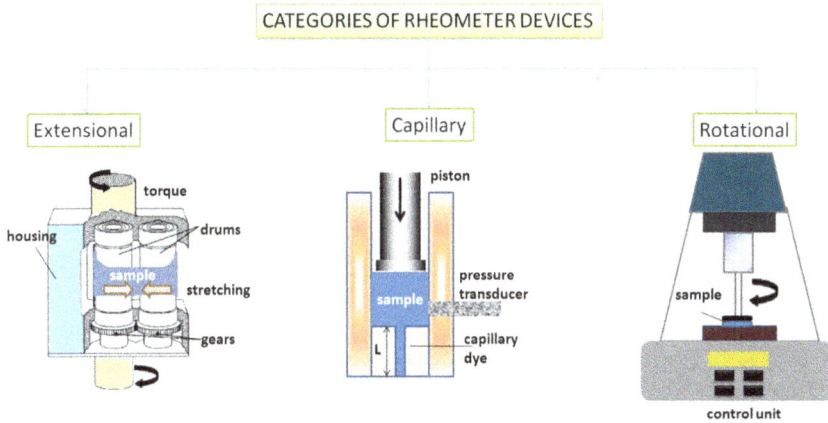

FIGURE 8.1 The illustration of the main categories of rheometer devices.

Fluid rheology experiments, made using rotational rheometers, can be divided in two types[27]:

- Rotational viscosity experiments: allow registering of the shear flow and the viscosity curves by presetting the value of strain or deformation rate. Such measurements provide data regarding the sample resistance to flow. During a strain-controlled test, the resulting deformation rate is obtained as the answer of the fluid, whereas during a deformation-controlled test, the resulting strain is achieved. Rotational measurements are often performed with cone–plate measuring system, since in this manner, the deformation rate is not impacted by the distance to the middle of the plate and hence is constant. On the other hand, the cone degree value determines a specific range of shear rate interval within the experiment can be done.

- Oscillation measurements: give information on elasticity, stickiness of the analyzed material. Here there are some subcategories of tests that can be made:

  - Amplitude sweep: the frequency is maintained constant, while the amplitude of the deformation/strain signal is enhanced step-wise from one measuring point to the next one. The diagram with strain (or shear stress) plotted on the x-axis and shear moduli on y-axis allows discerning the linear viscoelastic domain, that is, the point at which storage modulus is no longer constant because of the fluid structure disruption.

- Frequency sweep: serves to elucidate the time-dependent properties of a fluid in the non-destructive deformation interval. For such experiments, the shear frequency is ranged and strain or deformation amplitude is held constant. Larger frequencies can be applied to mimic fast motion on small timescales, but low frequencies are proper for describing slow motion on bigger timescales or at rest. At a deformation-controlled oscillatory test, the fluid is subjected to a periodic deformation and thus a periodic strain is generated, revealing displacement of phase δ relative to the preset deformation. The values of δ and the dissipation factor (tan δ) help to discern the nature of the studied fluid. So, tan $\delta<1$ is noted for the elastic samples and tan $\delta>1$ is depicting viscous fluids.

- Time sweep: is done at fixed value of the strain or deformation amplitude and constant angular velocity. The modifications in the fluid features over time are registered from the time-dependent behavior.

- Temperature sweep: the temperature ramp at fixed angular frequency and deformation is used. The temperature-dependent measurement is extracted from the dependence of shear moduli on temperature.

Based on the previously exposed problematic, a diagram can be constructed to highlight the principal properties that can be extracted from shear rheological investigations (Fig. 8.2).

For fluids that contain more than one solute, the rheological analysis is more complex and depends on the nature of the system components.[28-30] For instance, when dealing with reinforced fluids like polymers in solution or melt phase containing variable amounts of CNTs, the shear flow behavior reflects the mutual influence of each component on the overall rheological properties.

In dilute domain, where CNTs are apart from each other, they are interacting with polymer chains (molten samples) and possibly with solvent molecules (solution samples).[10] So, if excluding the Brownian motion of isolated and axisymmetric CNTs, one may remark a time periodicity of the angle of the axis of symmetry along the flow. The viscoelastic response of the macromolecular solution is far exceeding than that of the dispersed

CNTs. In the absence of polymer, the reduced viscosity is affected by the filler concentration, while the relaxation time is unchanged by the CNT concentration. Further the introduction of CNTs in the solvent determines a transition from isotropic to biphasic system and finally to a single liquid-crystalline state with randomly disposed domains.[31,32] It is possible to control the liquid-crystalline ordering via CNTs–solvent interactions, aspect of great significance for fiber spinning.

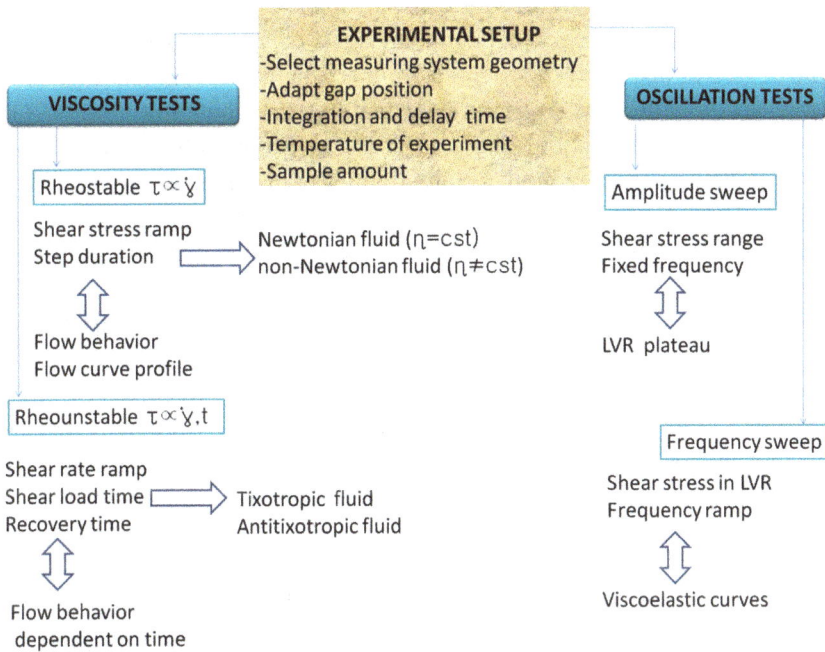

FIGURE 8.2 The scheme of the principal characteristics of fluids that can be achieved from shear rheological investigations.

The main issue of semi-dilute regime of CNTs in a system relies on overcoming of the strong inter-tube attractive forces that impede proper dispersion. For the situation of a polymer that displays no specific interactions with such fillers, the driving force for dispersion rise particularly from the gain in translational entropy, known to be very low for high molecular weight polymers and anisotropic nanoparticles and thus generates the de-mixing and weak dispersion.[33] In the situations which the polymer and CNTs display attractive interactions, rendering in polymer adsorption on to the fillers, the effective inter-CNTs attractions might be highly attractive at low polymer amounts and low to moderate adsorption strengths, while becomes short-range repulsive at large concentrations and elevated adsorption strength.[10] If the macromolecules are grafted onto CNTs, the excluded volume repulsion among the polymer chains that are tethered is prevalent and stabilizes the CNTs dispersion. Grafting or adsorption of small/large molecules on CNTs produces short-range repulsion thereby avoiding their aggregation and facilitating the dispersion. Such methods are useful at low and intermediate CNT concentrations, but at very high amounts, CNTs dispersion can be difficult. The degree of dispersion is often determined via the percolation threshold. More specifically, linear viscoelastic features are changing upon CNTs reinforcement from liquid-like to solid-like. Structurally, at this point of CNT percent, the network begins to appear in the matrix. Thus, for a polymer found in melt or liquid phase in the presence of small-amplitude linear oscillatory shear, in the small frequency (f) zone, the storage G' and loss modulus G", present a dependence as follows: $G' \sim f^2$ and $G'' \sim f^1$. In these conditions, the complex viscosity does not range with the variation of frequency, the sample depicting Newtonian behavior. Further introduction of CNTs in the matrix gradually changes the liquid-like dependency into a solid-like non-terminal behavior, and for this case, the complex viscosity at small frequencies diverges with $|\eta^*| \sim 1/f$. The evolution of the structural properties of the CNTs-based material at variable filler amount is shown to follow a particular sigmoidal dependence.[34] Beyond the rheological percolation threshold, inside the reinforced material, a continuous path is formed, spanning the matrix and consequently a sudden change in the physical properties occurs. More precisely, if discussing the electrical properties, one may remark that the insulating matrix is gradually becoming semiconducting and finally conducting, depending on the quantity of the embedded CNTs.

The effective aspect ratio of the filler and the percolation threshold can be viewed as interdependent factors which are inversely proportional. They can be determined from concentration-dependent linear viscoelastic characteristics.[35] The concentration ascribed to the percolation is parameter which depicts the onset of a percolation transition and is impacted by the system features. Generally, the structural/viscoelastic attributes of the CNTs composite, unlike the electrical ones, are affected by the rheological manifestation of the percolation, which in turn lacks imposing an absolute connectivity among the carbon fillers.[36] Hence, in the case of 3D isotropic/ random dispersion, the rigidity percolation foregoes the connectivity percolation or the electrical one is found to be bigger in regard to the geometrical threshold.

It is very important to highlight the fact that not only the reinforcement degree but also the state of dispersion determines the percolation threshold, together with the CNTs polydispersity and the local clustering.[10] The CNTs size polydispersity cannot be evaluated using simple experimental approaches. On the other hand, if a known volume fraction of chaotically disposed CNTs is placed in a matrix, the occurrence of the local clustering should produce an augmentation of the percolation threshold.

In the semi-dilute to concentrated regime, in the vicinity and beyond the percolation concentration, the shear flow behavior is ruled by the mesoscale superstructure. Also, in this range, beyond the typical time–temperature superposition, the CNTs composites display time–temperature–composition superposition—feature found for weakly attracting particles, like anisotropic carbon black fillers.[10] The linear viscoelastic reinforced fluids display an identical viscoelastic response in time and frequency domains. When the CNTs composite is filled in excess, the percolation superpositioning is no longer followed since there is a concurrence between the magnitudes of the matrix viscoelasticity and the fractal network and the variable nature of the filler network superstructure at concentrations related to the percolation point. Moreover, the introduction of larger quantity of CNTs in the matrix makes the dispersed anisotropic filler to adopt a nematic ordered structure so it impedes the analysis with fractals. At extremely high CNTs loading of the samples, the filler alignment in response to the handling could be an essential issue and disorientation kinetics in such nanocomposites have been shown to be extremely slow.[37]

In the semi-dilute regime, the nonlinear viscoelasticity must be discussed by highlighting the following aspects:

• Shear stress relaxation at large deformation.
• Steady shear properties.
• Flow generated alignments.

In the case of semi-dilute dispersions of CNTs included in a matrix, the stress relaxation behavior in reaction to a step shear strain depending on the applied strain amplitude can be categorized into three regimes:

- The zone of small deformations, where the sample has a response which is linear and the relaxation modulus is not affected by the strain amplitude. Here, the largest value for recording the linear properties is inversely varying with filler amount.
- The zone after the critical strain, where the stress relaxation displays a strain-softening with the appearance of the relaxation curve being maintained and enable to utilize the time-strain separability.
- The zone at highest deformations (very far from critical strain) where the relaxation spectrum cannot be preserved, while the time-strain superposability is invalid.

After exceeding the percolation point, the storage modulus is enhanced upon the CNTs reinforcement of the matrix. The same aspect is noted for the shear sensitivity of the structural elements, while the onset of non-linear behavior is described by a power law,[10] suggesting that the CNTs introduction in the matrix leads to stiffer and more fragile composite materials.[38,39] Similar behavior can be encountered for fractal networks, likes those of colloidal gels or flocculated silica spheres.[40] Another important aspect is that the shear sensitivity of the network reflected in the function, if it is scaled by the concentration shift factor, downfalls, and results single master curve. Based on this tendency of concentration shift factor, one may explain the modification of the "linear" viscosity caused by the dispersion of anisotropic fillers. Such notion can be used to depict the scaling of the non-linear deformation in CNTs composites, showing that the effective deformation of the CNTs suspension in the intermediate range of strain amplitudes can be "affine." Using mathematical approaches, it can be noted that the factorized non-linear memory function for self-similar CNTs network seems to be consistent with the concentration scaling of the composite function.[41] The local strain-dependent deformation can be valid for the fractal materials where small short-range interactions prevail; conversely, this fails for the composites characterized by dominant long-range interactions owing to H-bonding, ionic and macromolecular bridged gels.[10] The time–temperature–composition superposition indicates that beyond the percolation point, both the linear and non-linear viscoelasticity are impacted predominantly by the network superstructure. Overall, the non-linear viscoelastic regime could be imparted in two subregimes:

- The first one is linked to the reversible character of the network deformation, while the superposition principle is valid.

- The second one depicts the irreversible or permanent deformation of the sample, where the recovery process is very slow.

When dealing with steady shear properties and non-linear viscoelasticity, it is essential to mention that the CNTs reinforcements form elastic networks with yield stress. At compositions much beyond percolation, the use of steady shear produces a stress overshoot (separated from that of the matrix itself) which settles to a steady value at long times. The network superstructure, subjected to steady shear, begins to rearrange locally to house the displacement.[42] The stress enhancement is a consequence of the structural variations which are produced by the aggregation due to the collisions of the anisotropic filler clusters and the occurrence of novel bonds among them. At times that exceeding the one ascribed to the maximum value of stress, the fractal network is disrupted, lowering the effective stress supported by the network. Hence, the overshoot stress, or the maximum stress can be understood as being similar to a yield stress beyond which the network is able to flow. The time demanded to achieve the maximum stress was found to be less by the CNTs amount and more by the applied shear rate.

Another subject discussed in literature[10] is represented by the flow-induced structure of the CNTs dispersions, including the CNTs-directed hierarchical alignment of the macromolecules (from unit cell to lamellar arrangement).[43,44] For the situation of the CNTs-directed controlled crystallization, the alignment of polymer crystals can be accomplished when the filler is oriented by a shear or elongational flow fields or electrical fields.[44]

8.3 RHEOLOGY OF CNTs COMPOSITES AND ITS ROLE IN PRODUCT PROCESSING

Most literature reports dealing with rheological behavior was mainly studied for CNTs composites having polymer as matrix, such as epoxy resins,[17,45–47] poly(ethylene vinyl acetate),[48,49] poly(vinylidene fluoride),[50,51] polyamide,[52] polycarbonate,[53] polyethylene,[54–56] polypropylene,[57] and polystyrene.[58,59]

Pötschke et al.[34] performed a comprehensive rheological investigation of CNTs/polycarbonate composites prepared by diluting a masterbatch containing 15 wt.% of CNTs in a twin-screw extruder. Oscillatory experiments performed in LVE zone documented relevant changes in viscosity and shear moduli as a function of the system composition. Below 2 wt.% of CNTs in the sample, the flow behavior remains Newtonian alike to that of the used matrix. Upon filler introduction (>2 wt.% of CNTs) in the system, a great increase in the viscosity was observed particularly at low frequencies. Such differences between reinforced and pristine systems were hardly noted at higher frequencies owing to shear thinning behavior of the composites. Also, a concurrent enhancement in shear rheological moduli was observed, with an outstanding change of the frequency-dependence of the shear moduli in the low frequency zone. This produced the fadeaway of the homopolymer-like terminal behavior noticed for the pure polymer matrix. Based on this observation, it can be considered that 2 wt.% represents the rheological percolation threshold, which was further correlated with the electrical one.

Mirjalili et al.[60] showed that incorporation of CNTs in polymer resins such as epoxy produces an increase in the system viscosity. Hence, the processing of the loaded resins is challenging rendering poor composite performance.

Guadagno et al.[45] studied the impact of non-covalent functionalization of the CNTs and introduced such fillers in epoxy matrix. The rheological behavior of these liquid dispersions revealed some benefits that arise from avoiding agglomeration in the filler dispersion and this efficiently determines the lowering of viscosity of the reinforced epoxy material. In this manner, it is contrasted an important issue linked to the manufacturing processes of the CNTs-derived composites at reinforcement degrees beyond the electrical percolation point.

Yearsley and coworkers[47] investigated rheological properties of multiphase epoxy/CNTs/carbon black (CB) composites highlighting the rheological changes that are seen in the linear viscoelastic (LVE) domain. The CB and CNTs suspensions exhibit similar rheological behavior and the corresponding microstructures are found to be shear rate sensitive. A structure-dependent Maxwell-Voigt phenomenological model having certain yield stress was formulated, offering a good fit to the rheological data. Both systems present almost no differences in the values of the structure model parameters, even if the onset of rheology development took place at a concentration ranging with a decade. From the recorded data, one may conclude that the rheological changes are prevalently affected by aggregate dimensions and interactions, rather than the respective particulate and filament topography of the both filler microstructures. A similar analysis was performed by Sumfleth et al.,[17] which also examined the rheological characteristics of epoxy filled with CNTs/CB. As a result of the kindred mechanisms of dynamic agglomeration, in the cured state, there is no significant difference among the rheological and the electrical percolation threshold for the CNTs and CB samples. Because of this percolation coincidence, in uncured epoxy composites, the non-covalent interactions are insignificant. Moreover, after curing, the electrical percolation point is higher in regard to that of the uncured composite, since there is a great tendency of the CB and CNTs to produce conductive networks while curing. The curing conditions were found to be responsible for the dissimilarities between rheological and electrical percolation. In these conditions, the rheological percolation point should be viewed as an upper boundary for the electrical percolation threshold in the cured state. The occurrence of multifiller CB/CNTs networks in the matrix makes the registered flow behavior not so distinctive to that of the binary CNTs suspensions. On the other hand, the binary CB composites present a bigger percolation threshold. The observed differences among the binary CNTs suspension and the ternary CB/CNTs samples in elastic modulus at larger filler amounts turn out to be less than expected. Such synergistic effects in filler network appearance already exist in the epoxy suspension and become more obvious upon curing.

Hassanabadi et al.[48] prepared binary mixtures of CNTs/montmorillonite nanoclay, which were introduced in ethylene vinyl acetate (EVA) copolymer. The purpose of their work was to check if regardless of particle features they could notice a universal rheological behavior near the percolation. Therefore, rheological tests were made under small-amplitude oscillatory shear (SAOS), large amplitude oscillatory shear (LAOS), and transient shear step. SAOS recordings revealed that incorporation of the filler had a huge influence on the reptation relaxation time (τD) and less impact on the dynamics related to the Rouse relaxation time (τR). During the step shear transient experiments, it was noted a decrease of the critical shear rate for overshoot appearance as a result of chain confinement, whereas the occurrence of filler network strongly enhanced the level of stress overshoot. LAOS data indicated that formation of particle networks augmented the nonlinear parameters. In the rheological investigations, it was remarked that owing to the CNTs hollow structure, large surface area ascribed to its size and density, this filler displayed stronger effects in regard to nanoclay particles. Another important aspect is that, while the percolation point is dissimilar for CNTs and clay, both composites had analogous behavior at percolation, revealing a Rouse-dominated behavior.

As aforementioned, the non-linear oscillatory shear tests are highly useful for deeper investigation of the microstructure evolution in multiphase materials containing CNTs as reinforcements.[61-63] Kamkar et al.[50] employed non-linear rheology to examine the influence of secondary filler, namely, manganese dioxide nanowires (MnO_2NWs), on network structure of CNTs/poly(vinylidene fluoride) composites. Addition of the second filler in the system determined an improved state of dispersion of the CNTs in the polymer environment as also shown by microscopy techniques. The steady oscillatory experiments were made in the LVE domain to clarify the impact of the filler network on the flow conditions and viscoelastic characteristics. The transient response during steady shear flow emphasized that the stress overshoot of composites containing the two fillers magnified considerably in regard to the composites containing one type of filler. The latter can be ascribed to the microstructural changes in the analyzed multicomponent fluids. LAOS experiments were useful for clarification of the impact of MnO_2NWs on nonlinear viscoelastic properties of the composites. Based on the quantitative non-linear parameters, such as strain-stiffening ratio and shear-thickening ratio and also Lissajous-Bowditch graphs, one may notice that a more rigid filler network is achieved for the polymer loaded with binary fillers owing to the improved dispersion degree of CNTs. Considering this aspect, the composites with better dispersion of the fillers present a bigger number filler–filler and filler–polymer interaction sites. Hence, the rigid network of the multifiller polymer is more sensitive to deformation (e.g., input strain amplitude), which was revealed by the non-linear viscoelastic recorded data.

Polystyrene (PS)/CNTs composites are largely studied in literature due to their high applicative potential in electronics, membranes, packaging, and the automotive industry.[56,64-68] The common rheological behavior of such CNTs composites lies in the achievement of larger viscosity and shear moduli values upon continuous reinforcement of PS with CNTs. Similarly, to other CNTs/polymer systems, a deviation from Newtonian flow and a transition from liquid-like to solid-like behavior are remarked as the filler amount is higher in the matrix. As expected, this is attributed to the appearance of the filler network, which impacts the relaxation dynamics of the PS chains.[64-68] Amr and coworkers[64] reported a distinct rheological response for polymer composites loaded with nitric acid-treated CNTs. They noticed a decrease in the complex viscosity for the samples containing 0.1 and 0.5 wt.% of the treated CNTs, with respect to the pure polymer. Such uncommon behavior could be explained by the absence of inter-tube interactions and with the plasticizing action of the embedded particles at such low reinforcement percents. The importance of CNTs chemical modification with organic diazonium compounds on the rheological features of monodisperse PS was reported by Mitchell et al.[65] Linear viscoelastic recording on matrix and composite melts indicated a reduction of the percolation threshold upon the functionalization of CNTs since there is a bigger compatibility in regard to unmodified fillers. Extending rheological studies by varying the temperature and applying time–temperature principle led to new aspects related to terminal region. For PS containing modified CNTs, the elastic modulus displays frequency-independent behavior and also a finite yield stress. This determines a divergence in the complex viscosity and complex modulus data as a result of the interference of a percolated CNTs structure. Such aspects are remarked for the CNTs-derived systems but at larger filler amounts with respect to the fluids containing modified CNTs probably owing to the weaker particle dispersion in the latter case. Kota et al.[66] proposed a quantitative analysis of interpenetrating phase in the reinforced PS via melt rheology. For this purpose, the PS/CNTs was considered as an amalgamation of two phases: a continuous phase of polymer fluid doped with non-interacting CNTs and a continuous phase represented by the solid-like network of percolated CNTs. Using electrical percolation data they were able to evaluate the role of each phase to rheological characteristics of the composite fluid, allowing the achievement of the features of the continuous CNTs path and the isolation of the impact of non-interacting CNTs present in the macromolecular matrix. Then, the oscillatory data at variable frequencies indicated an interesting behavior related to "scaffold-like" microstructure that revealed a stick-slip friction mechanism at the interface of the percolated filler network at bigger shear frequencies. An interesting report that relates rheology data to processing of the PS/CNTs composites by extruding is that of McClory et al.[67] When placing the composite fluids in a twin-screw intermeshing co-rotational extruder, they observed an enhancement of the rheological functions (viscosity, elastic modulus, loss tangent) when augmenting the screw speed. In any case, at the applied speeds, the disentanglement of the primary CNTs agglomerates does not occur. The sample filled with 5 wt.% of CNTs is reaching the percolation threshold at all used screw speeds, showing that for a high CNTs content, a wide distribution of filler dispersion is attained regardless the magnitude of the screw speeds. The remarked arrangement of CNTs within the fluid PS medium is enough, from a rheological point-of-view, to change macromolecular dynamics.

The impact of distinct processing techniques on the rheological properties of the pristine/modified CNTs with phenyl propane ester and polystyrene functional groups was described by Faraguna et al.[68] To this end, they prepared some composites by melt blending and solution-mixing of two PS of distinct molecular weight and chemically functionalized CNTs. Linear viscoelastic data indicated that solution-mixed fluids displayed bigger complex viscosities and smaller percolation thresholds, comparatively to melt-mixed ones. A Kindred study was published by Kamkar and collaborators[69] for PS/CNTs materials, but here the analyses were conducted toward nonlinear rheological response with respect to the processing approaches. In the linear viscoelastic zone, solution-mixed samples presented a prevalent elastic response, whereas the melt-mixed materials displayed an overall viscous behavior. This finding, linked to the information extracted from the small critical strain amplitude for the onset of the non-linear domain, allowed emphasizing a stronger and denser interconnected network of filler in solution-mixed polymer specimens than in melt-mixed ones.

For a deeper understanding of the shear flow connection with the processing conditions, a case study will be presented based on the data from reference.[59] Typical shear thinning for PS solutions containing various amounts of CNTs was achieved especially at low shear rates. The dependence of the composite viscosity on the applied deformation is essential for determining the suitability of a certain deposition method. In some situations where the composite is very fluid, a good method for obtaining thin layers is spin coating. This method was previously applied for other polymers,[70] where the approach of Acrivos[71] was employed to relate the spin coating processing with shear flow features of non-Newtonian fluids. According to Acrivos,[71] the thickness for as-deposited wet polymer layer is linked to the fluid viscosity. During polymer layer formation, the fluid is under the influence of an inherent non-uniform shear field, which has a variable intensity from the center to the margins of the spinning disk. Consequently, this aspect produces variations in the uniformity of the wet film thickness. Figure 8.3 depicts the assessed wet film thickness against radial position for PS solution and its composites with CNTs processed by spin coating. The thickness of the CNTs-derived composite fluid and its corresponding matrix is impacted by the shear forces acting during processing which are reflected on the viscosity magnitude. The distribution of the wet composite layer thickness along the rotating disk radius is influenced by the rheological behavior. As indicated by Acrivos,[71] the shear thinning properties of the PS and its CNT-doped solutions determine a less uniform layer thickness. It can be noticed in Figure 8.3 that these fluids present a domain where they have a non-uniform wet layer thickness profile. The thickness value of these composites is disturbed in a proportional manner with viscosity variation during shearing. More precisely, the pristine PS solution, which is characterized by a larger domain of constant viscosity, has a bigger zone of uniform film thickness that decreases toward the margins of the spinning disk. Upon the addition of CNTs in the system, the viscosity becomes more and more sensitive to the shearing so the thickness of the composite layer is constant only to limited regions of the rotating disk.

In the cases where the composite solutions must be processed by electrospinning, it is widely known that the viscosity of the system should be very high in order to achieve defect-free fibers.[72] The solution conductive properties are also essential for this purpose. Therefore, the introduction of fillers like the CNTs would increase both viscosity and conductivity of the fluid so one may expect to attain fibers that lack morphological defects.

On the other hand, the pseudoplastic behavior of the most CNTs composite fluids is suitable for processing by tape casting technique. The variations of the viscosity below the blade can be correlated to those observed by means of rheology. According to the theory developed by Chou et al.,[73] there is a semi-empirical formula that connects the layer thickness to the solution viscosity. Based on this theory the film thickness is affected by two factors, namely, the shear-driven flow and the pressure-driven flow. The assessed thickness for the PS and PS/CNTs solutions is impacted by the height of the blade's edge, but also on coating velocity and rheological features of these fluids. Throughout the tape casting, the wet layer is sheared as a result of the surface tension of the solution. This is the reason why the changes in the system viscosity with the coating speed are of paramount importance. The dried film thickness reflects the effects of the solid content and the densification tendency of the PA and PS/CNTs fluid.

Figure 8.4 reveals that at low casting speeds, the shear flow characteristics are transposed in a decrease of the dry film thickness. The enhancement of the casting speed produces the diminishment of the specimen thickness. As indicated in literature,[73] the drag force becomes higher at larger velocities, and it is prevalent in regard to the pressure force. This generates the stretching of the fluid along the peeling belt. Considering these problems, one may notice that at casting speeds exceeding 7 mm/s, the shear-driven flow influences more the film thickness. In the 7–18 mm/s speed interval, the sample thickness becomes lower than 1200 μ. The CNTs loading of PS solution determines an increase of the viscosity which in turn slightly decreases the film thickness, as remarked in Figure 8.4.

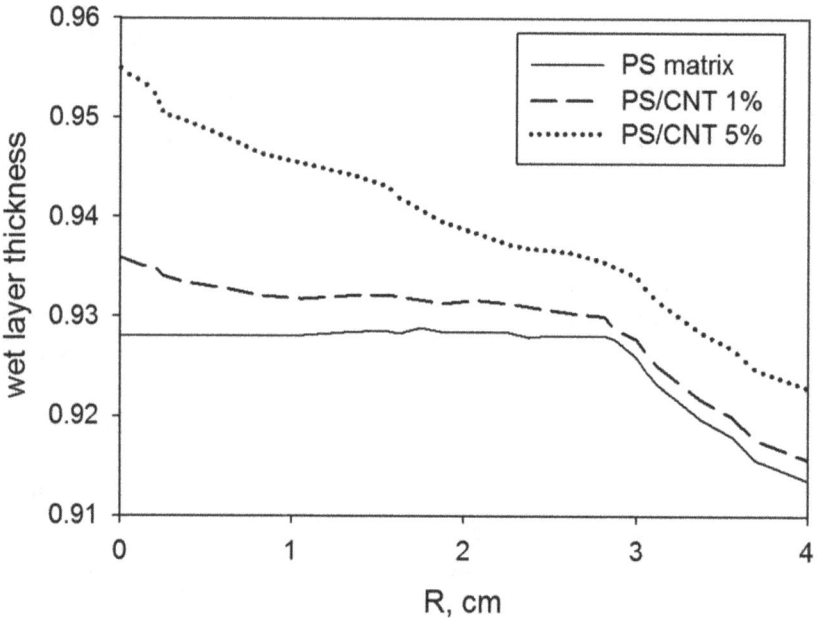

FIGURE 8.3   The assessed wet film thickness versus radial position for PS solution and its composites with CNTs processed by spin coating.

Spin coating deposition, electrospinning, and tape casting represent processing methods for polymer solutions and composite fluids that lie at the basis of many products, including membranes, dielectric layers or electrodes and so on. So, prior to product fabrication is important to perform a deep rheological analysis.

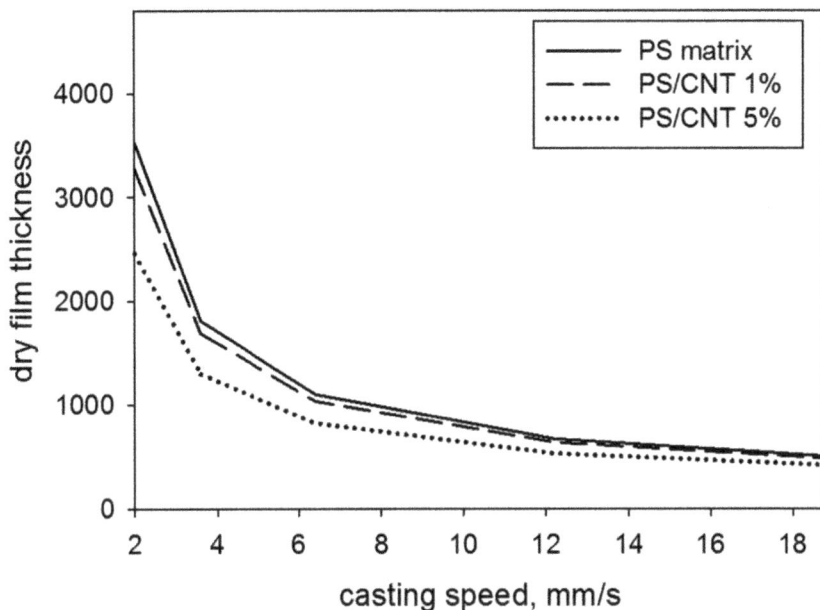

FIGURE 8.4   The assessed dry tape thickness versus casting speed for PS solution and its composites with CNTs processed by tape casting.

8.4 CONCLUSIONS

The CNTs-derived composites belong to a category of materials of high relevance in many applicative domains. Processing of such solutions or melts requires close analysis of the shear flow and viscoelastic properties. The establishment of the rheological behavior is essential for knowing the microstructure changes upon the addition of CNTs and their response to shear deformation. The dispersion state of the fillers and their interaction with the matrix is impacting the material viscoelastic characteristics. The rheological response enables the gaining of a knowledge on the processability characteristics of the filled fluids via spin coating deposition, electrospinning, tape casting. Introduction of CNTs in thermoplastics is reflected in certain changes in the macromolecular relaxation dynamics, which lead to changes in the low frequency viscosity, shear moduli, and loss tangent. Also, the CNTs are impacting the nonlinear viscoelastic responses of the polymer fluid, determining an anticipation of the transition toward the non-linear viscoelastic zone. As a function of the matrix features, particularly its macromolecular architecture and chemical peculiarities, the CNT-derived composites present specific rheological responses, highlighting the interaction that takes place in the material that results in distinct arrangements of loaded particles within the host matrix.

Generally, such anisotropic fillers determine a reduced effect in the big frequency interval and stronger at small frequencies, showing that the inserted CNTs can influence the macromolecular relaxation dynamics at length scales overcoming the entanglement distance. The latter aspect involves a minimal influence of the reinforcement agent on the polymer processability. The rheological response is quite essential for finding the adequate processing conditions of the CNTs composites from fluid state into uniform and stable coatings, films, or fiber-like materials.

KEYWORDS

- CNT polymer composites
- viscosity
- viscoelasticity
- percolation
- processing

REFERENCES

1. Anzar, N.; Hasan, R.; Tyagi, M.; Yadav, N.; Narang, J. Carbon Nanotube—a Review on Synthesis, Properties and Plethora of Applications in the Field of Biomedical Science. *Sens. Int.* **2020,** *1,* 100003.
2. Janas, D. Towards Monochiral Carbon Nanotubes: A Review of Progress in the Sorting of Single-Walled Carbon Nanotubes. *Mater. Chem. Front.* **2018,** *2,* 36–63.
3. Imani Yengejeh, S.; Kazemi, S. A.; Öchsner, A. Carbon Nanotubes as Reinforcement in Composites: A Review of the Analytical, Numerical and Experimental Approaches. *Comput. Mater. Sci.* **2017,** *136,* 85–101.
4. Qian, H.; Greenhalgh, E. S.; Shaffer, M. S. P.; Bismarck, A. Carbon Nanotube-Based Hierarchical Composites: A Review. *J. Mater. Chem.* **2010,** *20,* 4751.
5. Kumar, A.; Sharma, K.; Dixit, A. R. Carbon Nanotube- and Graphene-Reinforced Multiphase Polymeric Composites: Review on Their Properties and Applications. *J. Mater. Sci.* **2020,** *55,* 2682–2724.

6. Malkin Y. A.; Isayev, A. I. *Rheology Concepts, Methods, and Applications*; Elsevier: Toronto, 2012; p 528.

7. Song, Y. S.; Youn, J. R. Influence of Dispersion States of Carbon Nanotubes on Physical Properties of Epoxy Nanocomposites. *Carbon* **2005**, *43*, 1378–1385.

8. Li, J.; Ma, P. C.; Chow, W. S.; To, C. K.; Tang, B. Z.; Kim, J.-K. Correlations between Percolation Threshold, Dispersion State, and Aspect Ratio of Carbon Nanotubes. *Adv. Funct. Mater.* **2007**, *17*, 3207–3215.

9. Ma, A. W. K.; Yearsley, K. M.; Chinesta, F.; Mackley, M. R. A Review of the Microstructure and Rheology of Carbon Nanotube Suspensions. *Proc. Instit. Mech. Eng., Part N: J. Nanoeng. Nanosyst.* **2008**, *222*, 71–94.

10. Chatterjee, T.; Krishnamoorti, R. Rheology of Polymer Carbon Nanotubes Composites. *Soft Matter* **2013**, *9*, 9515.

11. Jang, S. H.; Kawashima, S.; Yin, H. Influence of Carbon Nanotube Clustering on Mechanical and Electrical Properties of Cement Pastes. *Materials (Basel)* **2016**, *9*, 220.

12. Zare, Y.; Rhee, K. Y. Study on the Effects of the Interphase Region on the Network Properties in Polymer Carbon Nanotube Nanocomposites. *Polymers* **2020**, *12*, 182.

13. Sinnott, S. B. Chemical Functionalization of Carbon Nanotubes. *J. Nanosci. Nanotechnol.* **2002**, *2*, 113–123.

14. Bilalis, P.; Katsigiannopoulos, D.; Avgeropoulos, A.; Sakellariou, G. Non-Covalent Functionalization of Carbon Nanotubes with Polymers. *RSC Adv.* **2014**, *4*, 2911–2934.

15. Kwon, G.; Heo, Y.; Shin, K.; Sung, B. J.; Electrical Percolation Networks of Carbon Nanotubes in a Shear Flow. *Phys. Rev. E*, **2012**, *85*, 011143.

16. Watt, M. R.; Gerhardt, R. A. Factors That Affect Network Formation in Carbon Nanotube Composites and Their Resultant Electrical Properties. *J. Compos. Sci.* **2020**, *4*, 100.

17. Sumfleth, J.; Buschhorn, S. T.; Schulte, K. Comparison of Rheological and Electrical Percolation Phenomena in Carbon Black and Carbon Nanotube Filled Epoxy Polymers. *J. Mater. Sci.* **2010**, *46*, 659–669.

18. Han, D.; Mei, H.; Xiao, S.; Dassios, K. G.; Cheng, L. A Review on the Processing Technologies of Carbon Nanotube/Silicon Carbide Composites. *J. Eur. Ceram. Soc.* **2018**, *38*, 3695–3708.

19. Mahanthesha, P.; Srinivasa, C. K.; Mohankumar, G. C. Processing and Characterization of Carbon Nanotubes Decorated with Pure Electroless Nickel and Their Magnetic Properties. *Procedia Mater. Sci.* **2014**, *5*, 883–890.

20. Grady, B. P. *Carbon Nanotube-Polymer Composites: Manufacture, Properties, and Applications*; Wiley: USA, 2011.

21. Weisenberger, M. C.; Andrews R.; Rantell T. Carbon Nanotube Polymer Composites: Recent Developments in Mechanical Properties. In *Physical Properties of Polymers Handbook*; Mark, J. E., Ed.; Springer: New York, 2007.

22. Macosko, C. W. *Rheology: Principles, Measurements, and Applications*; Wiley: USA, 1994.

23. Barnes, H. A. A Review of the Rheology of Filled Viscoelastic Systems. *Rheol. Rev.* **2003**, 1–36.

24. Ferry, J. D. *Viscoelasticity Properties of Polymers*; Wiley-Interscience: New York, 1980.

25. Wang, Y.; Su, B. W.; Gao, X. L. Study on Factors Influencing the Viscosity of Polyacrylamide Solution. *Adv. Mater. Res.* **2012**, *512–515*, 2439–2442.
26. Abraham, J.; Nair, S. T.; Maria, H. J.; George, S. C.; Thomas, S. Rheology of Polymer-Carbon Nanotube Composites. *Kirk-Othmer Encyclopedia Chem. Technol.* **2018**, 1–21.
27. Mezger, T. G. *The Rheology Handbook: For Users of Rotational and Oscillatory Rheometers*; Vicerntz Network: Germany, 2006.
28. Gupta, R. K. *Polymer and Composite Rheology*; CRC Press: USA, 2019.
29. Barzic, R. F.; Barzic, A. I.; Dumitrascu, G. Percolation Network Formation in Poly(4-Vinylpyridine)/Aluminum Nitride Nanocomposites: Rheological, Dielectric, and Thermal Investigations. *Polym. Compos.* **2013**, *35*, 1543–1552.
30. Kotsilkova, R. *Thermoset Nanocomposites for Engineering Applications*; Smithers Rapra Technology Limited: Shawbury, 2007.
31. Davis, V. A.; Ericson, L. M.; Parra-Vasquez, A. N. G.; Fan, H.; Wang, Y. H.; Prieto, V.; Longoria, J. A.; Ramesh, S.; Saini, R. K.; Kittrell, C.; Billups, W. E.; Adams, W. W.; Hauge, R. H.; Smalley, R. E.; Pasquali, M. Phase Behavior and Rheology of SWNTs in Superacids. *Macromolecules* **2004**, *37*, 154–160.
32. Davis, V. A.; Parra-Vasquez, A. N. G.; Green, M. J.; Rai, P. K.; Behabtu, N.; Prieto, V.; Booker, R. D.; Schmidt, J.; Kesselman, E.; Zhou, W.; Fan, H.; Adams, W. W.; Hauge, R. H.; Fischer, J. E.; Cohen, Y.; Talmon, Y.; Smalley, R. E.; Pasquali, M. True Solutions of Single-Walled Carbon Nanotubes for Assembly into Macroscopic Materials. *Nat. Nanotechnol.* **2009**, *4*, 830–834.
33. Larson, R. G. *The Structure and Rheology of Complex Fluids*; Oxford University Press: New York, 1999.
34. Pötschke, P.; Fornes, T. D.; Paul, D. R. Rheological Behavior of Multiwalled Carbon Nanotube/Polycarbonate Composites. *Polymer* **2002**, *43*, 3247–3255.
35. Chatterjee, T.; Yurekli, K.; Hadjiev V. G.; Krishnamoorti, R. Single-Walled Carbon Nanotube Dispersions in Poly(Ethylene Oxide). *Adv. Funct. Mater.* **2005**, *15*, 1832–1838.
36. Du, F.; Scogna, R. C.; Zhou, W.; Brand, S.; Fischer, J. E.; Winey, K. I. Nanotube Networks in Polymer Nanocomposites: Rheology and Electrical Conductivity. *Macromolecules* **2004**, *37*, 9048–9055.
37. Chatterjee, T.; Mitchell, A.; Hadjiev, V. G.; Krishnamoorti, R. Oriented Single-Walled Carbon Nanotubes–Poly(Ethylene Oxide) Nanocomposites. *Macromolecules* **2012**, *45*, 9357–9363.
38. Rahatekar, S. S.; Koziol, K. K.; Kline, S. R.; Hobbie, E. K.; Gilman, J. W.; Windle, A. H. Length-Dependent Mechanics of Carbon-Nanotube Networks. *Adv. Mater.* **2009**, *21*, 874–878.
39. Hobbie, E. K.; Fry, D. J. Rheology of Concentrated Carbon Nanotube Suspensions. *J. Chem. Phys.* **2007**, *126*, 124907.
40. Chen, M.; Russel, W. B. Characteristics of Flocculated Silica Dispersions. *J. Coll. Interf. Sci.* **1991**, *141*, 564–577.
41. Mongruel, A.; Cartault, M. Nonlinear Rheology of Styrene-Butadiene Rubber Filled with Carbon-Black or Silica Particles. *J. Rheol.* **2006**, *50*, 115–135.
42. Chatterjee, T.; Krishnamoorti, R. Steady Shear Response of Carbon Nanotube Networks Dispersed in Poly(Ethylene Oxide). *Macromolecules* **2008**, *41*, 5333–5338.
43. Minus, M. L.; Chae, H. G.; Kumar, S. Polyethylene Crystallization Nucleated by Carbon Nanotubes under Shear. *Acs Appl. Mater. Interf.* **2012**, *4*, 326–330.

44. Chatterjee, T.; Mitchell, C. A.; Hadjiev, V. G.; Krishnamoorti, R. Hierarchical Polymer–Nanotube Composites. *Adv. Mater.* **2007,** *19,* 3850–3853.

45. Guadagno, L.; Raimondo, M.; Vertuccio, L.; Naddeo, C.; Barra, G.; Longo, P.; Lamberti, P.; Spinelli, G.; Nobile, M. R. Morphological, Rheological and Electrical Properties of Composites Filled with Carbon Nanotubes Functionalized with 1-Pyrenebutyric Acid. *Compos. Part B* **2018,** *147,* 12–21.

46. Nobile, M. R.; Naddeo, C.; Raimondo, M.; Guadagno, L. Effect of functionalized Carbon Nanofillers on the Rheological Behavior of Structural Epoxy Resins. *AIP Conf. Proc.* **2019,** *2196,* 020027.

47. Yearsley, K. M.; Mackley, M. R.; Chinesta, F.; Leygue, A.; The Rheology of Multi-walled Carbon Nanotube and Carbon Black Suspensions. *J. Rheol.* **2012,** *56,* 1465.

48. Hassanabadi, H. M.; Wilhelm, M.; Rodrigue, D. A Rheological Criterion to Determine the Percolation Threshold in Polymer Nano-Composites. *Rheol. Acta* **2014,** *53,* 869–882.

49. Stan, F.; Stanciu, N. V.; Fetecau, C. Melt Rheological Properties of Ethylene-Vinyl Acetate/Multi-Walled Carbon Nanotube Composites. *Compos. Part B* **2017,** *110,* 20–31.

50. Kamkar, M.; Aliabadian, E.; Shayesteh Zeraati, A.; Sundararaj, U. Application of Nonlinear Rheology to Assess the Effect of Secondary Nanofiller on Network Structure of Hybrid Polymer Nanocomposites. *Phys. Fluids* **2018,** *30,* 023102.

51. Wu, D.; Wang, J.; Zhang, M.; Zhou, W. Rheology of Carbon Nanotubes–Filled Poly(Vinylidene Fluoride) Composites. *Ind. Eng. Chem. Res.* **2012,** *51,* 6705–6713.

52. Bai, J.; Goodridge, R. D.; Hague, R. J. M.; Song, M.; Okamoto, M. Influence of Carbon Nanotubes on the Rheology and dynamic mechanical properties of Polyamide-12 for Laser Sintering. *Polym. Test.* **2014,** *36,* 95–100.

53. Gao, X.; Isayev, A. I.; Yi, C. Ultrasonic Treatment of Polycarbonate/Carbon Nanotubes Composites. *Polymer* **2016,** *84,* 209–222.

54. Ahmad, A. A.; Al-Juhani, A. A.; Thomas, S.; De, S. K.; Atieh, M. A. Effect of Modified and Nonmodified Carbon Nanotubes on the Rheological Behavior of High Density Polyethylene Nanocomposite. *J. Nanomater.* **2013,** *2013,* 731860.

55. Feng, Y.; Zhou, R.; Zhang, P.; Sang, Q.; Dong, X.; Zhao, J.; Rheology Study of Chlorinated Polyethylene with Various Cl-Contents and Carbon Nanotube Composites. *J. Macromol. Sci. Part B,* **2010,** *49,* 57–65.

56. Arrigo, R.; Malucelli, G. Rheological Behavior of Polymer/Carbon Nanotube Composites: An Overview. *Materials* **2020,** *13,* 2771.

57. Girei, S. A.; Thomas, S. P.; Atieh, M. A.; Mezghani, K.; De, S. K.; Bandyopadhyay, S.; Al-Juhani, A. Effect of –COOH Functionalized Carbon Nanotubes on Mechanical, Dynamic Mechanical and Thermal Properties of Polypropylene Nanocomposites. *J. Thermoplast. Compos. Mater.* **2012,** *25,* 333–350.

58. Park, J. S.; An, J. H.; Jang, K. S.; Lee, S. J. Rheological and Electrical Properties of Polystyrene Nanocomposites via Incorporation of Polymer-Wrapped Carbon Nanotubes. *Korea-Australia Rheol. J.,* **2019,** *31,* 111–118.

59. Barzic, A. I. Percolation Effects in MCNT-Filled Polystyrene: Rheological, Optical, Adhesion and Conductive Investigations. *Mater. Plast.* **2021,** *1,* 10 pp.

60. Mirjalili, V.; Ashrafi, B.; Adhikari, K.; Hubert, P. Effect of the Single Walled Carbon Nanotube Content on Resin Flow. *Proc. 9th Int. Conf. Flow Proc. Compos. Mater. Montréal* **2008,** 1–9.

61. Hyun, K.; Wilhelm, M.; Klein, C. O.; Cho, K. S.; Nam, J. G.; Ahn, K. H.; Lee, S. J.; Ewoldt, R. H.; McKinley, G. H. A Review Of Nonlinear Oscillatory Shear Tests: Analysis and Application of Large Amplitude Oscillatory Shear (LAOS). *Progr. Polym. Sci.* **2011**, *36*, 1697–1753.

62. Lim, H. T.; Ahn, K. H. Nonlinear Viscoelasticity of Polymer Nanocomposites under Large Amplitude Oscillatory Shear Flow. *J. Rheol.* **2013**, *57*, 767.

63. Hyun, K.; Kim, W. A New Non-Linear Parameter Q from FT-Rheology under Nonlinear Dynamic Oscillatory Shear for Polymer Melts System. *Korea-Aust. Rheol. J.* **2011**, *23*, 227–235.

64. Amr, I. T.; Al-Amer, A.; Thomas, S.; Al-Harthi, M.; Girei, S. A.; Sougrat, R.; Atieh, M. A. Effect of Acid Treated Carbon Nanotubes on Mechanical, Rheological and Thermal Properties of Polystyrene Nanocomposites. *Compos. Part B* **2011**, *42*, 1554–1561.

65. Mitchell, C. A.; Bahr, J. L.; Arepalli, S.; Tour, J. M.; Krishnamoorti, R. Dispersion of Functionalized Carbon Nanotubes in Polystyrene. *Macomolecules* **2002**, *35*, 8825–8830.

66. Kota, A. K.; Cipriano, B. H.; Powell, D.; Raghavan, S. R.; Bruck, H. A. Quantitative Characterization of the Formation of an Interpenetrating Phase Composite in Polystyrene from the Percolation of Multiwalled Carbon Nanotubes. *Nanotechnology* **2007**, *18*, 505705.

67. McClory, C.; Pötschke, P.; McNally, T. Influence of Screw Speed on Electrical and Rheological Percolation of Melt-Mixed High-Impact Polystyrene/MWCNT Nano-composites. *Macromol. Mater. Eng.* **2011**, *296*, 59–69.

68. Faraguna, F.; Pötschke, P.; Pionteck, J. Preparation of Polystyrene Nanocomposites with Functionalized Carbon Nanotubes by Melt and Solution Mixing: Investigation of Dispersion, Melt Rheology, Electrical and Thermal Properties. *Polymer* **2017**, *132*, 325–341.

69. Kamkar, M.; Sultana, S. M. N.; Pawar, S. P.; Eshraghian, A.; Erfanian, E.; Sundararaj, E. The Key Role of Processing in Tuning Nonlinear Viscoelastic Properties and Micro-wave Absorption in CNT-Based Polymer Nanocomposites. *Mater. Today Commun.* **2020**, *24*, 101010.

70. Barzic, A. I.; Soroceanu, M.; Albu, R. M.; Ioanid, E. G.; Sacarescu, L.; Harabagiu, V. *Macromol. Res.* **2019**, *27*, 1210–1220.

71. Acrivos, A.; Shah, M. J.; Petersen, E. E. On the Flow of a Non-Newtonian Liquid on a Rotating Disk. *J. Appl. Phys.* **1960**, *31*, 963.

72. Chisca, S.; Barzic, A. I.; Sava, I.; Olaru, N.; Bruma, M.; Morphological and Rheological Insights on Polyimide Chain Entanglements for Electrospinning Produced Fibers. *J. Phys. Chem. B* **2012**, *116*, 9082–9088.

73. Chou, Y. T.; Ko, Y. T.; Yan, M. F. Fluid Flow Model for Ceramic Tape Casting. *J. Am. Ceram. Soc.* **1987**, *70*, C280–C282.

CHAPTER 9

THERMAL AND ELECTRICAL TRANSPORT IN CARBON NANOTUBES COMPOSITES

ANDREEA IRINA BARZIC

"Petru Poni" Institute of Macromolecular Chemistry,
Laboratory of Physical Chemistry of Polymers, 700487, Iasi, Romania

E-mail: irina_cosutchi@yahoo.com

ABSTRACT

Charge and phonon transport processes in CNTs-reinforced materials is an important topic in many fields, such as material science, nanotechnology, electronics, aeronautics, and so on. This chapter describes some basic aspects regarding thermal and electrical conductivity of CNTs-based composites, emphasizing the mechanisms that lie at the basis of these phenomena and also the factors that are impacting these physical properties. All these issues are very useful in understanding the phenomena governing the charge and phonon transport in multiphase systems and help to acquire insights on designing new materials with upgraded performance as demanded by the futuristic trends in high-tech devices that improve our daily life.

9.1 INTRODUCTION

Highly conductive reinforcement's agents have been largely studied because they have upgraded the performance of numerous materials in terms of thermal and electrical characteristics.[1] Among them, carbon-based fillers, such as graphene, carbon nanotubes (CNTs), carbon black and fullerene, seem to be the key to achieve progress and revolutionize most of the nowadays modern technologies.[2] The main benefit is that such fillers can be economically prepared into numerous shapes and dimensions from renewable resources, hence avoiding the damage of the natural environment, resulting from increasing pollution and greenhouse effect.[3,4]

CNTs have been discovered by Iijima[5] and since then many research works were published on this subject.[6–10] CNTs display a peculiar structure having an extended p-electron system, which render remarkable electronic and electrical features (1000–200,000 S/cm).[11] Another fact that has been established is that CNTs might present amphoteric behavior by exchanging negative charges with compounds that are electron donors or acceptors, resulting negatively or positively charged counterions.[12] Besides the CNTs charge transport outstanding abilities, they additionally exhibit good thermal conductivity which ranges between 2000 and 6000 W/m K at the temperature of 25°C.[13–15] The direct and quantitative evaluation of thermal conduction characteristics of individual CNTs is still difficult given the technological issues associated with the nanoscale experimental measurements.[16] Therefore, the thermal conductivity of CNTs found in most reports is mainly based on the theoretical simulations and estimations from indirect experiments,[17–19] leading to considerable scattered data.

CNTs are widely used to reinforce low conductive materials (especially polymers), opening novel perspectives for substituting the metallic components in devices from a variety of applications, power electronics, aeronautics, heat exchangers, electric motors, and generators. Fundamental knowledge in these domains on electrical and thermal transport could be of paramount importance to understand the design of multiphase materials that lie at the basis of many electronic components. This is also possible due to the multitude of additional advantages of CNT/polymer composites, namely, as light weight, corrosion strength, and facile processing. The existing interest to enhance the thermal/electrical conductivity of polymers is directed toward the selective loading of nanofillers with appropriate transport features. However, there are many factors that affect the electrical and thermal performance of the CNTs-derived composites:

- CNTs structure, topological defects, size, purification and graphitization, functionalization.

- The amount of added filler and its state of dispersion.
- CNTs interfacial compatibility with the host matrix.
- CNTs alignment induced during processing or produced by external fields.
- Matrix structural peculiarities, like crystallinity and thermal behavior.

The majority of the CNTs composites require the introduction of a specific percent of CNTs filler inside the matrix in order to attain a sufficient level of properties in regard to the initial polymer host. In other words, for obtaining an increase of the electrical/thermal conductivity with several orders of magnitude, the added CNTs should form a percolation network in the polymer environment. Another interesting aspect is that some studies show that for the same matrix doped with CNTs, the values of the conductivity and percolation point are not coinciding. The explanation for this might reside in the uncertainties related to type, degree of purification, size dispersity, aspect ratios, CNTs overlapping degree, crystalline orientation, straightness, and so on.[10,20,21] In any case, the lack of the CNTs surface reactivity restricts their practical use in composites preparation since it induces weak adhesion among the CNTs and polymer. Intrinsic van der Waals attraction between the fillers, in addition to their large surface area and aspect ratio, produces considerable agglomeration, thereby limiting the adequate transfer of their superior features to the polymer base material.[22,23] In the attempt to overcome such issues, several physical and chemical methodologies were developed.[10,13,24] Reviewing the reported approaches, one may find a guided tour via the toolbox for the dispersion of the fillers within polymer matrixes.[25–27] Other report speaks about the importance of the aspect ratio of the CNTs, their dispersibility, and the capacity to conglomerate on the parameters for the fabrication of composites with conductive features at low reinforcements.[10,28]

Considering this background, this chapter reviews certain relevant aspects concerning charge and heat transport in CNTs-containing composites, mainly having polymers as matrix. A brief description of the mechanisms lying at the basis of these phenomena is made and also the most important breakthroughs in these domains. Therefore, one may attain a reference to prepare the CNTs/polymer composites with desired thermal or electrical conductivity–physical properties that are paramount factors in numerous applications.

9.2 THERMAL TRANSPORT IN CNTs-BASED COMPOSITES

9.2.1 MECHANISMS OF THERMAL CONDUCTION

Heat transport implies the transfer of thermal energy from one zone to another by means of the energy carriers.[13,29–32] For instance, in solid phase, the heat transfer occurs via phonons, electrons, or even photons. Before discussing the thermal conductivity in composites, it is more meaningful to present some fundamental aspects regarding the mechanism in polymers (bad conductors) and crystalline materials (good conductors). Thermal conductivity can be estimated based on Debye eq 9.1:

$$k = \frac{C_p \upsilon l}{3} \tag{9.1}$$

where k is the thermal conductivity, C_p is the specific heat capacity, l is the phonon mean-free path and υ is the mean phonon velocity.

Generally, in good conductive materials having a perfectly crystalline, ordered structure, the mechanism of heat transport involves the following stages:

- Heat is reaching the first layer of the material.
- The atoms from the superficial layers begin to have vibrational energy.
- Heat is transmitted toward the next atomic layers at the same speed under the form of a wave.
- Heat diffuses in the material.
- Heat diffuses with a common vibrational mode (or a phonon) to the entire crystalline material.
- When heat arrives at the opposite surface of the material, it is partially transferred by conduction or radiation to the environs.

In the case of polymers, the main route of thermal energy transport is mainly assured by phonons (defined as quantized modes of vibration within the crystal lattice) and less by electrons since they lack mobility along the macromolecular chains. The problem with polymers is that they are prevalently amorphous, thus the term l is quite small owing to phonon scattering from various defects. This is why polymers are characterized by very low k values. An overall image on thermal conduction of polymers and their composites is provided in the several review papers.[13,29–32] It was evidenced that k parameter is affected by crystallinity, temperature, orientation of the polymer chains. Briefly, the mechanism of heat conduction in polymers can be depicted as follows[29]: one side of the material is in contact with heat source, so the heat is transmitted to the first layer of superficial atoms under the form of a vibration and then it is passed to the nearest atoms (but not with the same speed) and so on. So, heat conduction here is achieved by slow diffusion and not as a wave. Thermal energy propagation along the macromolecule could cause chaotic vibration and rotation of atoms, lowering k magnitude.

As in the case of other nonmetallic materials, CNTs have a thermal conductivity that is intermediated by phonons. The processes impacting the phonon transport in such fillers are including the scattering occurring at the boundary surface, amount of phonon active modes, the length of the free path for the phonons and inelastic Umklapp-scattering (explained as being "an anharmonic phonon–phonon or electron–phonon scattering process").[13,33–35] Heat transfer in CNTs is affected by many aspects, such as the atomic arrangement, filler size (i.e., diameter and length), the purity degree, the number of walls and defects. These issues are more deeply discussed in the review paper published by Han and Fina.[13]

Having in view the aforementioned elements, it is obvious that the low thermal conductivity of polymers can be raised by filling with conductive agents like CNTs. The mechanism of the thermal energy transport in CNTs-containing polymers is more complex. Such carbon-derived particles are known to display big specific surface area. When introduced in the polymer medium, several interfaces are produced, thereby generating phonon scattering and create interfacial thermal resistance. For this reason, the heat transfer in the composite system is diminished at filler–matrix interface. The mismatches between the reinforcement agent and the polymer determine phonon scattering at the interface, consequently the k magnitude is lowered. At compositions under the percolation point, the CNTs are not able to connect to each other to produce a thermal conduction pathway. In such situations, interfacial thermal resistance is considered to be the prevalent aspect that lowers thermal conduction of the composite. Functionalization of the fillers surface seems to be a good alternative to solve this problem. In a composite, CNTs act like an extremely conductive channel, whereas the adapted surface enables covalent and noncovalent linkage to the matrix host chains. The latter is expected to facilitate the phonon traveling from the fillers to the polymer and vice versa.[13] In some research works, researchers have presumed that the macromolecular chains of the matrix and the molecular chains on the surface of carbon-based filler are able to intertwine with each other and produce an interlayer. The latter might reduce the interfacial phonon scattering and diminish the interface thermal resistance.[36–39] If the amount of added CNTs is beyond the percolation threshold, the thermal energy in the composite mainly travels via the heat conduction pathway, considering the large thermal conductivity of inserted fillers. When reinforced material is brought in contact with the heat source, the thermal energy is transported fast via CNTs so the thermal conduction of the whole material is enhanced. Supplementing the thermal pathways and lowering the thermal resistance among the CNTs and the CNTs-polymer interface are key solutions for attaining composite materials characterized by high thermal conductivity.

Literature provides a general classification of thermally conductive polymer materials by accounting on three factors[30,40–43]:

- Symmetry of the structure: isotropic and anisotropic matrix materials.
- The level of conductivity: thermally/electrically conductive materials and thermally conductive/insulating materials.
- Fabrication method: materials with instrinsic conduction and reinforced thermally conductive materials.

Starting from this, Yang et al.[30] made a deep analysis regarding the heat conduction mechanism for the following categories of materials:

- The intrinsic thermally conductive materials: they are mainly crystalline structures, where the lattice vibration intermediates the heat transfer with constant speed, like the propagation of waves.[44]
- The filled polymeric systems with high conductivity: where the heat propagation implies thermally conductive path theory, thermally conductive percolation theory, and thermoelastic coefficient theory.[45]

9.2.2 CNTs COMPOSITES WITH THERMAL CONDUCTIVITY

In this section, some relevant reports regarding thermal conductivity of the CNTs/polymer composites are shortly presented. As summarized in the work of Han and Fina,[13] it was demonstrated that CNTs incorporation in polymers has a greater impact on the electrical conduction, while the thermal conductivity is not enlarged as expected. Indeed, there were noted certain improvements of k value, but they were very far from the theoretical thermal conduction of CNTs. This was particularly caused by the defects and phonon scattering in the composite material. Aliev and collaborators[46] have measured experimentally the thermal conduction in multiwalled CNTs and proved that filler bundling, tube–tube interconnections, dangling fiber ends, or misalignments are responsible for a dramatic reduction of the CNTs conduction from the theoretical values of about 3000 W/m K to the measured one of 50 W/m K. Even so, there are many studies that are devoted to analysis of the advantages of the CNTs loading in polymers for rising heat transfer. It can be easily deduced that with a larger amount of CNTs in the system, the thermal conduction would be greater. Nevertheless, despite extremely big reinforcement percent, the thermal conductivity of the composite has never attained the intrinsic value of the CNTs, nor was near to it. As an overall image, when inserting in a polymer around 30–50% wt CNTs, the thermal conductivity was enhanced by 10 times.[47]

The utilization of fillers characterized by biggest conduction is not always rendering to the highest thermal conductivity in the polymeric composite. In an attempt to improve the heat transport of polymers, the researchers have investigated a series of factors that limit the k magnitude, and when available, they developed solutions to overcome each particular issue, namely:

- The effect of interfaces: during reinforcement, many polymer/ particle interfacial areas appear, (interfacial resistance) and at percolation concentration, even more contact points between CNTs are formed (contact resistance). Theoretically, the acoustic mismatch of the phases determines the interfacial thermal barrier. Given the lack of the common vibration frequencies for the CNTs and the matrix, only small frequency phonon modes of filler are effective when nanotubes interact with a polymer through weak dispersion forces. Another element that causes interfacial resistance is the faulty physical contact between filler and polymer, which mainly is affected by the surface wettability. Moreover, the interface defects could emerge from internal stresses owing to the discrepancy in the filler and matrix thermal expansion coefficients. Good thermal contact (meaning small thermal resistance) between the filler and surrounding matrix might be in principle attained by close molecular contact and is ascribed to high efficiency in transferring heat from the CNTs to the polymer. This is not entirely ideal because of the different values of the mean-free path for phonons in each phase. The role of CNTs coupling with the host was presented for several types of CNTs, generally supposing that multiwalled CNTs are less sensitive to matrix coupling, because of the internal layer which lacks direct contact with the macromolecular medium. However, it was shown that the heat conduction contribution of inner walls of multiwalled CNTs seems to be insignificant.[48] As to the contact resistance, Shenogina et al.[49] have analyzed the contact between CNTs and proved that heat transport by filler–filler direct contact is not significant as a result of the weak Van der Waals forces binding the nanotubes, leading to an important thermal resistance and because of the small contact area. The contact resistance among such fillers was also examined by Zhong and Lukes,[50] indicating the diminishment in tube–tube resistance with longer CNTa, bigger overlap and lower spacing. Enhancement of the CNT–CNT contact is consequently mandatory to achieve more efficient heat transport among the fillers.[51] It can be stated that the resistance to the thermal energy transfer at the inter-CNT contact and the whole contact area could be considered paramount for heat exchange between CNTs surrounded by macromolecules. It was reported that dispersion and preferential arrangement of CNTs significantly impact the total contact area in a reinforced polymer material. A solution to reduce the thermal contact resistance in neighboring fillers is to subject them to chemical modification. Interactions of higher intensity than the Van der Waals forces might be advantageous for phonon transfer from one CNT to another.

• The dispersion degree of the filler: means separation of the individual nanoparticles, which is hard because of the small size and high aspect ratio, leading to the occurrence of CNTs bundles and aggregates.[52,53] A number of methods were formulated for the accomplishment of proper CNTs dispersion, such as functionalization, high power ultrasonication, and surfactant-assisted processing.[54–56] Despite the expectations, it appears that the solution in obtaining superior CNTs/ polymer composites for thermally conductive purposes relies in the capacity to control the aggregation of fillers within the polymeric medium to achieve an interconnecting network adequate for the desired heat transport. It is very hard to formulate general rules for the connection among the dispersion of CNTs and heat conduction in their composites. A reason for this is that literature lacks information about quantitative parameters to quantify dispersion and comparison of various results is difficult. Another issue is that dispersion is affected by functionalization, which in turn lowers filler heat transport features.[57,58] Consequently, the two effects cannot be investigated independently. In addition to these aspects, dispersion is influenced by the mixing energy imposed, but the latter also reduced heat transport of the fillers by inserting defects and/or shortening.[59] On the other hand, localization of thermally conductive paths was proved to be important for conductive purposes.[13] For enlarging the thermal contact area among the conductive nanotubes, localization of CNTs into co-continuous domain is highly desirable. So, achievement of conductive pathways by localizing CNTs or controlling their aggregation is beneficial for increasing k level. Localized aggregation of carbon fillers could be performed by colloidal-physics approaches using polymer emulsions or latexes. During reinforcement with CNTs, the polymers create excluded volume, resulting segregated network of nanoparticles. It could be possible to obtain a segregated network of CNTs and improve the thermal conductivity at very low loadings. Other approaches might be also employed to attain controlled segregation of the conductive particles, such as co-continuous immiscible polymer systems. They have shown to be suitable for building a percolation network for conduction purpose.[60] Co-continuous structures could be convenient to produce a conductive path in the material, selectively dispersing the fillers in one of the phases. Alternatively, impregnation of initially made conductive networks might be utilized to obtain composites with good control of conductive pathways by means of their volume.[13]

- The alignment of the fillers: This is possible owing to the anisotropic shape of the CNTs. This is excellent for the cases where the heat transport is desired to occur in a preferential direction. CNTs can be oriented within the polymer host by processing, CNTs arrays[61] or external force. The materials compounded by melt processing, CNTs mainly tend to orient along the flow direction. An anisotropic heat transport of the CNTs composites could be expected if all the nanotubes in the matrix are well aligned, which can be achieved through preformed CNTs arrays.[61] Preferential orientation of such anisotropic fillers along a known direction can be done by an external field, which could be of magnetic nature,[62] electric[63] or force fields (particularly stretching).[64–66] In contrast to the electrical conductivity, not so many reports on the thermal conductivity of CNTs/polymer materials having alignment filler under an external field can be found in literature. A theoretical study was carried out on thermal conduction of polystyrene/CNTs, where the fillers are either oriented radially or axial to the heat source,[67] showing that thermal energy travels in a larger amount along the CNTs' long axis. In Figure9.1 is illustrated another example of heat transport in a CNTs-loaded cellulose derivative at two reinforcements degrees, that is, 15% and 40%. At these filler concentrations, the system is surely beyond the percolation point. It is worth mentioning that temperature has some implications on the percolation threshold[68,69] since it enables a higher motion of the matrix chains and perturbs the filler network.

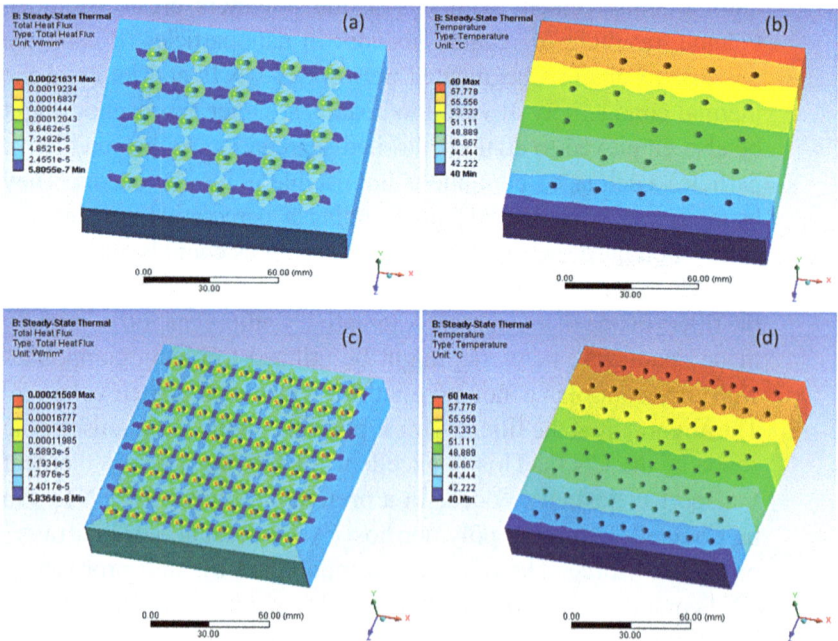

FIGURE 9.1 The heat flux (images from left) and the thermal gradient (images from right) for CNTs/cellulose derivative composite containing 15% (a,b) and 40% (c,d) oriented CNTs.

- The degree of crystallization of the polymer matrix: for an enhanced phononic conduction, it is mandatory to have geometrically regular and strong bonds, meaning that macromolecules are packed in crystal lattice structures. Such structural features in polymer host are favorably affecting the heat transport in the matrix phase at the interface between CNTs and polymer. In addition, in semicrystalline/CNTs composites, the filler might afford nucleation sites for polymers and speed up the crystal growth rate or even control its shape.[70] It was found by Liao et al.[71] that polymer crystallinity is influencing the state of dispersion of the CNTs. Polypropylene with various degrees of crystallinity was used in the investigation, and it was revealed that lower crystallinity of the matrix determines better dispersibility of the CNTs, but the heat conduction characteristics were not disclosed. Also, CNTs are not only impacting the nucleation, but it was discovered that the majority of the fillers are present in the amorphous phase in highly reinforced polymers. In matrix having big crystalline domains and increased crystallinity, this could outcome in the CNTs confinement in the amorphous part, where the large particle amount might be employed to enhance CNT–CNT thermal contact.

A relatively recent trend among the scientists dealing with reinforced polymers is the preparation of multifiller composites. So, besides the CNTs, the procedure deals with the insertion of two or more fillers with various shapes in the matrix. This seems to be a good approach in terms of conductive particle contact surface. Zhang et al.[72] have prepared insulating composites with heat conduction abilities by confining CNTs in phosphate glass (Pglass). The latter has a high interfacial tension with the polymer matrix, so powerful CNTs confinement is attained, thus enabling the appearance of conductive networks in the ternary composite. Su and coworkers[73] obtained composites by developing a novel strategy that led to three-dimensional staggered interconnected fillers that are vertically oriented hexagonal boron nitride (h-BN) platelets combined with nonaligned and dispersed NH_2-functionalized CNTs (aminated carbon nanotubes). Through the used methodology, the h-BN platelets surrounded by magnetic layer could be controlled by an external field, whereas CNT-NH_2 could not be controlled. The heat transport was improved, namely, at 30 wt.% h-BN being around 0.99 W/m·K. The thermal conductivity was almost the same through-plane with that measured in-plane of the multiphase material containing 26.43 wt.%

h-BN and 2 wt.% of CNT-NH$_2$. Yu et al.[74] have mixed graphite nanoplatelets (GNPs) and single-walled CNTs and introduced them in epoxy matrix, thus producing a highly efficient hybrid filler network with considerably lowered thermal contact resistance. Similar results were attained when using polypropylene as matrix in which CNTs and synthetic graphite were inserted.[75,76] Ghose et al.[77] have prepared ternary composites by blending suspensions of CNTs and aluminum nanofiller and loaded in poly(ethylene vinyl acetate) (EVA). It was found that the presence of aluminum particles determined the diminishment of the difference in the conduction in the axial and transverse directions. It could be assumed that this filler had the role of a bridging agent among the oriented conductive pathways in the composite system. Choi et al.[78] have compounded aluminum flakes and multiwalled CNTs fillers at 80% wt. to augment the polyacrylate's thermal conductivity from 0.50 to 1.67 W/m K. Heat conduction of the ternary composite was higher as the sheet thickness was larger. At a reinforcement level of 90 wt.%, the material loaded with aluminum powder of 13 μm displayed a superior conductivity that when filling with 3 μm powder. Moreover, when reinforcing with mixture of two powders, a synergistic impact on the thermal conductivity was noted. Morphological studies revealed that the state of CNTs dispersion in the polyacrylate/Al-flake system was showing agglomeration of the carbon-based fillers.

Having in view this background, it seems that the adequate exploitation of the thermal conductivity and aspect ratio of CNTs demands a close and deep understanding of the control of the micro- and nanostructure of the reinforced material. Many parameters are acknowledged to have an essential role in the conductivity performance, including interfacial resistance, filler distribution, dispersion, and degree of orientation. In any case, these parameters, most of the times are interrelated and certain issues appear in regard to their independent discussion. Therefore, the field of thermally conductive materials still has more aspects to be elucidated prior to attaining the desired performance of the composite materials.

9.3   ELECTRICAL TRANSPORT IN CNTs-BASED COMPOSITES

9.3.1   MECHANISMS OF ELECTRICAL CONDUCTION

The mechanisms of electrical conduction in the polymer composites were detailed in the work published by Min et al.[10] CNTs are believed to be covalently integrated within the matrix and they are contributing to the cross-linked structure and should not be viewed as a separate component.[79] At sufficient loading the percolation threshold is attained and the filler network is present in the material, there is achieved a critical minimum distance among the anisotropic shaped particles. In this case, the electron conduction is favored via a "hopping" or "tunneling" mechanism. In order to produce larger bulk electrical conductivity in the composite, some reports revealed that the addition of graphite with other conductive particles like CNTs lead to good results owing to the tridimensional conductive networks.[80] The occurrence of the nanoparticle network[81] or creation of a continuous electron path[82,83] is thought to be the principal motivation of the electron transport via tunneling or electron hopping that happens along the CNTs interconnects. Seidel et al.[84] proposed a micromechanics model to estimate the consequence of the effects of electron hopping and the appearance of conductive networks on the electrical conductivity of the CNTs-doped polymers. Their approach involves composite cylinders model as a nanoscale relevant volume element. In the latter, the process of electron hopping is inserted under the form of a continuum interphase layer. This leads to a distinct percolation concentration attributed to the electron hopping. The comparison of the model data with the experimental ones shows that there is a good concordance, hence, the model adequately describes the effect of the electron hopping and filler bundling on the material's conductivity and percolation point. There are studies that presume that the small percolation concentrations remarked in experiments are the result of the nanoscale phenomena (i.e., electron hopping),[85,86] while several research articles are more concentrated on dispersion arguments, pointing out that the small percolation concentrations are attributed to occurrence of the conducting networks which enable a large amount of electrons to flow through the loaded polymer.[87,88] However, Seidel et al.[84] concluded that for multiwalled CNTs/polymer systems, the presence of conductive networks is the major cause for the significant increases in conductivity noted at very low filler concentrations. Moreover, for single-walled CNTs systems, they consider that both the conductive networks and electron hopping generate huge increases in the conductivity of the loaded polymer at few percents of the CNTs. They explain that conducting networks' appearance is seen as the prevalent factor at concentrations far under the percolation limit ascribed to electron hopping process. Ounaies and coworkers[87] have shown that there are nonlinear I–V relationships during reinforcement with CNTs in excess of 0.1 vol.%. Such non-Ohmic behavior is considered to be determined by a tunneling mechanism. Conduction might take place via electron hopping from filler to the neighboring one when they are brought sufficiently close, but not always is required to physically touch each other.[10] Conductive paths are present inside the composites owing to quantum tunneling effects, where the length among the conductive particles is suitable for electron hopping occurrence. Percolation theory supposes that paths are composed of nanoparticles in direct contact. However, for quantum tunneling, there is a contact resistance or potential barriers inside the conductive routes that negatively influences the level of the composite conductivity.[89] It is rational to expect that at larger reinforcement levels with conductive inclusions like CNTs it is possible to attain a secondary threshold, where the fillers are found in direct contact, removing the contact resistance effects. Therefore, electrical conductivity of the CNTs/ polymer should be closer to that of the filler.

9.3.2   CNTs COMPOSITES WITH ELECTRICAL CONDUCTIVITY

The literature is abundant in research articles dealing with various aspects on electrical conduction in polymer composites. Some of them are focused on theoretical evaluation by using common models used for thermal conductivity or other calculation formulas,[13,90,91] while others combine them with experimental results.[2,8,9,92,–94] A large interval of values for conductivity and percolation point of CNTs/polymer materials have been published in the literature during the past years, as a function of the processing approach, polymer matrix features and nanotube type. This section of the chapter is presenting a brief and comprehensive source for polymer/CNTs research, emphasizing, if possible, fundamental aspects that relate the CNTs or polymer structure with the electrical performance of the composite.

Mora et al.[92] prepared CNTs/polylactic acid (PLA) and CNTs/high density polyethylene (HDPE) by complex manufacturing procedure, combining shear-induced melt blending and 3D printing. They performed conductivity measurements and modeling of 3D-printed composites. Small percolation thresholds were achieved based on performed experiments as 0.23 and 0.18 vol.% of CNTs for CNTs/PLA and CNTs/HDPE systems, respectively. The "micromechanics-based two-parameter agglomeration model" was conceived to foretell the electrical conductivity of the

CNTs-based composites while accounting for microstructure characteristics. It was also demonstrated that the two agglomeration parameters are employed to characterize segregated structures, wherein CNTs are limited to particular locations within the polymer host. This is the first model which takes into consideration the segregation of CNTs in the polymer when dealing with electrical conductivity. The close values of theoretical and measured data indicate the adequacy of the proposed approach. In addition, it was revealed that this model leads to reliable results when applied for elastomeric and thermoplastic composites, offering solutions for the design of electroconductive materials.

Aguilar et al.[93] has studied the impact of CNTs clustering on the electrical conductivity of reinforced polysulfone. To this end, they prepared two sets of samples having 0.05–0.75% filler: one that contain uniformly disposed CNTs and another with agglomerated CNTs at the microscale. The percolation concentration was detected at 0.11% and 0.068% w/w for the uniformly dispersed and agglomerated samples, respectively. A higher level of conductivity with two to four orders of magnitude for the composite containing nondispersed CNTs in regard to uniformly dispersed materials in the upper limit of the percolation point (0.1–0.3% w/w) was found. The enhanced conductivity of the agglomerated state can be accounted for the larger filler-to-filler contact after the percolating network has appeared which enables the electron transfer easier.

Spitalsky et al.[94] showed that in order to accomplish a dramatic enhancement of conduction features in a composite, one should pay attention to polymer properties and synthesis method, aspect ratio of CNTs, disruption of CNTs agglomerates, degree of distribution, and orientation. All these are affecting the level of conductivity, the percolation threshold, and the critical exponent.

Bhatia et al.[9] presented a critical review devoted to the current state-of-the-art on low temperature charge transport in CNTs-based composites, highlighting the various conduction mechanisms which were identified. At the percolation point and beyond, there is a great temperature dependence of conductivity. The charge transport in CNTs systems should augment the conductivity as a result of one-dimensional nature of the filler even at the percolation, but several experimental papers deviate appreciably from this idea. When the composite is found in the vicinity of percolation and displays homogeneous filler distribution, the material exhibits smaller temperature influence on the conductivity. This finding clarifies that uniform dispersion of the filler in the matrix without excess ultrasonication is essential to boost the charge transport in the CNTs composite. More efforts are still required to entirely quantify how certain parameters could impact the conduction mechanisms of the reinforced polymers.

Doh et al.[8] studied the influence of CNTs chirality and length on the electrical charge transport accounting for quantum tunneling resistance in loaded polymers via Monte Carlo computing. The nonoriented spatial positioning of CNTs was realized with a one-dimensional line segment and the periodic boundary settings in a 2D volume element. Intersection spots among each anisotropic filler were computed to achieve the connectivity lists of the connected network path. The approach has included the intrinsic resistance of the filler and the inter-CNTs tunneling resistance. The assessed data were compared with the numerical results generating a good fit in many cases. The relevance of including tunneling effects was more obvious for short filler strands with bigger diameters. For longer CNTs-derived networks, the tunneling effects have less impact on conduction. It was also pointed out that taking the stochastics of filler length distributions produces a lower percolation point, which is adequate for a realistic scenario. Moreover, the proportion of armchair and zig–zag structures in the filler might be accurately evaluated by considering the distinct cumulative amounts of the two CNTs structures in the formed network. The geometrical model along with its Monte Carlo simulation is very useful for understanding how to obtain a conductive material by evaluation of the proportion of metallic and semiconducting CNTs required.

Dal Lago[2] incorporated in polymer blends several kinds of carbon fillers (CNT, carbon black, graphene) and analyzed their effect on charge transport level. Recording the electrical conductivity, it was indicated that volume resistivity was reduced when the CNTs and carbon black percents were augmented, even if the addition of melt-blended graphene lacks impact on this feature. For the latter, solution blending was used to better disperse graphene, thus lowering the percolation concentration as a function of the solvent evaporation technique. They have remarked a gradual enhancement in all of the dielectric performance of the reinforced materials, in terms of loss factor, with temperature and the percentage of the filler.

Bairan et al.[95] have incorporated multifillers, such as graphene, carbon black, and CNTs, in pollypropylene in order to increase the charge transport in the composite and also to enhance the mechanical performance. It was demonstrated that few percents (6 wt.%) of the CNTs in the system induced greater conductivity of about 158.32 S/cm, while the density and shore hardness were 1.64 g/cm³ and 81.5 (SH), respectively. Moreover, the optimum level of flexural strength accomplished was 29.86 MPa at CNTs loading of 5 wt.%.

Souier et al.[96] investigated the possibility to improve charge carriers transport in CNTs included in the epoxy polymer. For intensification of the electrical conductivity and consequent shielding features, they have vertically aligned the filler arrays within the matrix. They prepared a CNTs "forest" by catalytic chemical vapor deposition, the thickness being around 1.4 mm. The experimental used method relied on "Current Sensing Atomic Force Microscopy" (CS-AFM), where the AFM tips are covered with platinum and are brought in contact with sample surface having protruding CNTs. In the same time, the back of the composite is metalized with silver to finalize the circuit. This can be equivalent with miniaturization of the two-probe method. The recorded images reveal high spatial resolution and indicate resistive zones (dark spots) ascribed to the epoxy matrix, and conductive domains (colored spots) attributed to the CNTs. The measurements were done by applying contact forces ranging from 6 to 75 nN. Electrical transport seems to be impacted by wetting characteristics of CNTs–metal interface, and the resistance at point junctions that scale with the dimension of interconnecting filler.

9.4   CONCLUSIONS

The remarkable physical properties reported for CNTs, added to their low density, make this new form of carbon ideal for polymer reinforcement. There is a growing interest in studying CNTs-based composites for enhanced thermal and/or electrical conductivity. Even at very low loading levels, there are noted outstanding changes in the conductive abilities of the material. However, it should highlight the relevance of the factors which impact the thermal and electrical properties. To attain the most extraordinary properties at the lowest amounts of included CNTs, some strategies should be expressed in regard to the formulation and processing technologies, besides the chemical approaches involved in the structural design of the composite. It is essential to give more attention to the interfacial compatibility between the filler and the matrix, since by controlling this, one may achieve proper dispersion of CNTs within the macromolecular host and render the desired level of conductivity. Various results are found in the literature even for similar CNTs because there are some factors that determine the conductive features of CNTs these include chirality, length distribution, state of aggregation, ends and sidewall defect, diameter size, and number of walls. Functionalization of CNTs is good for their proper dispersion in the matrix, but it should be made carefully as in turn, it lowers the intrinsic conductive properties of the fillers.

The thermal conductivity of CNTs-containing polymers has generally been discussed by analogy with electrical conductivity. However considerable variations exit between the thermal and electrical conductivities that arise mainly from the following two aspects: (1) the CNTs/matrix conductivity ratio is of many orders of magnitude lower for the thermal conductivity and (2) the differences between the electronic and the phononic conduction mechanisms. These arguments should be highly considered when judging the electrical and thermal transport characteristics of a polymer composite.

In the end, the extensive application of the conductive loaded polymers has not been fully developed, giving new investigation opportunities to the research community. It is trusted that in the near future, such multiphase materials could help in the production of the electrical wires, the antistatic instruments and also in upgrading the performance of certain current products used in power electronics, electric generators, and so on.

KEYWORDS

- **CNT fillers**
- **composites**
- **phonon transport**
- **charge transport**
- **interface**

REFERENCES

1. Gulrez, S. K. H.; Ali Mohsin, M. E.; Shaikh, H.; Anis, A.; Pulose, A. M.; Yadav, M. K.; Qua, E. H. P.; Al-Zahrani, S. M. A Review on Electrically Conductive Polypropylene and Polyethylene. *Polym. Compos.* **2013,** *35,* 900–914.

2. Dal Lago, E.; Cagnin, E.; Boaretti, C.; Roso, M.; Lorenzetti, A.; Modesti, M. Influence of Different Carbon-Based Fillers on Electrical and Mechanical Properties of a PC/ABS Blend. *Polymers (Basel)* **2019**, *12*, 29.

3. Srivastava, S. K.; Mishra, Y. K. Nanocarbon Reinforced Rubber Nanocomposites: Detailed Insights about Mechanical, Dynamical Mechanical Properties, Payne, and Mullin Effects. *Nanomaterials (Basel)* **2018**, *8*, 945.

4. Ong, Y. T.; Ahmad, A. L.; Zein, S. H. S.; Tan, S. H. A Review on Carbon Nanotubes in an Environmental Protection and Green Engineering Perspective. *Braz. J. Chem. Eng.* **2010**, *27*, 227–242.

5. Iijima, S. Helical Microtubules of Graphitic Carbon. *Nature* **1991**, *354*, 56–58.

6. Sundaram, R. M.; Sekiguchi, A.; Sekiya, M.; Yamada, T.; Hata, K. Copper/Carbon Nanotube Composites: Research Trends and Outlook. *R. Soc. Open Sci.* **2018**, *5*, 180814.

7. Kumanek, B.; Janas, D. Thermal Conductivity of Carbon Nanotube Networks: A Review. *J. Mater. Sci.* **2019**, *54*, 7397–7427.

8. Doh, J.; Park, S.-I.; Yang, Q.; Raghavan, N. The Effect of Carbon Nanotube Chirality on the Electrical Conductivity of Polymer Nanocomposites Considering Tunneling Resistance. *Nanotechnology* **2019**, *30*, 465701.

9. Bhatia, R.; Kumari, K.; Rani, R.; Suri, A.; Pahuja, U.; Singh, D. A Critical Review of Experimental Results on Low Temperature Charge Transport in Carbon Nanotubes Based Composites. *Rev. Phys.* **2018**, *3*, 15–25.

10. Min, C.; Shen, X.; Shi, Z.; Chen, L.; Xu, Z. The Electrical Properties and Conducting Mechanisms of Carbon Nanotube/Polymer Nanocomposites: A Review. *Polym.-Plast. Technol. Eng.* **2010**, *49*, 1172–1181.

11. Ebbesen, T. W.; Lezec, H. J.; Hiura, H.; Bennett, J. W.; Ghaem, H. F.; Thio, T. Electrical Conductivity of Individual Carbon Nanotubes. *Nature* **1996**, *382*, 54–56.

12. Maity, A.; Ray, S. S.; Pillai, S. K. Morphology and Electrical Conductivity of Poly(N-vinylcarbazole)=Carbon Nanotubes Nanocomposite Synthesized by Solid State Polymerization. *Macromol. Rapid Commun.* **2007**, *28*, 2224–2229.

13. Han, Z.; Fina, A. Thermal Conductivity of Carbon Nanotubes and Their Polymer Nanocomposites: A Review. *Progr. Polym. Sci.* **2011**, *36*, 914–944.

14. Wypych, G. *Handbook of Fillers: Physical Properties of Fillers and Filled Materials*; ChemTec Publishing: Toronto, 2000.

15. Fischer, J. E. Carbon Nanotubes: Structure and Properties. In *Carbon Nanomaterials*; Gogotsi, Y., Ed.; Taylor and Francis: New York, 2006; pp 51–58.

16. Xie, H.; Cai, A.; Wang, X. Thermal Diffusivity and Conductivity Of Multiwalled Carbon Nanotube Arrays. *Phys Lett. A* **2007**, *369*, 120–123.

17. Nan, C. W.; Shi, Z.; Lin, Y. A Simple Model for Thermal Conductivity of Carbon Nanotube-Based Composites. *Chem. Phys. Lett.* **2003**, *375*, 666–669.

18. Grujicic, M.; Cao, G.; Gersten, B. Atomic-Scale Computations of the Lattice Contribution to Thermal Conductivity of Single-Walled Carbon Nanotubes. *Mater. Sci. Eng. B*, **2004**, *107*, 204–216.

19. Hepplestone, S. P.; Ciavarella, A. M.; Janke, C.; Srivastava, G. P. Size and Temperature Dependence of the Specific Heat Capacity of Carbon Nanotubes. *Surf. Sci.* **2006**, *600*, 3633–3636.

20. Majidian, M.; Grimaldi, C.; Forró, L.; Magrez, A. Role of the Particle Size Polydispersity in the Electrical Conductivity of Carbon Nanotube-Epoxy Composites. *Sci. Rep.* **2017,** *7,* 12553.

21. Dasari, A.; Yu, Z. Z.; Mai, Y. W. Electrically Conductive and Super-Tough Polyamide-Based Nanocomposites. *Polymer* **2009,** *50,* 4112–4121.

22. Coleman, J. N.; Curran, S.; Dalton, A. B.; Davey, A. P.; McCarthy, B.; Blau, W.; Barklie, R. C.; Percolation-Dominated Conductivity in a Conjugated-Polymer-Carbon-Nanotube Composite. *Phys. Rev. B* **1998,** *58,* 7492–7495.

23. Geng, Y.; Liu, M. Y.; Li, J.; Shi, X. M.; Kim, J. K. Effects of Surfactant Treatment on Mechanical and Electrical Properties of CNT-Epoxy Nanocomposites. *Compos. Part A* **2008,** *39,* 1876–1883.

24. Singh, I. V.; Tanaka, M.; Endo, M. Effect of Interface on the Thermal Conductivity of Carbon Nanotube Composites. *Int. J. Therm. Sci.* **2007,** *46,* 842–847.

25. Grossiord, N.; Loos, J.; Regev, O.; Koning, C. E. Toolbox for Dispersing Carbon Nanotubes into Polymers to Get Conductive Nanocomposites. *Chem. Mater.* **2006,** *18,* 1089–1099.

26. Manzetti, S.; Gabriel, J. C. P. Methods for Dispersing Carbon Nanotubes for Nanotechnology Applications: Liquid Nanocrystals, Suspensions, Polyelectrolytes, Colloids and Organization Control. *Int. Nano Lett.* **2019,** *9,* 31–49.

27. Vaisman, L.; Wagner, H. D.; Marom, G. The Role of Surfactants in Dispersion of Carbon Nanotubes. *Adv. Colloid Interf. Sci.* **2006,** *128,* 37–46.

28. Gojny, F. H.; Wichmann, M. H. G.; Fiedler, B.; Kinloch, I. A.; Bauhofer, W.; Windle, A. H.; Schulte, K. Evaluation and Identification of Electrical and Thermal Conduction Mechanisms in Carbon Nanotube=Epoxy Composites. *Polymer* **2006,** *47,* 2036–2045.

29. Burger, N.; Laachachi, A.; Ferriol, M.; Lutz, M.; Toniazzo, V.; Ruch, D. Review of Thermal Conductivity in Composites: Mechanisms, Parameters and Theory. *Progr. Polym. Sci.* **2016,** *61,* 1–28.

30. Yang, X.; Liang, C.; Ma, T.; Guo, Y.; Kong, J.; Gu, J.; Chen, M.; Zhu, J. A Review on Thermally Conductive Polymeric Composites: Classification, Measurement, Model and Equations, Mechanism and Fabrication Methods. *Adv. Compos. Hybrid Mater.* **2018,** *1,* 207–230.

31. Li, A.; Zhang, C.; Zhang, Y. F. Thermal Conductivity of Graphene-Polymercomposites: Mechanisms, Properties, and Applications. *Polymers* **2017,** *9,* 437.

32. Hu, J.; Huang, Y.; Yao, Y.; Pan, G.; Sun, J.; Zeng, X.; Sun, R.; Xu, J. B.; Song, B.; Wong, C. P. Polymer Composite with Improved Thermal Conductivity by Constructing a Hierarchically Ordered Three-Dimensional Interconnected Network of BN. *ACS Appl. Mater. Interf.* **2017,** *9,* 13544–13553.

33. Maultzsch, J.; Reich, S.; Thomsen, C.; Dobardzic, E.; Miloevic, I.; Damnjanovic, M. Phonon Dispersion of Carbon Nanotubes. *Solid State Commun.* **2002,** *121,* 471–474.

34. Lindsay, L.; Broido, D. A.; Mingo, N. Lattice Thermal Conductivity of Single-Walled Carbon Nanotubes: Beyond the Relaxation Time Approximation and Phonon-Phonon Scattering Selection Rules. *Phys. Rev. B* **2009,** *80,* 125407.

35. Ishii, H.; Kobayashi, N.; Hirose, K. Electron–Phonon Coupling Effect on Quantum Transport in Carbon Nanotubes Using Time-Dependent Wave-Packet Approach. *Phys. E* **2007,** *40,* 249–252.

36. Yang, S.-Y.; Ma, C.-C. M.; Teng, C.-C.; Huang, Y.-W.; Liao, S.-H.; Huang, Y.-L.; Lee, T. M.; Tien, H. W.; Chiou, K.-C. Effect of Functionalized Carbon Nanotubes on the Thermal Conductivity of Epoxy Composites. *Carbon* **2010**, *48*, 592–603.

37. Hu, Z.; Li, N.; Li, J.; Zhang, C.; Song, Y.; Li, X.; Wu, G.; Xie, F. Huang, Y. Facile Preparation of Poly(P-Phenylene Benzobisoxazole)/Graphene Composite Films via One-Pot in Situ Polymerization. *Polymer* **2015**, *71*, 8–14.

38. Xiao, W.; Luo, X.; Ma, P.; Zhai, X.; Fan, T.; Li, X. Structure Factors of Carbon Nanotubes on the Thermal Conductivity of Carbon Nanotube/Epoxy Composites. *AIP Adv.* **2018**, *8*, 035107.

39. Chao, M.; Li, Y.; Wu, G.; Zhou, Z.; Yan, L. Functionalized Multiwalled Carbon Nanotube-Reinforced Polyimide Composite Films with Enhanced Mechanical and Thermal Properties. *Int. J. Polym. Sci.* **2019**, *2019*, 1–12.

40. Qian, X.; Gu, X. K.; Dresselhaus, M. S.; Yane, R. G.; Anisotropic Tuning of graphite thermal conductivity by lithium intercalation. *J. Phys. Chem. Lett.* **2016**, *7*, 4744–4750.

41. King, J. A.; Via, M. D.; Caspary, J. A.; Jubinski, M. M.; Miskioglu, I.; Mills, O. P.; Bogucki, G. R.; Electrical and Thermal Conductivity and Resins. *J. Appl. Polym. Sci.* **2010**, *118*, 2512–2520.

42. Lee, J. H.; Koh, C. Y.; Singer, J. P.; Jeon, S. J.; Maldovan, M.; Stein, O.; Thomas, E. L. 25th Anniversary Article: Ordered Polymer Structures for the Engineering of Photons and Phonons. *Adv. Mater.* **2014**, *26*, 532–568.

43. Zhang, T.; Luo, T. F. Role of Chain Morphology and Stiffness in Thermal Conductivity of Amorphous Polymers. *J. Phys. Chem. B* **2016**, *120*, 803–812.

44. Choy, C. Thermal Conductivity of Polymers. *Polymer*, **1977**, *18*, 984–1004.

45. Shen, X.; Wang, Z.; Wu, Y.; Liu, X.; He, Y. B.; Kim, J. K. Multilayer Graphene Enables Higher Efficiency in Improving Thermal Conductivities of Graphene/Epoxy Composites. *Nano Lett.* **2016**, 16, 3585–3593.

46. Aliev, A. E.; Lima, M. H.; Silverman, E. M.; Baughman, R. H. Thermal Conductivity of Multi-Walled Carbon Nanotube Sheets: Radiation Losses and Quenching of Phonon Modes. *Nanotechnology* **2009**, *21*, 035709.

47. Gaska, K.; Rybak, A.; Kapusta, C.; Sekula, R.; Siwek, A. Enhanced Thermal Conductivity of Epoxy-Matrix Composites with Hybrid Fillers. *Polym. Adv. Technol.* **2014**, *26*, 26–31.

48. Bagchi, B.; Nomura S. On the Effective Thermal Conductivity of Carbon Nanotube Reinforced Polymer Composites. *Compos. Sci. Technol.* **2006**, *66*, 1703–1712.

49. Shenogina, N.; Shenogin, S.; Xue, L.; Keblinski, P. On the Lack of Thermal Percolation in Carbon Nanotube Composites. *Appl. Phys. Lett.* **2005**, *87*, 133106.

50. Zhong, H.; Lukes, J. R.; Interfacial Thermal Resistance between Carbon Nanotubes: Molecular Dynamic Simulations and Analytical Thermal Modeling. *Phys. Rev. B* **2006**, *74*, 125403.

51. Pradhan, N. R.; Duan, H.; Liang, J.; Iannacchione, G. S. The Specific Heat and Effective Thermal Conductivity of Composites Containing Single-Wall and Multi-Wall Carbon Nanotubes. *Nanotechnology* **2009**, *20*, 245705.

52. Endo, M.; Hayashi, T.; Kim, Y. A. Large-Scale Production of Carbon Nanotubes and Their Applications. *Pure Appl. Chem.* **2006**, *78*, 1703–1713.

53. Pramanik, C.; Gissinger, J. R.; Kumar, S.; Heinz, H. Carbon Nanotube Dispersion in Solvents and Polymer Solutions: Mechanisms, Assembly, and Preferences. *ACS Nano* **2017,** *11,* 12805–12816.

54. Xie, X. L.; Mai, Y. W.; Zhou, X. P. Dispersion and Alignment of Carbon Nanotubes in Polymer Matrix: A Review. *Mater. Sci. Eng. R* **2005,** *49,* 89–112.

55. Paredes, J. I.; Burghard, M. Dispersions of Individual Single-Walled Carbon Nanotubes of High Length. *Langmuir* **2004,** *20,* 5149–5152.

56. Ke, G.; Guan, W. C.; Tang, C. Y.; Hu, Z.; Guan, W. J.; Zeng, D. L.; Deng, F. Covalent Modification of Multi-Walled Carbon Nanotubes with a Low Molecular Weight Chitosan. *Chin. Chem. Lett.* **2007,** *18,* 361–364.

57. Shen, J.; Hu, Y.; Qin, C.; Li, C.; Ye, M. Dispersion Behavior of Single-Walled Carbon Nanotubes by Grafting of Amphiphilic Block Copolymer. *Compos. A* **2008,** *39,* 1679–1683.

58. Spitalsky, Z.; Matejka L.; Slouf, M.; Konyushenko, E. N.; Kovárová, J.; Zemek, J.; Kotek, J.; Modification of Carbon Nanotubes and Its Effect on Properties of Carbon Nanotube/Epoxy Nanocomposites. *Polym. Compos.* **2009,** *30,* 1378–1387.

59. Hong, J.; Lee, J.; Hong, C. K.; Shim, S. E. Effect of Dispersion State of Carbon Nanotube on the Thermal Conductivity of Poly(Dimethyl Siloxane) Composites. *Curr. Appl. Phys.* **2010,** *10,* 359–363.

60. Bose, S.; Bhattacharyya, A. R.; Kulkarni, A. R.; Pöschke, P. Electrical, Rheological and Morphological Studies in Co-Continuous Blends of Polyamide 6 and Acrylonitrile–Butadiene–Styrene with Multiwall Carbon Nanotubes Prepared by Melt Blending. *Compos. Sci. Technol.* **2009,** *69,* 365–372.

61. Shaikh, S.; Li, L.; Lafdi, K.; Huie, J. Thermal Conductivity of an Aligned Carbon Nanotube Array. *Carbon* **2007,** *45,* 2608–2613.

62. Garmestani, H.; Al-Haik, M. S.; Dahmen, K.; Tannenbaum, R.; Li, D.; Sablin, S. S.; Hussaini, M. Y. Polymer-Mediated Alignment of Carbon Nanotubes under High Magnetic Fields. *Adv. Mater.* **2003,** *15,* 1918–1921.

63. Dai, J.; Wang, Q.; Li, W.; Wei, Z.; Xu, G. Properties of Well Aligned SWNT Modified Poly (Methyl Methacrylate) Nanocomposites. *Mater. Lett.* **2007,** *61,* 27–29.

64. Tang, Z.; Huang, Q.; Liu, Y.; Chen, Y.; Guo, B.; Zhang, L. Uniaxial Stretching-Induced Alignment of Carbon Nanotubes in Cross-Linked Elastomer Enabled by Dynamic Cross-Link Reshuffling. *ACS Macro Lett.* **2019,** *8,* 1575–1581.

65. Goh, P. S.; Ismail, A. F.; Ng, B. C. Directional Alignment of Carbon Nanotubes in Polymer Matrices: Contemporary Approaches and Future Advances. *Composites Part A* **2014,** *56,* 103–126.

66. Nam, T. H.; Goto, K.; Nakayama, H.; Oshima, K.; Premalal, V.; Shimamura, Y.; Inoue, Y.; Naito, K.; Kobayashi, S. Effects of Stretching on Mechanical Properties of Aligned Multi-Walled Carbon Nanotube/Epoxy Composites. *Composites Part A* **2014,** *64,* 194–202.

67. Barzic, A. I.; Barzic, R. F. Thermal Conduction in Polystyrene/Carbon Nanotubes: Effects of Nanofiller Orientation and Percolation Process. *Rev. Roum. Chim.* **2015,** *60,* 803–807.

68. Barzic, A. I. Temperature Implications on the Rheological Percolation Threshold in Poly(4-Vinylpyridine)/Barium Titanate Nanocomposites. *Rev. Roum. Chim.* **2014,** *59,* 515–519.

69. Barzic, A. I. Percolation Effects in MCNT-Filled Polystyrene: Rheological, Optical, Adhesion and Conductive Investigations. *Mater. Plast.* **2021**, *1*, 10 pp.

70. Bhattacharyya, A. R.; Sreekumar, T. V.; Liu, T.; Kumar, S.; Ericson, L. M.; Hauge, R. H.; Smalley, R. E. Crystallization and Orientation Studies in Polypropylene/Single Wall Carbon Nanotube Composite. *Polymer* **2003**, *44*, 2373–237.

71. Liao, S. H.; Yen, C. Y.; Weng, C. C.; Lin, Y. F.; Ma, C. M.; Yang, C. H.; Tsai, M. C.; Yen, M. Y.; Hsiao, M. C.; Lee, S. J.; Xie, X. F.; Hsiao, Y. H. Preparation and Properties of Carbon Nanotube/Polypropylene Nanocomposite Bipolar Plates for Polymer Electrolyte Membrane Fuel Cells. *J. Power Sourc.* **2008**, *185*, 1225–1232.

72. Zhang, L.; Li, X.; Deng, H.; Jing, Y.; Fu, Q. Enhanced Thermal Conductivity and Electrical Insulation Properties of Polymer Composites via Constructing Pglass/ CNTs Confined Hybrid Fillers. *Composites Part A*, **2018**, *115*, 1–7.

73. Su, Z.; Wang, H.; He, J.; Guo, Y.; Qu, Q.; Tian, X. Fabrication of Thermal Conductivity Enhanced Polymer Composites by Constructing an Oriented Three-Dimensional Staggered Interconnected Network of Boron Nitride Platelets and Carbon Nanotubes. *ACS Appl. Mater. Interf.* **2018**, *10*, 36342–36351.

74. Yu, A.; Ramesh, P.; Sun, X.; Bekyarova, E.; Itkis, M. E.; Haddon, R. C. Enhanced Thermal Conductivity in a Hybrid Graphite Nanoplatelet–Carbon Nanotube Filler for Epoxy Composites. *Adv. Mater.* **2008**, *20*, 4740–4744.

75. King, J. A.; Johnson, B. A.; Via, M. D.; Ciarkowski, C. J. Effects of Carbon Fillers in Thermally Conductive Polypropylene Based Resins. *Polym. Compos.* **2010**, *31*, 497–506.

76. King, J. A.; Gaxiola, D. L.; Johnson, B. A.; Keith, J. M. Thermal Conductivity of Carbon-Filled Polypropylene-Based Resins. *J. Compos. Mater.* **2010**, *44*, 839–855.

77. Ghose, S.; Watson, K. A.; Working, D. C.; Connell, J. W.; Smith, Jr, J. G.; Sun, Y. P. Thermal Conductivity of Ethylene Vinyl Acetate Copolymer/Nanofiller Blends. *Compos. Sci. Technol.* **2008**, *68*, 1843–1853.

78. Choi, S. W.; Yoon, K. H.; Jeong, S. S. Morphology and Thermal Conductivity of Polyacrylate Composites Containing Aluminum/Multi-Walled Carbon Nanotubes. *Composites Part A* **2013**, *45*, 1–5.

79. Zhu, J.; Peng, H. Q.; Rodriguez-Macias, F.; Margrave, J. L.; Khabashesku, V. N.; Imam, A. M.; Lozano, K.; Barrera, E. V. Reinforcing Epoxy Polymer Composites through Covalent Integration of Functionalized Nanotubes. *Adv. Funct. Mater.* **2004**, *14*, 643–648.

80. Yang, Y. K.; Xie, X. L.; Wu, J. G.; Yang, Z. F.; Wang, X. T.; Mai, Y. W. Multiwalled Carbon Nanotubes Functionalized by Hyperbranched Poly(Urea-Urethane)s by a One-Pot Polycondensation. *Macromol. Rapid Commun.* **2006**, *27*, 1695–1701.

81. Kota, A. K.; Cipriano, B. H.; Duesterberg, M. K.; Gershon, A. L.; Powell, D.; Raghavan, S. R.; Bruck, H. A. Electrical and Rheological Percolation in Polystyrene=MWCNT Nanocomposites. *Macromolecules* **2007**, *40*, 7400–7406.

82. Zainal, N. F. A.; Azira, A. A.; Nik, S. F.; Rusop, M. The Electrical and Optical Properties of PMMA=MWCNTs Nanocomposite Thin Films. *Nanosci. Nanotechnol.* **2009**, *1136*, 750–754.

83. Kymakis, E.; Alexandou, I.; Amaratunga, G. A. J. Single-Walled Carbon Nanotube-Polymer Composites: Electrical, Optical and Structural Investigation. *Synth. Met.* **2002**, *127*, 59–62.

84. Seidel, G. D.; Lagoudas, D. C. A Micromechanics Model for the Electrical Conductivity of Nanotube-Polymer Nanocomposites. *J. Compos. Mater.* **2009,** *43,* 917–941.

85. Du, F. M.; Scogna, R. C.; Zhou, W.; Brand, S.; Fischer, J. E.; Winey, K. I. Nanotube Networks in Polymer Nanocomposites: Rheology and Electrical Conductivity. *Macromolecules* **2004,** *37,* 9048–9055.

86. Smith, R. C.; Carey, J. D.; Murphy, R. J.; Blau, W. J.; Coleman, J. N.; Silva, S. R. P. Charge Transport Effects in Field Emission from Carbon Nanotube-Polymer Composites. *Appl. Phys. Lett.* **2005,** *87,* 263105.

87. Ounaies, Z.; Park, C.; Wise, K. E.; Siochi, E. J.; Harrison, J. S. Electrical Properties of Single Wall Carbon Nanotube Reinforced Polyimide Composites. *Compos. Sci. Technol.* **2003,** *63,* 1637–1646.

88. Wu, S. H.; Masaharu, I.; Natsuki, T.; Ni, Q. Q. Electrical Conduction and Percolation Behavior of Carbon Nanotubes-UPR Nanocomposites. *J. Reinf. Plast. Compos.* **2006,** *25,* 1957–1966.

89. Kilbride, B. E.; Coleman, J. N.; Fraysse, J.; Fournet, P.; Cadek, M.; Drury, A.; Hutzler, S.; Roth, S.; Blau, W. J. Experimental Observation of Scaling Laws for Alternating Current and Direct Current Conductivity in Polymer-Carbon Nanotube Composite Thin Films. *J. Appl. Phys.* **2002,** *92,* 4024–4030.

90. Barzic, R. F.; Barzic, A. I.; Dumitrascu, G. Percolation Network Formation in Poly(4-Vinylpyridine)/Aluminum Nitride Nanocomposites: Rheological, Dielectric, and Thermal Investigations. *Polym. Compos.* **2013,** *35,* 1543–1552.

91. Zare, R. Calculation of the Electrical Conductivity of Polymer Nanocomposites Assuming the Interphase Layer Surrounding Carbon Nanotubes. *Polym.* **2020,** *12,* 404.

92. Mora, A.; Verma, P.; Kumar, S. Electrical Conductivity of CNT/Polymer Composites: 3D Printing, Measurements and Modeling. *Compos. Part B: Eng.* **2020,** *183,* 107600.

93. Aguilar, J. O.; Bautista-Quijano, J. R.; Aviles, F. Influence of Carbon Nanotube Clustering on the Electrical Conductivity of Polymer Composite Films. *Express Polym. Lett.* **2010,** *4,* 292–299.

94. Spitalsky, Z.; Tasis, D.; Papagelis, K.; Galiotis, C. Carbon Nanotube–Polymer Composites: Chemistry, Processing, Mechanical and Electrical Properties. *Progr. Polym. Sci.* **2010,** *35,* 357–401.

95. Bairan, A.; Selamat, M. Z.; Sahadan, S. N.; Malingam, S. D.; Mohamad, N. Effect of Carbon Nanotubes Loading in Multifiller Polymer Composite as Bipolar Plate for PEM Fuel Cell. *Procedia Chem.* **2016,** *19,* 91–97.

96. Souier, T.; Maragliano, C.; Stefancich, M.; Chiesa, M. How to Achieve High Electrical Conductivity in Aligned Carbon Nanotube Polymer Composites. *Carbon* **2013,** *64,* 150–157.

CHAPTER 10

FTIR SPECTROSCOPY FOR CARBON NANOTUBE-BASED NANOMATERIALS IN BIOMEDICAL APPLICATIONS

MIOARA DROBOTA[1*], MARIA ANDREEA LUNGAN[2] and IULIAN RADU[3]

[1]*"Petru Poni" Institute of Macromolecular Chemistry, Aleea Grigore Ghica Voda 41A, Iasi, 700487, Romania*

[2]*Sara Pharm Solutions, Calea Rahovei 266-268, Electromagnetica Business Park, Bucureşti, 050912*

[3]*Department of Surgery, University of Medicine and Pharmacy "Grigore T. Popa" Iasi, Romania; Regional Institute of Oncology, I-st Surgical Oncology, Iasi, Romania*

Corresponding author. E-mail: miamiara@icmpp.ro

ABSTRACT

FTIR spectroscopy (Fourier-transform infrared spectroscopy) provides a versatile technique of identification and characterization of varied chemical structures to obtain information from biological to composite materials.

Due to their unique properties, such as chemical and physical properties, including structural diversity, thermal, optical, mechanical, electrical properties, carbon-based nanomaterials have attracted great attention for biomedical applications.

This manuscript presents some recent studies for various biomedical applications, in this paper, we present the FTIR vibrational spectroscopy for the characterization of different materials such as carbon-based nanomaterials, including carbon nanotubes, graphene oxide for biomedical application.

10.1   INTRODUCTION

Fourier-transform infrared spectroscopy is one of the noninvasive methods for characterizing compounds and biomaterials used in the biomedical fields. FTIR spectroscopy is a fast alternative that is applied to an industrial frame.[1]

Significant attention was granted to carbon in recent decades. This attention has been aroused by the different shapes and dimensions it can generate, because of its capacity given by the variable hybridization of carbon atomic orbitals that can ultimately generate intelligent materials.[2] These carbon-based structures may be found in different forms and involved in different structures. These structures in combination with various polymers can be assembled into different membranes, and are used in various applications. Biomedical applications are currently of great interest especially for the tissue engineering.[3]

Tissue engineering is the field that aims, through various promising ways, to regenerate and heal denatured tissues. This field of tissue engineering is based on three factors that must be taken into account: cells, tissues, and their interaction with external biomedium. Tissue is one of the most important elements in tissue engineering.[4,5] The extracellular matrix (ECM) in tissue engineering must simulate and be an appropriate environment for cell survival and control the behavior and functions of the cell.[4,6] The purpose of tissue engineering is to regenerate the tissue, so the scaffolding must be biodegradable to be able to be removed from the body medium during the gradual growth of the newly regenerated tissue. The scaffolding should have mechanical strength[7] and the architecture should be suitable for cell attachment and migration.[7-9] Carbon (C) is found almost everywhere, being the building block of all living systems and the most important material used in many applications.[8] carbon-based nanostructured materials at nanoscale have been used in various applications. Due to its four valence electrons, the carbon may hybridize in many ways and generate a great variety of structures. Functionalized carbon nanotubes (CNTs) can thus be found in two forms of single-walled CNTs (SWCNTs) and multiwalled CNTs (MWCNTs) that are assimilated in biomaterials for characteristic enhancements and scaffold functions.[9,10] The CNTs have some characteristics, such as high length-to-diameter ratio, high surface-to-volume ratio, low density, high mechanical flexibility, electrical and thermal conductivity.[11,12] The functionalization of CNTs with functional groups such as carboxyl (COOH) leads to an increase in their polarity and thus induces hydrophilicity and biocompatibility, respectively a more uniform distribution, ensuring to requested connection with the materials matrix.[10] Polymers with hydrophilic character such as gelatin and with hydrophobic character such as polycaprolactone (PCL)[13] can lead to composites with remarkable properties in combination with CNTs for tissue engineering. Gelatin is a facile biopolymer that is highly biodegradable with inadequate mechanical properties and PCL is a linear semicrystalline polymer with remarkable rheological, biocompatible, and non-toxic properties. These two polymers can be applied together in one composite in the field of tissue engineering to regenerate different tissue.[14,15] The combination of carbon-based structures and PCL, gelatin or even collagen in composites[14] may be applicable as tissue-engineering scaffolds which influences the biocompatibility, hydrophilicity, and the corresponding mechanical properties.[16] Due to the properties induced by the interaction of the CNTs with the matrix of a material, they allow the application as reinforcement in low-density composite materials, such as polymer-based composites.[15] The amount of MWCNTs functionalized with carboxylic functional groups in various molar ratios with different combination of polymers leads to composite constructs with clearly superior physical and mechanical properties that has applications in the field of tissue engineering.[17]

In this chapter, some important composites with carbon-based structures are presented, which have applications in biomedical field, they are characterized using Fourier-transform infrared (FTIR) spectroscopy.

10.2 FOURIER-TRANSFORM INFRARED SPECTROSCOPY AND CARBON NANOTUBES

The infrared (IR) spectroscopy is a spectroscopic analytical method, a technique which is used to characterize different structures based on the interaction of the infrared radiation with matter.[1]

FTIR spectrometry combines the interferometer with an older mathematical principle. The Fourier transform converts the information obtained from an interferometer (interferogram) when the radiation passes through the investigated sample and is finally sent to the detector. The information is materialized by means of a spectrum.

Therefore, the FTIR spectrum in infrared spectroscopy is a diagram, that represents a response from a detector device. We will find on the y-axis the values as percentage of the transmittance as function of wavenumber (cm^{-1}) (represented on x-axis). The response indicates an interaction between infrared radiation and a sample. There is a mathematical relationship between Absorbance and Transmittance according the following relation:

$$A = \log 1/T$$

where A is the absorbance and T is the transmittance (%T/100)

The mid-infrared region (MIR) is the commonly used range in chemical analysis and it presents the wide spectral range between 4000 and 500 cm^{-1}. For the group frequency region the spectrum occurs between 4000 and 1300 cm^{-1} (2.5–8.0 μm^{-1}) and for fingerprint region, the spectral range occurs between 1300 and 500 cm^{-1} (8.0–20 μm).

As mentioned before, carbon, due its four valence electrons, may generate a great variety of forms, such as linear chains, planar sheets and tetrahedral structures, used as part of some carbon-based nanostructured materials, and developed for various applications.[18] Among them, different types of CNTs can induce different properties to the studied nanomaterial and the FTIR spectrum will also have certain vibration changes. In addition, the nanotubes can have single wall or multiple walls: superimposed (scrolled) or cylindrical, with the nanotubes ends closed or opened, with or without branches. The morphology of the nanotubes can also be different, straight or curved (bent), depending on the width of the inner wall. The presence of structural defects, impurities, adsorbed molecules, and network defects should also be taken into consideration.

FTIR is used for the functionalized CNTs investigation, which could also estimate any defect that may occur. The investigation of the functionalization of the nanotubes will be presented in FTIR spectrum as a change in vibration. This could be a change in the morphological properties, and last but not least, a change in their toxicity. In FTIR spectra, the vibrations appear at 1600 and 3450 cm^{-1}, these being assigned vO-H according to Osswald et al., and that at 1445 cm^{-1} according to Misra et al., which is assigned for MWCNTs.[19,20]

The size of MWCNTs present an important influence in vibrations. When MWCNTs present a disordered arrangement in disordered amorphous carbon, a peak appears around 1340 cm^{-1} as well at 3000 cm^{-1} assigned CH_x.[21] A single peak at 1584 cm^{-1} is characteristic to SWCNTs in their ordered structure[21] and for functionalized MWCNTs another vibration appears at 1725 cm^{-1} (COOH groups).[22]

The functionalized SWCNTs-COOH compared with the original SWCNTs can be assigned as defect sites. These defects are the result of the removal of additives used in the oxidation process. It has been found that these defective nanotubes are much more reactive than those without defects. If SWCNTs are functionalized and have a tangential or radial arrangement along the length of the nanotube axis, in the FTIR spectrum, one vibration can be observed at 1590 cm^{-1} and another at 1350 cm^{-1}, which are assigned to the amorphous carbon.[23]

Rike Yudianti et al. treated the MWCNTs to introduce new functionality to the CNTs, which was achieved using a chemical oxidation with HNO_3, causing structural chemical changes that were evaluated in the FTIR spectrum, in the MIR region. In the FTIR spectrum from Figure 10.1 we can see the presence of the two forms: the pristine MWCNTs and oxidized-MWCNTs.[24]

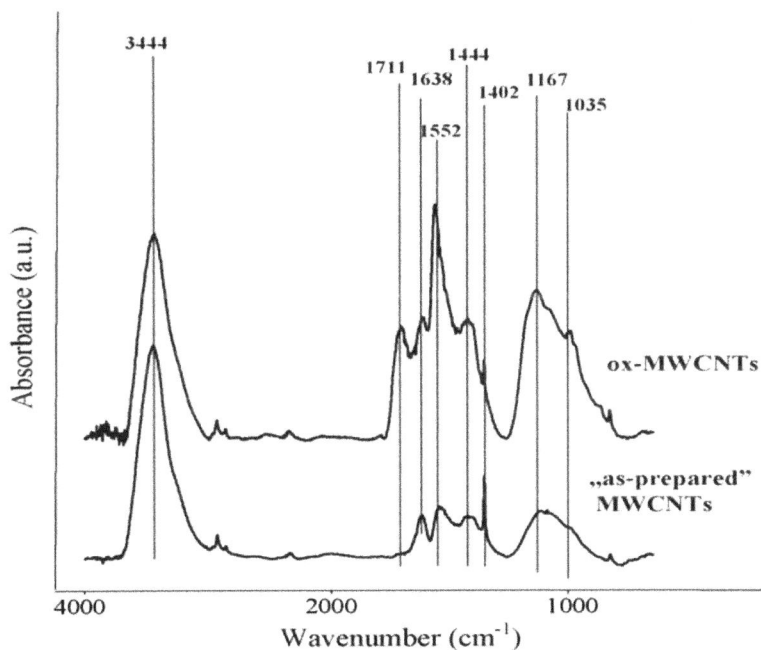

FIGURE 10.1   MWCNTs and MWCNTs-functionalized FTIR spectra.

Source: Reprinted with permission from Ref. [24]. © 2020 Elsevier.

After functionalization of nanotubes, the vibrations appear at 2800–2900 cm^{-1} corresponding to vasCH and vsCH, attributed to the alkyl chain and the hexagonal structure on MWCNTs which were confirmed in the range 1532–1560 cm^{-1}, like a band which is attributed to the carbon double bond $vC = C$.[25] A decrease in vibration attributed to $vC = C$ indicates an oxidation at MWCNTs surface and the appearance of the band at 1753 cm^{-1} could be characterized by the carbonyl such as carboxyl group.[26,27] This behavior shows a large number of hexagonal carbon.

The vibration band at 1550 cm^{-1} could indicate the vibration from unsaturated aromatic structure assigned due to $vC = C$ bonds. The bands at 1444 cm^{-1} assigned for OH (OH in-plane deformation) and from C–O moiety (e.g., C–O–C groups oxides of structural, oxygen bridges) are present in the FTIR spectra which confirmed the MWCNTs.

After functionalization, in spectrum, we detect at 3444 cm^{-1} a vibration attributed for OH, which was a decrease, and added another peak at 3190 cm^{-1}, which confirmed the presence of OH groups at the carbon surface. The signals appeared at 1711 and 1638 cm^{-1} are characteristic for carbonyl vibration from carboxyl groups (–COOH) and of ketone/quinone groups.[28] In the region 1250–950 cm^{-1}, the bands are increased due the amounts of hydrated surface oxides (O–H deformation and C–O stretching and aromatic carboxylic acids). The major differences between SWCNTs and MWCNTs are observed (Fig. 10.2) in the spectrum for SWCNTs compared with MWCNTs, the appearance of a very wide band which is much more intense has been observed for SWCNTs at 3300–3500 cm^{-1} (OH groups in this carbon sample) and another band in the range of 800–1200 cm^{-1} (stretching vibration attributed C-OH) comparative to MWCNTs.[29]

These bands are related to the oxygen–hydrogen and carbon–oxygen bonds in the CNTs structure. We can conclude that SWCNTs contains chemical functionalities, such as the –OH and –COOH carboxyl groups. Thus, SWCNTs have more hydrophilic surface, while MWCNTs appear to be hydrophobic. The hydrophilic and hydrophobic nature of single and multiple wall CNTs was observed and confirmed after their immersion in water.[29] SWCNTs are evenly dispersed throughout the water volume, while MWNTs accumulate on the surface.

10.3 CNTs–GELATIN COMPOSITE

High molecular weight polypeptides obtained by collagen hydrolysis lead to the formation of gelatin. It is widely used in the medical field, in the composition of dermal regeneration dressings, it is the most common as a material for obtaining biocompatible materials with controlled drug delivery.[29,30]

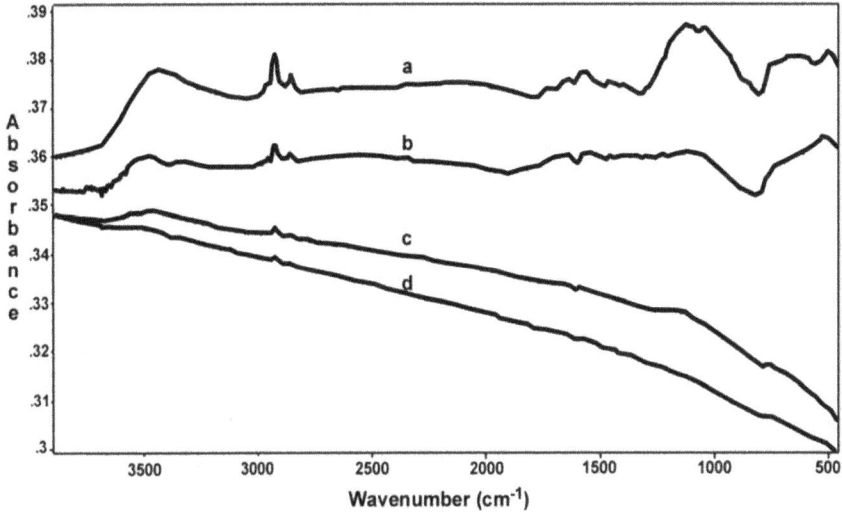

FIGURE 10.2   SWCNTs (a: after baseline corection; c: befor baseline) and MWCNTs (b: after baseline correction; c: befor baseline) FTIR spectra.

Source: Reprinted with permission from Ref. [29]. © 2021 Elsevier.

Gelatin due to its weak mechanical strength has limitations, the structure needs to be strengthened to be used as a material in practical applications.[31]

In biological applications, CNTs with single and multiwalled walls, due to their electrical properties can be used as a bioelectronic nanomaterial[34] in immobilizing biological species and proteins[35] with both SWCNTs and MWCNTs as a potential biosensor.[36]

From gelatin structure, the vibration of amide A occurs in the range of 3169–3366 cm^{-1}. This vibration is represented by the stretching vibration vNH induced by the hydrogen linking mode.

When the involved amine group is free, the vibration occurs in the range 3400–3440 cm^{-1}, while its position is shifted to lower frequencies of the vibration attributed to vNH when the amine group is involved in hydrogen bonds, meaning it is not free.

Amino groups from gelatin, when are free, are a consequence of a pronounced hydrolysis, as a result of obtaining gelatin. Amide B vibration presents a range between 3128 and 3190 cm^{-1}, representing stretch vibrations from vNH which are implied in gelatin groups. The vibration of amide I is a vibration which presents contribution of the stretch vibration vC = O. Muyongaa et al.[37] reported the vibrations for the amide I band between 1600 and 1700 cm^{-1} from gelatin extracted from the skin. Amide II vibration presents a complex vibration, mainly consisted of the deformation vibration δ(NH) and v(CN) stretching vibration in the amide groups.[38,39] CNTs with different shapes can be used in polymeric systems, due to the fact that they have a high capacity to tolerate large deformations (have good mechanical properties)[29–32] without being damaged.[33]

Amide II vibration is presented in the range of 1560–1500 cm^{-1} reported by Yakimets et al.[39] The decrease in conformation of the triple helix is much more sensitive to hydration[41] than the changes occurring in the secondary structure conformation.[42] The Amide III band is a vibration occurring between 1234 and 1237 cm^{-1}. Vibration's intensity at lower wavenumber at 887 and 920 cm^{-1} appears due to gelatin obtaining process for those vibrations most likely representing vCN corresponding to the primary amines.[43]

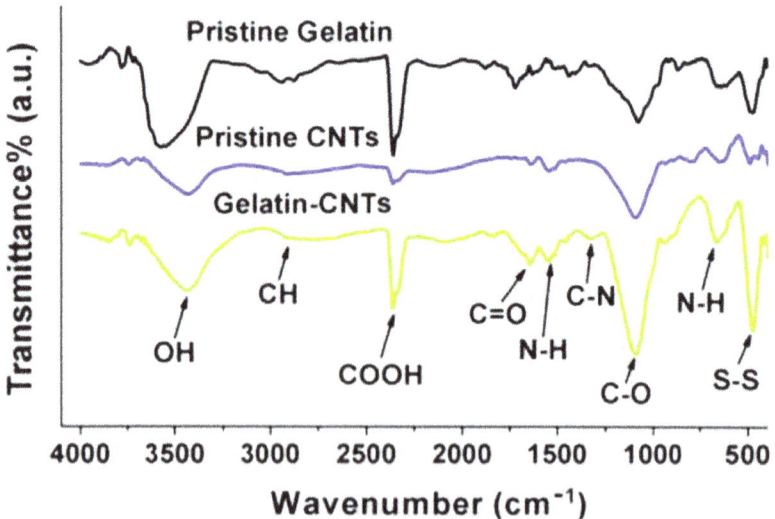

FIGURE 10.3 The FTIR spectra for gelatin, pristine CNTs, and gelatin–CNTs.

Source: Reprinted with permission from Ref. [45]. © 2021 Elsevier.

FTIR spectra of pure gelatin, pure CNTs, and gelatin–CNTs are presented in Figure 10.3.[44] All spectra for pure gelatin and gelatin–CNT show strong peaks at 2359, 1084, and 472 cm^{-1}, representing hydrogen binding from COOH groups,[45] vCO stretching vibrations[46] and S–S vibrations[47] of the gelatin component.

The peaks located at 2930, 1720, 1550, and 1320 cm^{-1} in the gelatin–CNTs composite spectrum due to vCH stretching vibrations, carbonyl group v(C = O),[48] δNH bending vibration,[49] and vCN stretching vibrations can be observed, and they confirm the presence of gelatin–CNTs composites.

10.4 CNTs–COLLAGEN COMPOSITE

Collagen is a biopolymer with excellent properties and contains the most part of the proteins from human body.[49] This biomaterial has a significant contribution in the connection between tissues, in the skin, tendons, cornea and bones, and membranes.

Figure 10.4[50] shows the FTIR spectra for collagen and collagen–MWCNTs composites and illustrates the influence of MWCNTs in the interaction with collagen molecules. The amide A at 3300 cm^{-1} is present in the FTIR spectrum and is due to the Fermi resonance of vN-H and the overtone of amide II.[51–53] These vibrations from amide A (stretching vibrations of the N-H group) are at higher wavenumbers for other proteins, usually occurring in range 3400–3440 cm^{-1}. Due to the formation of hydrogen bonds, the position of the band is shifted to a wavenumber less than 3300 cm^{-1}.[51] This behavior to the formation of hydrogen bonds is specific to amino acids, such as proline and hydroxyproline, in hydrogen linkages, which differentiates its vibrations from other proteins. The band located at 3078 cm^{-1} is attributed to amide B and the vibration present at 1629 cm^{-1} are attributed to amide I (80% stretch vCO) and the vibration located at 1543 cm^{-1} are attributed to amide II (40% stretch vC—N, 60% δN –H bend).[51]

This indicates that the addition of MWCNTs affects the CH$_3$ stretching vibrations of collagen molecules as shown in Figure 10.5.[53,54] When a higher concentration of CNTs is added, the secondary structure of the collagen, which can be identified by amide II, is not affected, only vCH$_3$ from the collagen molecules are affected. Thus, an intensification of the 2920 and 3070 cm^{-1} signals is observed. The wavenumbers of the amide I,

II, and III are directly related to the collagen structure. The vibration for amide I in the region 1628–1631 cm^{-1} is assigned to $\nu C = O$, this being a vibration along the polypeptide axis. In the structure of the peptide, we find in the interval of 1550–1600 cm^{-1}, the vibration of amide II located at 1546 cm^{-1}. The interval between 1235 and 1238 cm^{-1} is attributed to amide III. The latter corresponds to the vibrations of the pyrrolidine ring of proline and hydroxyproline. The FTIR results indicate that the structure of the collagen/MWCNTs composite has a well-defined secondary structure, not being significantly influenced by the presence of CNTs.

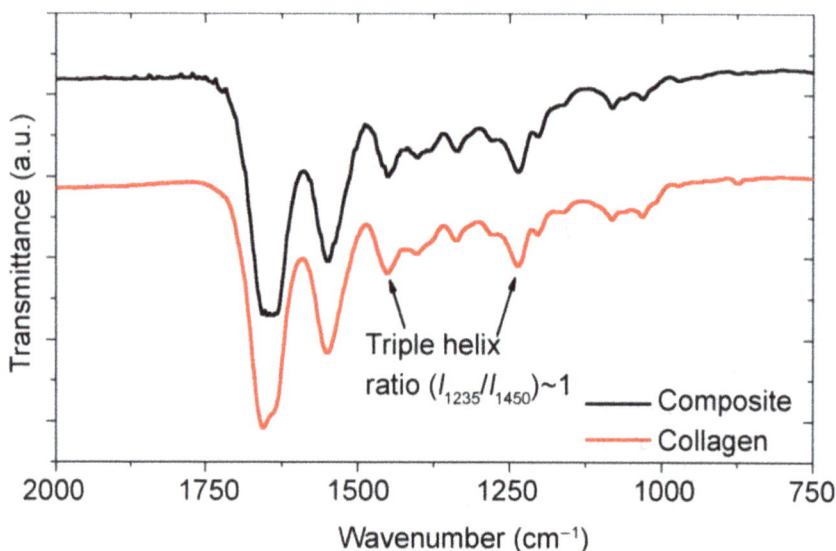

FIGURE 10.4   The FTIR spectrafor collagen and collagen–MWCNTs composite.

Source: Reprinted with permission from Ref. [50]. © 2021 Springer.

10.5   CNTs–PCL COMPOSITE

PCL is a polyester. The FTIR spectrum of pristine PCL shows characteristic signals in the range 2900–2800 cm^{-1} assigned to the CH stretching, at 2944 and 2865 cm^{-1} for vas(CH_2) and vs(CH_2), another signal between 1720 and 1750 cm^{-1} for carbonyl stretching ν ($C = O$) at 1725 cm^{-1} from ester groups, C=O-O and C-O stretching.[55–57]

   FTIR spectra of PCL shows stretching vibrations in fibers located at ~ 1293 cm^{-1} for C-O and C-C (for the crystalline phase),[58] vibrations for

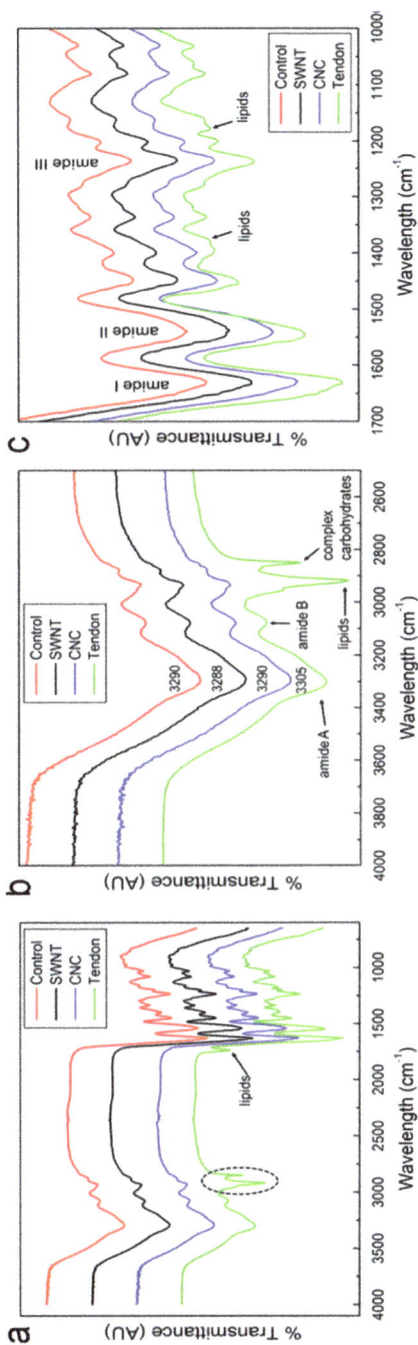

FIGURE 10.5 FTIR spectra for collagen fibers and collagen–MWCNTs in different ranges: (a) 1000–4000 cm^{-1}, (b) 2500–4000 cm^{-1}, and (c)1000–1700 cm^{-1}.

Source: Reprinted with permission from Ref. [54]. © 2021 Elsevier.

vasC-O-C at $\sim$ 1240 and stretching vibrations at 1191 cm^{-1} for ν(OC – O). For the fiber, the vibrations for vs(C-O-C) was at 1159 cm^{-1}, but in the film, this vibration was found at 1170 cm^{-1}. According to Hudecki et al., the FTIR spectra from Figure 10.6[59] for the pristine PCL show some characteristic signals. In some cases, a small signal will appear at 3440 cm^{-1} attributed to νOH which may confirm the presence of a certain adsorbed moisture or terminal groups remained from the initiator used to open the caprolactone cyclic ring.[60]

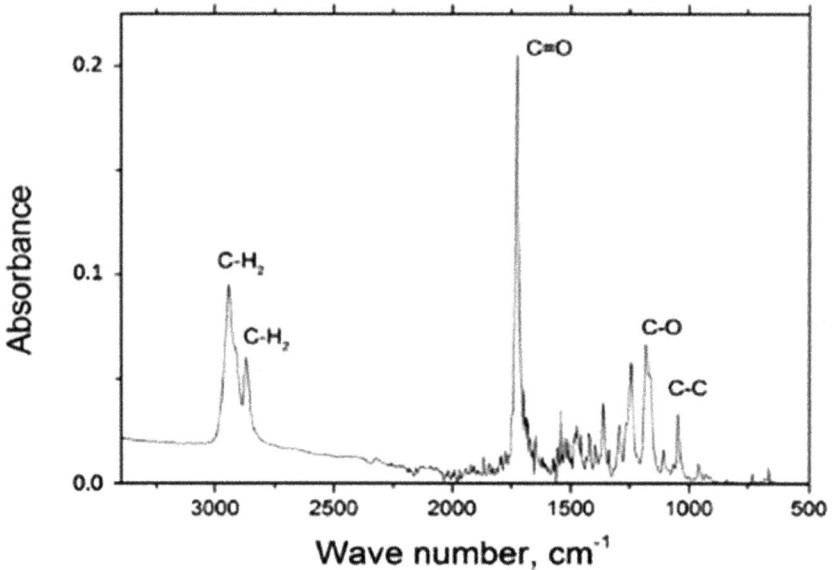

FIGURE 10.6   FTIR spectrum of PCL polymer.

Source: Reprinted with permission from Ref. [59]. © 2021 MDPI.

In composite PCL–nanomaterials like carbon nanofiber (PCL–CNF) with different amount of carbon, a small vibration appears at 1470 cm^{-1} corresponding to νC=C aromatic ring. For the PCL–nanographite (PCL–NG) nanocomposite, these signals are moved to a lower wavenumber in IR.[61] The presence of increasing content of NG is confirmed by the appearance of absorption band at 1675, 1090, and 3500 cm^{-1} attributed to C=C, C-O-C, and OH groups. The shift of these peaks to lower wavenumbers in the FTIR spectra of PCL–NG compared with those obtained for PCL polymer indicates a certain interaction that is most likely caused by the formation of hydrogen bonds between NG and PCL in the created nanocomposites. These carbon (CNF and NG) composites cause a decrease in the crystallinity compared with PCL justified by the decrease of vibrations for carbonyl stretching $v(C = O)$ at 1725 cm^{-1} and at ~ 1293 cm^{-1}.

The presence of gelatin in various proportions, in different molar ratios with PCL, appears in the FTIR spectra[61] and is indicated by the presence of the characteristic bands respectively of amide I and amide II. Vibration of Amide I at 1650 (C = O stretching) was assigned to both random coil and α-helix conformation of gelatin,[61,62] while amide II band is present at 1540 cm^{-1} (mixed vibration of δ(N-H) bending and v(C-N) stretching). When the content of gelatin is increased, the peak at 1750 cm^{-1} intensifies due to the increase in the number of carbonyl groups.[63,64]

According to Ghasemi-Mobarakeh et al.[65] the characteristic vibrations of PCL and gelatin were confirmed for PCL/gelatin nanofibers and a slight change to lower wavenumbers occurred. This phenomenon is further an evidence of the existence of interactions between the ester groups of PCL with the amide groups of gelatin and graphene.

For the graphene spectrum, characteristic peaks were observed. The peak for OH is present at 3431 cm^{-1}, stretching vibration for CH arises at 2920 and 2851 cm^{-1}, stretching vibration for CO occurs at 1728 cm^{-1}, and the peak for v(C=C) arises at 1631 cm^{-1}. After graphene incorporation in PCL/gelatin composite, the characteristic hydroxyl band has broadened indicating that there is possibility of an interaction between graphene and NH from gelatin molecules. In the FTIR spectrum, this appears only as a shoulder not as a well-defined band, this being only due to the formation of hydrogen bonds. All the bands for both composites PCL/gelatin and PCL/gelatin/graphene present a similar location and form.[64,66]

10.6 APPLICATIONS

The CNTs, defined like smart nanomaterials in biomedical applications, were used due to their characteristic and unique properties in the regeneration and the reconstruction of an adequate tissue. These were evaluated like an ideal structure which could sustain and stimulate the growth in different tissues.[67,71] The CNTs imply all the demands, including the excellent electrical conductivity, to obtain the materials and their development for the neuronal regeneration. Lovan și colab.[68] report that some CNTs possess the best surface which induces the dendritic and cellular adhesion. Increasing the neural circuits will determine the high neural network connectivity. Mazzatenta and collaborators[69] developed a SWCNTs-neuron integrated CNT system.[70,71]

Detailed analysis of the scanning electron microscopy at larger magnifications suggested the presence of some close interactions between cell membranes and SWCNTs at the level of neural processes and cell surfaces. The highly oriented design of SWCNTs-based interfaces led to the conclusion that neurons can grow and develop functional circuits on SWCNTs surfaces, indicating the overall biocompatibility of purified SWCNTs. Scanning electron microscopy observations of the neuron/SWCNT contacts suggested the adhesion of the neuronal cells. However, the records seem to exclude a long-term neurotoxicity of these interactions, because the detected modeled spontaneous activity is ubiquitous in the cultured neural circuits. Electrical coupling between neurons and SWCNTs is a relevant issue so that the experimental results, reinforced by modeling, support the idea that any resistive coupling between biomembranes and SWCNTs cannot be qualitatively distinguished from a coupling between SWCNT and patch pipette through the patch-seal to the ground. Currents by SWCNTs stimulation do not conclusively demonstrate a resistive coupling between SWCNTs and neurons. Indeed, by detecting synaptic responses, evoked and provoked in neurons, one can evaluate the effectiveness of the SWCNTs-neuron interaction. Such a phenomenological observation suggests that SWCNT-coated substrates could provide the best mechanical coupling between artificial devices and neuronal tissue.

In different works are presented that the CNT incorporation in biopolymer-based hydrogels could promote the growth of cardiomyocytes in domains like cardiac tissue engineering or cardiac patches.[72,73]

The composite like graphene incorporated in nanofiber mats shows increased conductivity and could also find application in different areas, such as electrical stimulation in tissue engineering as a biosensor. This could be considered as an indication for stimulating cell proliferation and differentiation.[74]

According to Su et al.,[75] in the therapies procedures, black scars appear on the mice skin injected with graphene oxide (GO), nano-GO (PNG), and nano-GO-RGD peptides (PNG-RGD), with a decrease in tumors that reached to 380% indicating limited efficacy of GO as a photothermal agent in GO-injected mice. The mice bearing tumors were injected intratumorally with PNG or PNG-RGD. This confirmed that PNG can effectively suppress the tumor growth that has been observed during tumor cell progression. Mice were euthanized without any treatment after 16 days. The tumor size reached 420%. The effects of the PNG-RGD agent developed on the basis of a RGD conjugated with PNG are selectively for target brain cancer cells. The photothermal conversion shows the performance of PNG–RGD and contributes to an increased therapeutic efficacy compared with the tumor that showed a high recrudescence in mice injected with the same amount of GO when they were exposed to laser irradiation.

CNTs are used in magnetic resonance imaging (MRI) as contrast agents, because it has been established that CNTs doped with metal impurities could be used as MRI contrast agents (taking into consideration that in this technique the image could be detected. without a depth limit).[75] In fact, several studies have been conducted to evaluate the potential of using CNTs. The results of Li et al.[76] suggest that SWCNTs synthesized with an amphiphilic gadolinium (III) chelate using stearic acid have potential as highly efficacious MRI-NIR imaging contrast agents. The functionalized nanotubes MWCNTs were intramuscularly injected and the results clearly demonstrated that MWCNTs could function as both positive and negative paramagnetic contrast agents.

The new PCL/gelatin nanofiber mats incorporated with graphene were prepared by electrospinning technique. The nanofibers nanocomposite could be observed in the resulting mats with higher hydrophilicity due to the presence of oxygen-containing groups attached to the structure of graphene. No cytotoxicity in the presence of graphene was observed. Also, there were more cells proliferated on graphene-doped PCL/gelatin nanofibers than those of PCL/gelatin resulting from the biological properties of graphene. The excellent properties of biocompatibility of the nanofibrous mat are a consequence of the improvement in hydrophilicity of the PCL/gelatin/graphene biomimetic nanofiber scaffold which could be considered as an upgrade in tissue-engineering biomedical application.[77]

The hydrophobic interaction between MWCNTs and gelatin forms composites, so that CNTs inhibit solubility and swelling capacity. The incorporation of the MWCNTs into gelatin also causes a significant increase in the tensile strength by decreasing the elongation at breaking and an increase in the Young's modulus of composite films.[78] The significant increase of Young's modulus of the composite could be due to the that nanotubes are well dispersed in the gelatin matrix and there is a strong adhesion between that two components. MWCNTs particles located in the gelatin matrix increase the opacity of the gelatin film, most likely due to the light scattering effect of CNTs. Mohammad et al. confirmed that gelatin/MWCNTs/Au composite films can be used to quickly detect the bacteria in biological fluids, both Gram-positive and Gram-negative types.[78] These properties suggest a high potential of gelatinous nanocomposites/gold-particles/MWCNTs in biomedical applications.[79] The nanocomposites could facilitate the interaction of anticancer drug to targeted cancer cells. This makes the nanocomposites ideal candidates for the enhancement of targeting efficiency in drug delivery, providing a new perspective for cancer chemotherapy.[80]

The FT-IR confirms[76] the covalent reaction between MWCNT–NH$_2$ and collagen. The MWCNT–NH$_2$ were covalently grafted to the surface of the neat collagen.

The combination of CNTs–collagen opens, in fact, a new perspective in the nanomaterials self-assembly for biomedical applications, especially based on hemostatic effect. The hydrophilicity is an important characteristic property of the hemostatic biomaterials. Thus, the hydrophilicity of MWCNT–NH$_2$/collagen composites increases with the increase of MWCNT–NH$_2$ concentration in composites, so the water absorption increases. Due to the increase in the concentration of the carboxyl groups and amine groups, the water absorption of the composites increased. Thus, the time required for the blood to coagulate outside the body is an important indicator of a coagulant for the activity of hemostatic materials. The addition of MWCNT–NH$_2$ decreases the blood clotting time, indicating that MWCNT–NH$_2$/collagen has significant pro-coagulant activity. Moreover, the coagulation time was significantly reduced with the increase of the amount of MWCNT–NH$_2$ in the composite.

FTIR spectrum was also reported by Sun et al.[81] In their study, when the SWCNTs were incorporated with collagen hydrogels, it determined the alignment and assembly of the cells, allowing the formation of the engineered cardiac functional tissues with greater synchronous contraction potential in the newly obtained tissues, thus increasing the heart rate.[82,83]

In the collagen molecule, there are hydrogen bonds due the assembled structure, which is implicitly the α-helix structure. In FTIR spectrum for amide A, the signal was moved to lower wavenumber, this is due to the structure of MWCNTs which contain the unsaturated links. That behavior could be due to the formed hydrogen bonds.[38]

Generally, the FTIR results indicate the secondary structure in collagen and in the collagen–MWCNTs composite, more precisely of the well-defined and stabilized α-helix conformation. The composite films formed from PCL–MWCNTs were investigated by FTIR,[84] and can be used in biomedical application such as neuronal networks, because it allows us to obtain optimum roughness, involved in the anchoring of the neuronal cells which are going to grow on these composite.[74] In this case, the studied cytocompatibility is superior compared with another composite formed with nanotubes carbon.

The structure of the gelatin–MWCNTs composite confirmed by FTIR[85] will be used for drug delivery. The composite from both component, gelatin and MWCNTs-functionalized will lead to the formation of microspheres, and also could be used as nanofillers due to the biocompatibility properties.[86]

10.7   CONCLUSIONS

FTIR spectroscopy is a nondestructive and versatile method with which functional groups can be characterized and identified depending on the vibration mode specific to those groups. Single-walled and multiwalled CNTs are less common in FTIR spectra because their nanometric size makes them difficult to be detected. In fact, if the aim is to obtain composites, nanotubes will be used in a functionalized form.

In the FTIR spectra of the composites obtained based on gelatin and CNTs, displacements of the polymer characteristic bands will be found or in some cases, certain bands of low intensity looking like shoulders will be found. This behavior is the result of interactions between the two components of the composite, namely, the polymer and the nanotubes.

The signals of the gelatin–CNTs composite in the FTIR spectrum show displacements of the wavenumbers in the lower region of the gelatin characteristic bands, the nanotubes being very well covered by this polymer. According to the FTIR spectra of collagen–CNTs composites, it can be said that CNTs do not affect the chemical composition or structure of collagen during fiber formation, do not react chemically with collagen molecules and/or affect the formation of fibrils. When the N-H group present in the collagen molecules interacts with the π bonds of MWCNTs, the peak of amide A moves at lower frequencies. When a higher concentration of CNTs is added, the secondary structure of the collagen, which can be identified by amide II, is not affected, only νCH_3 from the collagen molecules are affected. Thus, an intensification of the 2920 and 3070 cm^{-1} signal, respectively, is observed. In the composite of PCL nanomaterials, such as carbon nanofiber (PCL–CNF) with different amounts of carbon, a small vibration occurs at 1470 cm^{-1} corresponding to the aromatic ring $\nu C = C$ that is found in the structure of the nanotubes.

Due to the polyaromatic structure of CNTs and their unique properties, the nanotubes are able to conjugate with a wide variety of therapeutic molecules (drugs, proteins, antibodies, DNA, enzymes, etc.) used in cancer therapy. Also, these CNTs have been used in biomedicine to bind antineoplastic drugs and antibiotics or other biomolecule bonds (genes, antibodies, proteins, vaccines, DNA, biosensors, cells, etc.) that have applications in gene therapy, immunotherapy, tissue regeneration, and diagnosis of various ailments.

ACKNOWLEDGMENT

The financial support of European Social Fund for Regional Development, Competitiveness Operational Program Axis 1—Project "Petru Poni Institute of Macromolecular Chemistry—Interdisciplinary Pol for Smart Specialization through Research and Innovation and Technology Transfer in Bio(nano)polymeric Materials and (Eco)Technology" InoMatPol (ID P_36_570, Contract 142/10.10.2016, cod MySMIS: 107464) is gratefully acknowledged.

KEYWORDS

- carbon nanotubes
- collagen
- composite
- FTIR
- gelatin
- poly(ε-caprolactone)

REFERENCES

1. Movasaghi, Z.; Rehman, S.; Rehman, I. Fourier Transform Infrared (FTIR) Spectroscopy of Biological Tissues. *Appl. Spectrosc. Rev.* **2008**, *43*, 134–179.
2. Dwivedi, N.; Dubey, K. C.; Shukla, R. K. Structural and Optical Studies of Multi-Walled Carbon Nanotubes. *Mater. Today: Proc.* **2020**, *29*, 872–875.
3. Stuart, B. *Infrared Spectroscopy: Fundamentals and Applications*; John Wiley & Sons, Ltd ISBNs: 0-470-85427-8 (HB); 0-470-85428-6 (PB). 2004.
4. Lavik, E.; Langer, R. Tissue Engineering: Current State and Perspectives. *Appl. Microbiol. Biotechnol.* **2004**, *65*, 1–8.
5. Ferraris, S.; Spriano, S. Antibacterial Titanium Surfaces in for Medical Implants. *Mater. Sci. Eng. C.* **2016**, *61*, 965–978.
6. Pan, Y.; Dong, S.; Hao, Y.; Zhao, Xie. Demineralized Bone Matrix Gelatin as Scaffold for Tissue Engineering. *Afr. J. Microbiol. Res.* **2010**, *4*, 865–870.
7. Sandeep, K.; Ruma, R.; Neeraj, D.; Tankeshwarab, K.; Kim, K-H. Carbon Nanotubes: A Novel Material for Multifaceted Applications in Human Healthcare. *Chem. Soc. Rev.* **2017**, *46*, 158–196.
8. Kasoju, N; Bhonde, R. R.; Bora, U. Fabrication of a Novel Micro–Nanofibrous Nonwoven Scaffold with Antheraea Assama Silk Fibroin for Use in Tissue Engineering. *Mater. Lett.* **2009**, *63*, 2466–2469.
9. Kouklin, N.; Tzolov, M.; Straus, D.; Yin, A.; Xu, J. M. Infrared Absorption Properties of Carbon Nanotubes Synthesized by Chemical Vapor Deposition. *Appl. Phys. Lett.* **2004**, *85*, 4463–4465.
10. Dai, H. Carbon Nanotubes: Synthesis, Integration, and Properties. *Acc Chem Res.* **2002**, *35*, 1035–1044.
11. Hernandez, I.; Kumar, A.; Joddar, B. A Bioactive Hydrogel and 3D Printed Polycaprolactone System for Bone Tissue Engineering. *Gels* **2017**, *3*, 26. doi:10.3390/gels3030026
12. Laurencin, C. T.; Nair, L. S. *Nanotechnology and Regenerative Engineering: The Scaffold*, 2nd ed.; CRC Press: Boka Raton, 2015.
13. Gorodzha, S. N.; Surmeneva, M. A.; Surmenev, R. A. Fabrication and Characterization of Polycaprolactone Cross-Linked and Highly-Aligned 3-D Artificial Scaffolds for Bone Tissue Regeneration via Electrospinning Technology. *IOP Conf. Ser.: Mater. Sci. Eng.* **2015**, *98*, 012024.
14. Hirata, E.; Uo, M.; Takita, H.; Akasaka, T.; Watari, F.; Yokoyama, A. Multiwalled CNT Coating of 3D Collagen Scaffolds for Bone Tissue Engineering. *Carbon* **2011**, *49*, 3284–3291.
15. Zadehnajar, P.; Akbari, B.; Karbasi, S.; Mirmusavi, M. H.; Preparation and Characterization of Poly ε-Caprolactone-Gelatin/Multi-Walled Carbon Nanotubes Electrospun Scaffolds for Cartilage Tissue Engineering Applications. *Int. J. Polym. Mater.* **2020**, *5*, 326–337.
16. Tan, W.; Twomey, J.; Guo, D.; Madhavan, K.; Li, M. Evaluation of Nanostructural, Mechanical, and Biological Properties of Collagen–Nanotube Composites. *IEEE Trans. Nanobiosci.* **2010**, *9*, 111–120.
17. Juan, W.; Petefish, J. W.; Hillier, A. C; Schneide, I. C. Epitaxially Grown Collagen Fibrils Reveal Diversity in Contact Guidance Behavior among Cancer Cells. *Langmuir* **2015**, *31*, 307–314.

18. Pulikkathara, M. X.; Kuznetsov, O. V.; Khabashesku, V. N. Sidewall Covalent Functionalization of Single Wall Carbon Nanotubes through Reactions of Fluoronanotubes with Urea, Guanidine, and Thiourea. *Chem Mater* **2008**, *20*, 2685–2695.

19. Osswald, S.; Havel, M.; Gogotsi, Y. Monitoring Oxidation of Multiwalled Carbon Nanotubes by Raman Spectroscopy. *J. Raman. Spectrosc.* **2007**, *38*, 728–736.

20. Misra, A.; Tyagi, P. K.; Rai, P.; Misra, D. S. FTIR Spectroscopy of Multiwalled Carbon Nanotubes: A Simple Approach to Study the Nitrogen Doping. *J. Nanosci. Nanotechnol.* **2007**, *7*, 1820–1823.

21. Jeong, K.; Furtado, C. A.; Xiaoming, L.; Chen, G.; Eklund, P. C. Raman and IR Spectroscopy of Chemically Processed Single-Walled Carbon Nanotubes. *J. Am. Chem. Soc.* **2005**, *127*, 15437–15445.

22. Garcia-Vida, F. J.; Pitarke, J. M.; Pendry, J. B. Effective Medium Theory of the Optical Properties of Aligned Carbon Nanotubes. *Phys. Rev. Lett.* **1997**, *78*, 4289–4292.

23. Hamon, M. A.; Hu, H.; Bhowmik, P.; Niyogi, S.; Zhao, B.; Itkis, M. E.; Haddon, R. C. End-Group and Defect Analysis of Soluble Single-Walled Carbon Nanotubes. *Chem. Phys. Lett.* **2001**, *347*, 8–12.

24. Stobinski, L.; Lesiak, B.; Kövér, L.; Tóth, J.; Biniak, S.; Trykowski, G.; Judek, J. Multiwall Carbon Nanotubes Purification and Oxidation by Nitric Acid Studied by the FTIR and Electron Spectroscopy Methods. *J. Alloys Compd.* **2010**, *501*, 77–84.

25. Vesali, N.M; Khodadadi, A. A.; Mortazavi, Y.; Alizadeh, O. S.;Pourfayaz, F.; Mosadegh S. S. Functionalization of Carbon Nanotubes Using Nitric Acid Oxidation and DBD Plasma. *World Acad. Sci. Eng. Technol.* **2009**, *49*, 177–179.

26. Chen, S.; Shen, W.; Wu, G.; Chen, D.; Jiang, M. A New Approach to Functionalization of Single Walled Carbon Nanotube with Both Alkyl and Carbonyl Groups. *Chem. Phys. Lett.* **2005**, *402*, 302–317.

27. Lee, S. Y.; Park, S. J. Hydrogen Adsorption of Acid-Treated Multi-Walled Carbon Nanotubes at Low Temperature. *Bull. Korean Chem. Soc.* **2010**, *31*, 1596–1600.

28. Kastner, J.; Pichler, T.; Kuzmany, H.; Curran, S.; Blau, W.; Weldon, D. N.; Delamesiere, M.; Draper, S.; Zandbergend, H. Resonance Raman and Infrared Spectroscopy of Carbon Nanotubes. *Chem. Phys. Lett.* **1994**, *221*, 53–58.

29. Fraczek, A.; Menaszek, E.; Paluszkiewicz, C.; Blazewicz, M. Comparative in Vivo Biocompatibility Study of Single and Multi-Wall Carbon Nanotubes. *Acta Biomater.* **2008**, *4*, 1593–1602.

30. Cao, Y.; Zhou, Y. M.; Shan, Y.; Ju, H. X.; Xue, X. J. Preparation and Characterization of Grafted Collagen-Multiwalled Carbon Nanotubes Composites. *J. Nanosci. Nanotechnol.* **2007**, *7*, 447–451.

31. Jing, Z.; Wu, Y.; Su, W.; Tian, M.; Jiang, W.; Cao, L.; Zhao, L.; Zhao, Z. Carbon Nanotube Reinforced Collagen/Hydroxyapatite Scaffolds Improve Bone Tissue Formation in Vitro and in Vivo. *Ann. Biomed. Eng.* **2017**, *45*, 2075–2087.

32. Tan, W.; Twomey, J.; Guo, D.; Madhavan, K.; Li, M. Evaluation of Nanostructural, Mechanical, and Biological Properties of Collagen-Nanotube Composites. *IEEE Trans. Nanobiosci.* **2010**, *9*, 111–120.

33. Andrews, R.; Weisenberger, M. Carbon Nanotube Polymer Composites. *Curr. Opin. Solid State Mater. Sci.* **2004**, *8*, 31–37.

34. Yan, B.; Gremlich, H. U.; Moss, S.; Coppola, G. M.; Sun, Q.; Liu, L. A Comparison of Various FTIR and FT Raman Methods: Applications in the Reaction Optimization Stage of Combinatorial Chemistry. *J. Comb. Chem.* **1999,** *1*, 46–54.
35. Pal, K.; Banthia, A. K.; Majumdar, D. K. Preparation and Characterization of Polyvinyl Alcohol—Gelatin Hydrogel Membranes for Biomedical Applications. *AAPS Pharm. Sci. Tech.* **2007,** *8*, E1–E5.
36. Mombeshora, E. T.; Simoyi, R.; Nyamori, V. O.; Ndungu, P. Multiwalled Carbon Nanotube-Titania Nanocomposites: Understanding Nano-Structural Parameters and Functionality in Dye-Sensitized Solar Cells. *S. Afr. J. Chem.* **2015,** *68*, 153–164.
37. Muyongaa, J. H.; Colec, C. G. B.; Duodu, K. G. Characterisation of Acid Soluble Collagen from Skins of Young and Adult Nile Perch. *Food Hydrocoll.* **2004,** *18*, 581–592.
38. Bandekar, J. Amide Modes and Protein Conformation. *Biochim. Biophys. Acta Protein Struct. Mol. Enzymol.* **1992,** *1120*, 123–143.
39. Lavialle, F.; Adams, R. G.; Levin, I. W. Infrared Spectroscopic Study of the Secondary Structure of Melittin in Water, 2-Chloroethanol, and Phospholipid Bilayer Dispersions. *Biochemistry* **1982,** *21*, 2305–2312.
40. Yakimets, I.; Wellner, N.; Smith, A. C.; Wilson, H. R.; Farhat, I.; Mitchell, R. H. Mechanical Properties with Respect to Water Content of Gelatin Films in Glassy State. *J. Polym.* **2005,** *46*, 12577–12585.
41. Wellner, N.; Belton, P. S.; Tatham, A. S. Fourier Transform IR Spectroscopic Study of Hydration-Induced Structure Changes in the Solid State of Omega-Gliadins. *Biochem. J.* **1996,** *319*, 741–747.
42. Friess, W.; Lee, G. Basic Thermoanalytical Studies of Insoluble Collagen Matrices. *Biomaterials* **1996,** *17*, 2289–2294.
43. Téllez, C. A.; Felcman, J.; De Moraes Silva, A. Fourier Transform Infrared and Raman Spectra of N-Di-Isopropylphosphorylguanidine (DPG). *Spectrochim. Acta. A* **2000,** *56*, 1563–1574.
44. Dhibar, S.; Das, C. K. Silver Nanoparticles Decorated Polyaniline/Multiwalled Carbon Nanotubes Nanocomposites for High-Performance Supercapacitor Electrode. *Ind. Eng. Chem. Res.* **2014,** *53*, 3495–3508.
45. Li, Y.; Li, R.; Fu, X.; Wang, Y.; Zhong, W. H. Bio-surfactant for Defect Control: Multifunctional Gelatin Coated MWCNTs for Conductive Epoxy Nanocomposites. *Compos. Sci. Technol.* **2018,** *159*, 216–224.
46. Massoumi, B.; Hosseinzadeh, M.; Jaymand, M. Electrically Conductive Nanocomposite Adhesives Based on Epoxy or Chloroprene Containing Polyaniline, and Carbon Nanotubes. *J. Mater. Sci. Mater. Electron.* **2015,** *26*, 6057–6067.
47. Shoulders, M. D.; Raines, R. T. Collagen Structure and Stability. *Annu. Rev. Biochem.* **2009,** *78*, 929–958.
48. Fujigaya, T.; Nakashima, N. Non-Covalent Polymer Wrapping of Carbon Nanotubes and the Role of Wrapped Polymers as Functional Dispersants. *Sci. Technol. Adv. Mater.* **2015,** *16*, 1–21.
49. Gelsea, K.; Poschl, E.; Aigner, T. Collagens—Structure, Function, and Biosynthesis. *Adv. Drug Deliv. Rev.* **2003,** *55*, 1531–1546.

50. Da Silva, E. E.; Della Colleta, H. H. M.; Ferlauto, A. S.; Moreira, R. L.; Resende, R. R.; Oliveira, S.; Kitten, G. T.; Lacerda, R. G. Nanostructured 3-d Collagen/Nanotube Biocomposites for Future Bone Regeneration Scaffolds. *Nano Res.* **2009**, *2*, 462–473.

51. Brodsky, B.; Doyle, E. G.; Elkan, B.; Blout, R. Infrared Spectroscopy of Collagen and Collagen-Like Polypeptides. *Biopolymers* **1975**, *14*, 937–957.

52. Barth, A.; Zscherp, C. What Vibrations Tell Us about Proteins. *Quart. Rev. Biophys.* **2002**, *35*, 369–430.

53. Giraud-Guille, M. M.; Besseau, L.; Chopin, C.; Durand, P. D. Herbage Structural Aspects of Fish Skin Collagen Which Forms Ordered Arrays via Liquid Crystalline States. *Biomaterials* **2000**, *21*, 899–906.

54. Green, E. C.; Zhang, Y. H.; Li, H.; Minus, L. M. Gel-Spinning of Mimetic Collagen and Collagen/Nano-Carbon Fibers: Understanding Multi-Scale Influences on Molecular Ordering and Fibril Alignment. *J. Mech. Behav. Biomed. Mater.* **2017**, *65*, 552–564.

55. Jing, Z.; Wu, Y.; Su, W.; Tian, M.; Jiang, W.; Cao, L.; Zhao, L.; Zhao, Z. Carbon Nanotube Reinforced Collagen/Hydroxyapatite Scaffolds Improve Bone Tissue Formation in Vitro and in Vivo, *Ann. Biomed. Eng.* **2017**, *45*, 2075–2087.

56. Ferreira, P.; Santos, P.; Alves, P.; Carvalho, M. P.; de Sá KD, Miguel SP, Correia, I. J. e Coimbra. Photocrosslinkable Electrospun Fiber Meshes for Tissue Engineering Applications. *Eur. Polym. J.* **2017**, *97*, 210–219.

57. Jiang, Y. C.; Jiang, L.; Huang, A.; Wang, X. F.; Li, Q.; Turng, L. S. Electrospun Polycaprolactone/Gelatin Composites with Enhanced Cell. Matrix Interactions as Blood Vessel Endothelial Layer Scaffolds. *Mater. Sci. Eng. C.* **2017**, *71*, 901–908.

58. Lim, M. M.; Sultana, N. Comparison on in Vitro Degradation of Polycaprolactone and Polycaprolactone/Gelatin Nanofibrous Scaffold. *Malaysian J. Anal. Sci.* **2017**, *21*, 627–632.

59. Hudecki, A.; Łyko-Morawska, D.; Likus, W.; Skonieczna, M.; Markowski, J.; Wilk, R.; Kolano-Burian, A.; Maziarz, W.; Adamska, J.; Łos, M. J. Composite Nanofibers Containing Multiwall Carbon Nanotubes as Biodegradable Membranes in Reconstructive Medicine. *Nanomaterials* **2019**, *9*, 63. doi:10.3390/nano9010063.

60. Gopinathan, J.; Quigley, A. F.; Bhattacharyya, A.; Padhye, R.; Robert, M. I.; Kapsa, R.; Nayak, R. A.; Shanks, S. H. Preparation, Characterisation, and in Vitro Evaluation of Electrically Conducting Poly(ε-Caprolactone)-Based Nanocomposite Scaffolds Using PC12 Cells. *J. Biomed. Mater. Res. A.* **2016**, *104A*, 853–865.

61. Dawei, L.; Weiming, C.; Binbin, S.; Haoxuan, L.; Tong, W.; Qinfei, K.; Chen, H.; Hany, EI-H.; Al-Deyab, S. S.; Xiumei, M. A Comparison of Nanoscale and Multiscale PCL/Gelatin Scaffolds Prepared by Disc-Electrospinning. *Colloid Surf. B.* **2016**, *146*, 632–641.

62. Kushwaha, S. K. S.; Ghoshal, S.; Rai, A. K.; Singh, S. Carbon Nanotubes as a Novel Drug Delivery System for Anticancer Therapy: A Review. *Braz. J. Pharm. Sci.* **2013**, *49*, 630–643.

63. Bakhshali, M.; Ramezani, M.; Jaymand, M.; Ahmadinejad, M.; Bakhshali, M.; Multi-Walled Carbon Nanotubes-g-Poly(Ethylene Glycol)b-Poly(ε-Caprolactone) Synthesis, Characterization, and Properties. *J. Polym. Res.* **2015**, *22*, 214. doi: 10.1007/s10965-015-0863-7

64. Lim, Y. C.; Johnson, J.; Fei, Z.; Wu, Y.; Farson, D. F.; Lannutti, J. J.; Choi, H. W.; Le, J. L. Micropatterning and Characterization of Electrospun Poly(ε-Caprolactone)/

Gelatin Nanofiber Tissue Scaffolds by Femtosecond Laser Ablation for Tissue Engineering Applications. *Biotechnol. Bioeng.* **2011**, *108*, 116–126.

65. Ghasemi-Mobarakeh, L.; Prabhakaran, M. P.; Morshed, M. et al. Electrospun Poly-ε caprolactone)/Gelatin Nanofibrous Scaffolds for Nerve Tissue Engineering. *Biomaterials* **2008**, *29*, 4532–4539.

66. Heidari, M.; Bahrami, H.; Ranjbar-Mohammadi, M. Fabrication, Optimization and Characterization of Electrospun Poly(Caprolactone)/Gelatin/Graphene Nanofibrous Mats. *Mater. Sci. Eng. C.* **2017**, *78*, 218–229.

67. Yang, Y.; Chawla, A.; Zhang, J.; Esa, A.; Jang, H. L.; Khademhosseini, A. Applications of Nanotechnology for Regenerative Medicine; Healing Tissues at the Nanoscale. *Principles of Regenerative Medicine*, **2019**.

68. Lovat, V.; Pantarotto, D.; Lagostena, L.; Cacciari, B.; Grandolfo, M.; Righi, M.; Spalluto, G.; Prato, M.; Ballerini, L. Carbon Nanotube Substrates Boost Neuronal Electrical Signaling. *Nano Lett.* **2005**, *5*, 1107–1110.

69. Mazzatenta, A.; Giugliano, M.; Campidelli, S.; Gambazzi, L.; Businaro, L.; Markram, H.; Prato, M.; Ballerini, L. Interfacing Neurons with Carbon Nanotubes: Electrical Signal Transfer and Synaptic Stimulation in Cultured Brain Circuits. *J. Neurosci.* **2007**, *27*, 6931–6936.

70. Ren, J.; Xu, Q.; Chen, X.; Li, W.; Guo, K.; Zhao, Y.; Wang, Q.; Zhang, Z.; Peng, H.; Li, Y. G. Superaligned Carbon Nanotubes Guide Oriented Cell Growth and Promote Electrophysiological Homogeneity for Synthetic Cardiac Tissues. *Adv. Mater.* **2017**, *29*, 1702713. doi.org/10.1002/adma.201702713.

71. Silva, E.; Vasconcellos, L. M. R.; Bruno, V. M. R.; Dos Santos D. S.; Campana-Filho, S. P.; Marciano F. R.; Webster, T. J.; Lobo, A. O. PDLLA Honeycomb-Like Scaffolds with a High Loading of Superhydrophilic Graphene/Multi-Walled Carbon Nanotubes Promote Osteoblast in Vitro Functions and Guided in Vivo Bone Regeneration. *Mater. Sci. Eng. C* **2017**, *73*, 31–39.

72. Yu, H.; Zhao, H.; Huang, C.; Du, Y. Mechanically and Electrically Enhanced CNT– Collagen Hydrogels as Potential Scaffolds for Engineered Cardiac Constructs. *ACS Biomater. Sci. Eng.* **2017**, *3*, 3017–3021.

73. Pok, S.; Vitale, F.; Eichmann, S. L.; Benavides, O. M.; Pasquali, M.; Jacot, J. G. Biocompatible Carbon Nanotube–Chitosan Scaffold Matching the Electrical Conductivity of the Heart. *ACS Nano* **2014**, *8*, 9822–9832.

74. Wang, L.; Wu, Y. B.; Hu, T. L.; Ma, P. X.;. Guo, B. L. Aligned Conductive Core-Shell Biomimetic Scaffolds Based on Nanofiber Yarns/Hydrogel for Enhanced 3D Neurite Outgrowth Alignment and Elongation. *Acta Biomater.* **2019**, *96*, 175–187.

75. Su, S.; Wang, J.; Vargas, E.; Wei, J.; Martínez-Zaguilán, R.; Souad, R. S.; Pantoya, M. L.; Wang, S.; Chaudhuri, J.; Qi, J. Porphyrin Immobilized Nanographene Oxide for Enhanced and Targeted Photothermal Therapy of Brain Cancer. *ACS Biomater. Sci. Eng.* **2016**, *8*, 1357–1366. doi:10.1021/acsbiomaterials.6b00290

76. Gheith, M. K.; Sinani, V. A.; Wicksted, J. P.; Matts, R. L.; Kotov, N. A. Single-Walled Carbon Nanotube Polyelectrolyte Multilayers and Freestanding Films as a Biocompatible Platform for Neuroprosthetic Implants. *Adv. Mater.* **2005**, *17*, 2663–2667.

77. Heidari, M.; Bahrami, H.; Ranjbar-Mohammadi, M. Fabrication, Optimization and Characterization of Electrospun Poly(Caprolactone)/Gelatin/Graphene Nanofibrous Mats. *Mater. Sci. Eng. C.* **2017**, *78*, 218–229.

78. Mohammad, H. R. Z.; Seifi, M.; Shoja, S. A Facile Method toward Potentially Next-Generation Bacteria Detectors Using Polymer/MWCNT/Au Nanocomposite Films: A Possibility to Detecting Ability through the Shift in Resonance Frequency. *Mater. Res. Express.* **2019,** *6*, 045004. doi.org/10.1088/2053-1591/aaf15c.

79. Zhang, J. J.; Gu, M. M.; Zheng, T. T.; Zhu, J. J. Synthesis of Gelatin-Stabilized Gold Nanoparticles and Assembly of Carboxylic Single-Walled Carbon Nanotubes/Au Composites for Cytosensing and Drug Uptake. *Anal. Chem.* **2009,** *81*, 6641–6648.

80. Negri, V.; Pacheco-Torres, J.; Calle, D.; López-Larrubia, P. Carbon Nanotubes in Biomedicine. *Top Curr. Chem.* **2020,** *378*, 15. doi: 10.1007/s41061-019-0278-8.

81. Sun, H.; Zhou, J.; Huang, Z.; Qu, L.; Lin, N.; Liang, C.; Dai, R.; Tang, L.; Tian, F. Carbon Nanotube Incorporated Collagen Hydrogels Improve Cell Alignment and the Performance of Cardiac Constructs. *Int. J. Nanomed.* **2017,** *12*, 3109–3120.

82. Graupner, R.; Abraham, J.; Wunderlich, D.; Vencelová, A.; Lauffer, P.; Röhrl, J.; Hundhausen, M.; Ley, L.; Hirsch, A. Nucleophilic–Alkylation–Reoxidation: A Functionalization Sequence for Single-Wall Carbon Nanotubes. *J. Am. Chem. Soc.* **2006,** *128*, 6683–6689.

83. Roshanbinfar, K; Mohammadi, Z.; Sheikh-Mahdi, M. A.; Dehghan, M. M.; Oommen, O. P.; Hilborn, J.; Engel, F. B. Carbon Nanotube Doped Pericardial Matrix-Derived Electroconductive Biohybrid Hydrogel for Cardiac Tissue Engineering. *Biomater. Sci.* **2019,** *7*, 3906–3917.

84. Zeng, H.; Gao, C.; Yan, D. Poly(e-Caprolactone)-Functionalized Carbon Nanotubes and Their Biodegradation Properties. *Adv. Funct. Mater.* **2006,** *16*, 812–818.

85. Mawhinney, D. B.; Naumenko,V.; Kuznetsova, A.; Yates, J. T. Infrared Spectral Evidence for the Etching of Carbon Nanotubes: Ozone Oxidation at 298 K. *J. Am. Chem. Soc.* **2000,** *122*, 2383–2384.

86. Li, D.; Li, S.; Liu, J.; Zhana, L.; Wang, P.; Zhu, Hongshui; Wei, J. Surface Modification of Carbon Nanotube with Gelatin via Mussel Inspired Method. *Mater. Sci. Eng. C* **2020,** *112*, 110887. doi.org/10.1016/j.msec.2020.110887.

CHAPTER 11

CARBON NANOTUBES-BASED COMPOSITE MATERIALS FOR ELECTROMAGNETIC SHIELDING APPLICATIONS

ADRIAN GHEMES*, GABRIEL ABABEI, GEORGE STOIAN, LUIZA BUDEANU-RACILA, NICOLETA LUPU, and HORIA CHIRIAC

National Institute of Research and Development for Technical Physics, Iasi, Romania

Corresponding author. E-mail: aghemes@phys.iasi.ro

ABSTRACT

Electromagnetic interference between electronic devices becomes a major problem nowadays due to tremendous increase of their use in our daily life. To avoid such interferences, metal-based materials are used to fabricate enclosures capable of minimizing the radiation escape. Anyway, the use of metals as electromagnetic radiation shielding materials has certain disadvantages like weight increase and small corrosion resistance. Recently, conductive polymer composites became a better choice due to their lightweight, corrosion resistance, and flexibility in fabrication.

In this chapter, we first briefly discuss about the electromagnetic radiation shielding using conventional metal-based shields. Then, we make an overview on conductive polymer composites applied for electromagnetic shields. In the next section, we introduce a new hybrid composite material based on multiwalled carbon nanotubes (MWCNTs) and soft magnetic CoFeB nanoparticles with applications for electromagnetic interference shielding. High aspect ratio MWCNTs have been synthesized by chloride-mediated chemical vapor deposition method. Then, a composite material consisting of an epoxy resin, MWCNTs and CoFeB nanoparticles was prepared and its electromagnetic radiation shielding capability was tested. The new material showed good absorption properties in the microwave frequency domain. Few attempts to improve the shielding effectiveness of these new composite materials are also presented and discussed.

11.1   INTRODUCTION

In our daily life, the usage of electrical and electronic devices increases more and more, each device being a new emitter of electromagnetic radiation. The radiation escape from a certain device may interfere with the devices nearby and this leads to abnormal operation of those devices. This is the so-called electromagnetic interference (EMI) and became a major problem nowadays. Few examples of EMIs are disturbance of radio and TV signals from low altitude flying aircrafts, noises in microphones from a cell phone, electronic devices interference with navigation signals during takeoff and landing of aircrafts. The radiation emitted by GSM mobile phones for example, can be extremely harmful to any implantable electronic device in the human body.[1]

EMI shielding represents a combination of factors meant to limit or minimize the radiation escape from a certain device or to block the harmful radiation to enter into an electronic device. The EMI shielding can be realized by reflection and absorption of the electromagnetic radiation by a shielding material. An electromagnetic radiation shielding material should possess either mobile charge carriers (electrons or holes) or electric and/or magnetic dipoles which interact with the electric (E) and magnetic (H) vectors of the incident electromagnetic radiation. Usually, metals and alloys are known to be the best EMI-shielding materials. Thus, metal-based materials, such as sheet metal, metal screen, or metal foam are used to fabricate enclosures which are capable to minimize the radiation escape from a certain device. However, the use of metals as shielding materials has certain disadvantages. The increase in the mass of the final device and small corrosion resistance being the major drawbacks. Therefore, other alternatives had to be found.

In the last two decades, conductive polymer composites became a better choice due to their lightweight, corrosion resistance, flexibility in fabrication and versatility of the shapes. In general, most of the polymers exhibit poor electrical conductivity, but this can be improved by mixing with different conductive fillers. The electrically conductive fillers might be either low-melting point alloys like bismuth or antimony, metals like copper, aluminum, zinc, nickel, stainless steel or carbon-based materials, such as carbon black, carbon fibers or powder-like carbon nanotubes (CNTs).[2] These fillers are very well dispersed into a polymer matrix and then the electromagnetic radiation shielding capability of the composite is tested either in reflection or absorption mode. Although this method proved its suitability and some promising results were obtained,[3] still their shielding effectiveness (SE) is limited by the conductivity of the filler material, maximum loading capacity, and percolation threshold. The percolation threshold strongly depends on the aspect ratio of the conductive filler dispersed into matrix, that is, higher aspect ratio of the filler decreases the percolation threshold.

After their first report about 30 years ago,[4] CNTs, because of their unique characteristics, attracted the interest of the scientific community. Among these special properties, their small mass density and huge electrical conductivity made CNTs very promising in order to obtain a lightweight composite material with good electromagnetic shielding efficiency. Many research groups around the world tried to turn the huge potential of both single-walled CNTs (SWCNTs) and multiwalled CNTs (MWCNTs) into efficient electromagnetic shielding material.

Thus, SWCNTs/polyurethane composites were prepared by Liu et al.[5] and they obtained a maximum absorption value of 22 dB at a frequency of 8.8GHz for 5 wt.% SWCNTs loading, while Li et al.[6] obtained shielding efficiency up to 20 dB at 1.5 GHz by using SWCNTs/epoxy composites with 15% SWCNTs loading. Singh et al.[7] used high aspect ratio MWCNTs to prepare reinforced low-density polyethylene (LDPE) composites. They obtained average SE values of −22.4 dB for 10% MWCNTs–LPDE composites in Ku-band frequency range (12.4–18 GHz). Gupta et al.[8] synthesized poly(trimethylene terephthalate)/multiwalled CNT composites with various amounts of MWCNTs. They observed SE values of −36 to −42 dB in Ku-band frequency range using 20–30 μm long and 20–40 nm in diameter MWCNTs.

Other groups have tried to use hybrid combinations of CNTs and magnetic nanoparticles in order to increase the SE, still obtaining lightweight composite materials as compared with their bulk metal counterparts. For example, NiO, Fe_3O_4, Ag, $NiCo_2$ and core-shell $Fe@Fe_3O_4$ nanoparticles[9–12] were interspersed with CNTs to improve the electromagnetic wave absorption characteristics of as-prepared composite materials.

In this work, we propose to use high aspect ratio MWCNTs in combination with soft magnetic CoFeB nanoparticles in order to obtain a new composite material for electromagnetic radiation shielding. The new material will benefit from the high electrical conductivity of the MWCNTs and superior magnetic properties of the CoFeB nanoparticles to enhance its EMI shielding capability.

11.2   EXPERIMENATAL METHODS AND MATERIALS

11.2.1   SYNTHESIS OF MULTIWALLED CARBON NANOTUBES

MWCNTs were synthesized by a chloride-mediated chemical vapor deposition (CVD) method reported by Inoue in 2008.[13] The experimental setup used for MWCNTs synthesis was built in our laboratory and is presented in Figure 11.1.

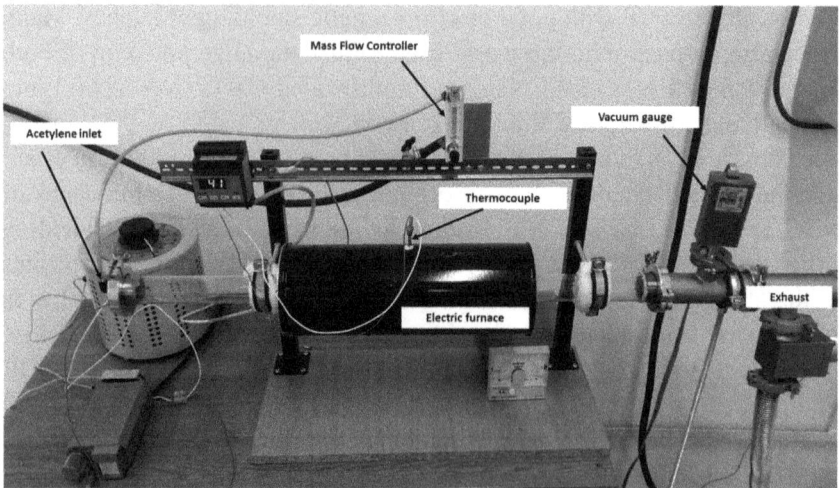

FIGURE 11.1   Chemical vapor deposition equipment used for MWCNTs synthesis.

We used anhydrous iron chloride ($FeCl_2$) powder 99.5% (Alfa Aesar) as a source of Fe which plays the role of catalyst and acetylene (C_2H_2) as carbon source. Iron chloride powder and substrates were introduced into a quartz tube furnace which was evacuated to a pressure of about 10^{-3} Torr using a rotary pump. Then the furnace was heated to a temperature of 850°C and C_2H_2 was introduced into the growth area at a flow rate of 200 cm³/min. Then, the MWCNTs synthesis started and the gas flow was maintained for different durations, from few minutes up to 1 h, at a constant pressure of 10 Torr. The substrates we used were either fused quartz plates or oxidized silicon wafers as shown in Figure 11.2.

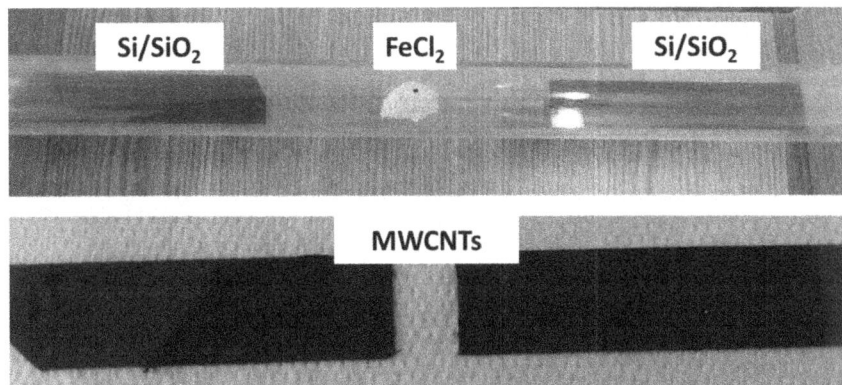

FIGURE 11.2 Si/SiO$_2$ substrates and FeCl$_2$ powder (up) and MWCNTs on substrates (down).

11.2.2 CoFeB NANOPARTICLES SYNTHESIS METHOD

CoFeB nanoparticles were prepared by the chemical reduction of cobalt chloride hexahydrate (CoCl$_2$·6H$_2$O) and iron sulfate heptahydrate (FeSO$_4$·7H$_2$O) salts in aqueous solution of sodium borohydride (NaBH$_4$) which is used as boron source and reducing agent.[14] Additionally, poly-vinylpyrolidone has been used as dispersant and dimension controlling agent. The pH of the solution during synthesis was maintained above 12.6.

After synthesis, the CoFeB nanoparticles were separated by centrifuga-tion at 8000 min^{-1} for 10 min. Finally, as-synthesized nanoparticles were thoroughly washed with distilled water and alcohol followed by drying in vacuum at room temperature.

By using a Fe/Co ration in solution of 0.2, CoFeB nanoparticles were obtained with diameters of around 30 nm at a synthesis temperature of 10°C. The spectrophotometric measurements indicated a composition as $Co_{78.6\pm0.2}Fe_{8.31\pm0.2}B_{10.58\pm0.6}$ (wt.%), irrespective of the temperature during the nanoparticles synthesis process.

11.2.3 PREPARATION OF COMPOSITE MATERIALS

We have prepared two kinds of CNTs-based composite materials. First, as-grown MWCNTs were dispersed by ultrasonication in water, dried, and densified by ethanol in order to obtain a MWCNTs paper, also known as Bucky paper. Then, pieces of 5 × 20 mm were cut and stacked between epoxy resin layers in order to obtain composite materials with 1, 2, and 3 MWCNTs paper layers as shown in Figure 11.3.

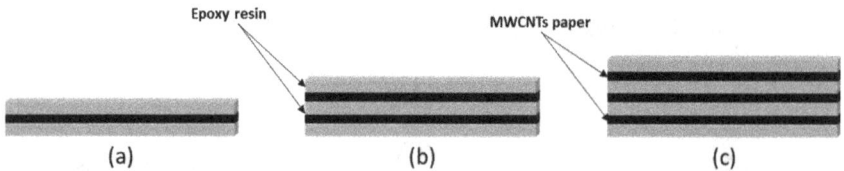

FIGURE 11.3 Schematic representation of MWCNTs/epoxy resin composite material with one MWCNTs layer (a), two MWCNTs layers (b), and three MWCNTs layers (c).

The test samples were parallelepipeds with a length of 22 mm, width of 10 mm, and thickness of 5 mm. The transmission coefficient S_{21} (dB) of the samples was determined using a WR90 type X-band microwave guide connected to a PNAL5230 Agilent Vector Network Analyzer (VNA) after a SOLT (short-open load-through) calibration of the setup, in the frequency range from 8.2 to 12 GHz.

Second, for the preparation of hybrid composite material, MWCNTs and CoFeB nanoparticles were mixed by ultrasonication for 1 h with one component of a four-component transparent epoxy resin (EPON 812), to assure homogeneity. Then, the other three components were added and the mixture was stirred for 10 min. Finally, the material was pressed in a cylindrical Teflon® mold and introduced for 1 h into an oven at a temperature of 60°C for polymerization of the resin.

We obtained toroidal-shaped samples with outer diameter Φ_{out} = 7 mm, inner diameter Φ_{in} = 3 mm, and thickness t = 3 mm which were used for measurements in the microwave frequency range. For optical microscope observations of the material's structure, we used thin slices of 0.4 mm which were cut from a parallelepiped.

High frequency measurements were done using the 7 mm coaxial transmission line method in the frequency range from 1 to 8 GHz by connecting the coaxial cell between the emission and reception ports of vector network analyzer (Agilent VNA N5203A) as shown in Figure 11.4.

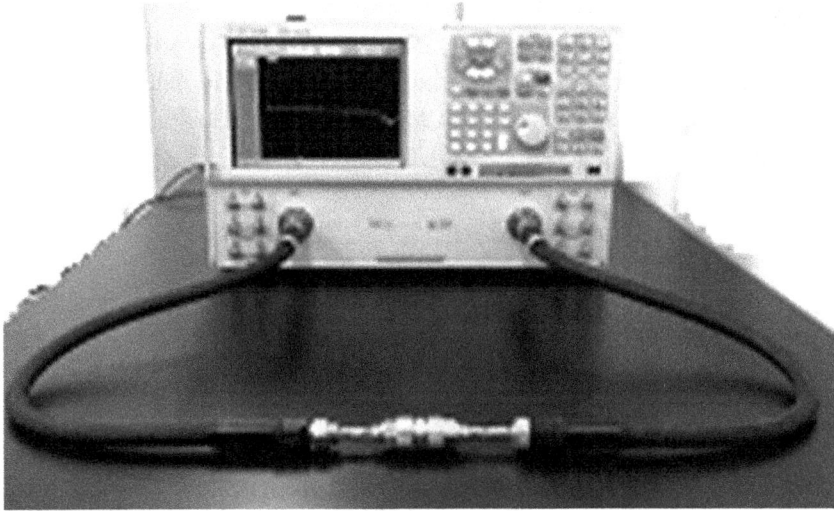

FIGURE 11.4   Experimental setup for high frequency measurements.

The microwave absorption characteristics of the samples were evaluated by measuring the transmission S21 (dB) and reflection S11 (dB) coefficients respectively of the toroidal samples.

11.3   RESULTS AND DISCUSSION

11.3.1   CHARACTERIZATION OF AS-GROWN MULTIWALLED CARBON NANOTUBES

By using the experimental procedure described in Section 11.2.1, we obtained vertically aligned MWCNTs with length up to 300 µm for 20 min growth. As shown in Figure 11.5, MWCNTs are little wavy from bottom to tip, indicating that individual CNTs grow at different rates.

As-grown MWCNTs were characterized in terms of purity by energy-dispersive X-ray spectroscopy (EDX). The measurements were taken in three spots at the bottom, center, and top of the array. As we can easily observe from Table 11.1, very small amount of Fe impurities is left into the CNTs array, less than 1% in all the areas.

The MWCNTs characteristics were further investigated by transmission electron microscopy (TEM) analysis. As illustrated in Figure 11.6, inside individual nanotubes, we may find small metallic parts which are remaining mainly at the roots of the nanotubes. These are iron species (iron oxides, and iron carbides) which are entering the inner hole of the nanotubes at the initial stage of the growth.

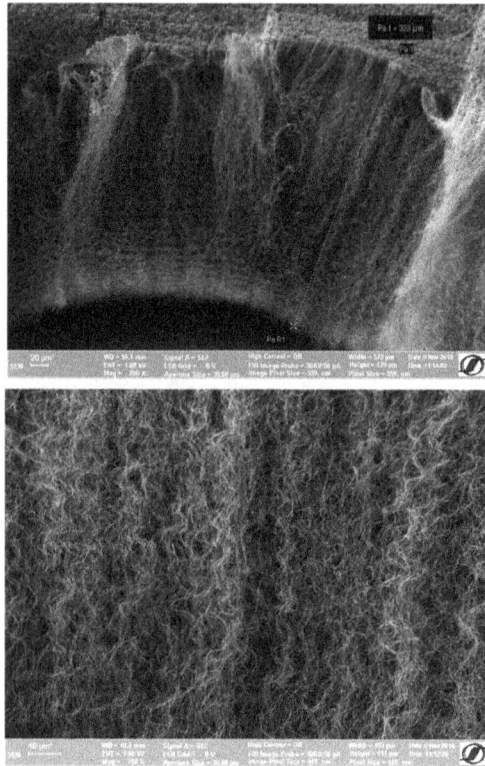

FIGURE 11.5   FE-SEM images of a vertically aligned MWCNTs array.

TABLE 11.1   EDX Analysis in Three Points along MWCNT Arrays.

Spectrum	Position	C (at.%)	Fe (at.%)
Spectrum 1	Bottom	99.54	0.46
Spectrum 2	Center	99.35	0.65
Spectrum 3	Top	99.29	0.71
Mean		99.39	0.61
Std. deviation		0.13	0.13
Max.		99.54	0.71
Min.		99.29	0.46

The inset in Figure 11.6 shows a HRTEM image of an individual CNT. The outer diameter of as-grown CNTs is about 30 nm while the inner hole has a diameter of 6 nm. These nanotubes consist in more than 30 walls. Also, from the HRTEM image, we can observe a "clean" surface indicating a very small amount of amorphous carbon being deposited on the surface of MWCNTs.

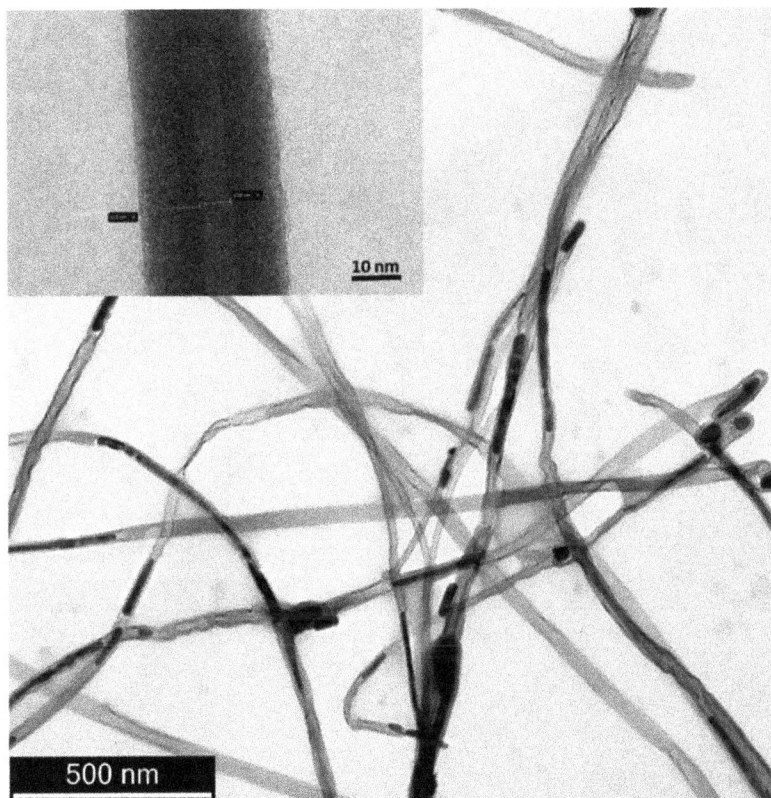

FIGURE 11.6   TEM image of as-grown multiwalled carbon nanotubes.

11.3.2   CHARACTERIZATION OF MWCNTS/CoFeB COMPOSITE MATERIAL

In order to check the dispersion and homogeneity of the of the hybrid composite material, we cut 0.4 mm thin slices and we performed optical microscope observations. We found a good distribution of the CoFeB nanoparticles within the polymeric matrix, while MWCNTs are locally agglomerated as shown in Figure 11.7.

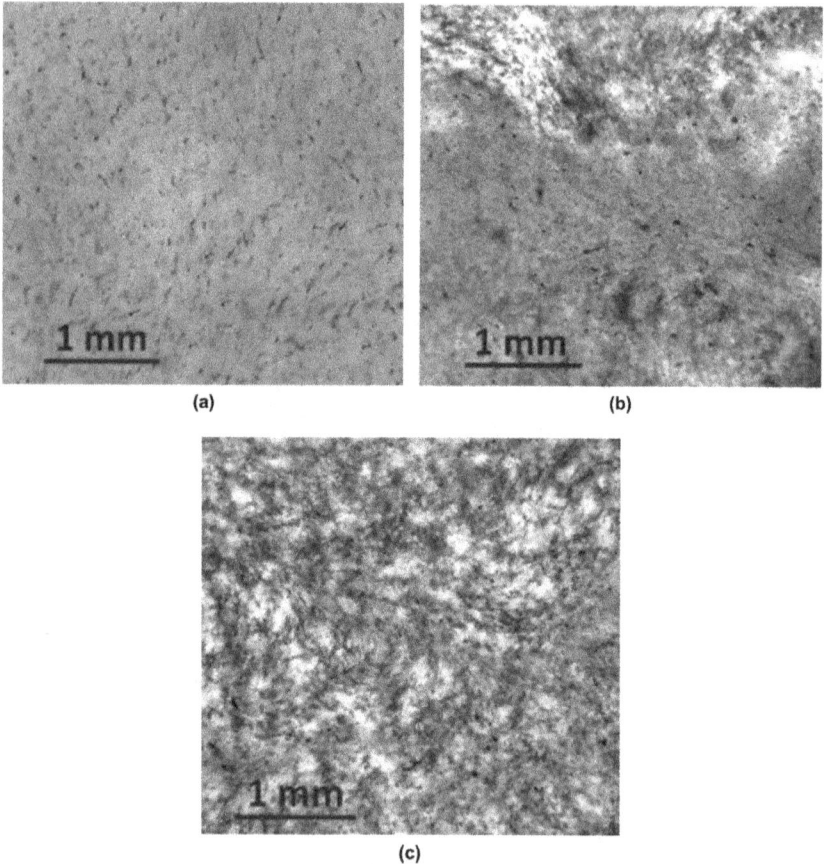

(a)   (b)

(c)

FIGURE 11.7   Optical microscope images of CoFeB nanoparticles (a), MWCNTs (b), and MWCNTs/CoFeB nanoparticles (c) dispersed in epoxy resin.

The local agglomeration of CNTs appears because we have not used any surfactants during dispersion. In general, it is very difficult to obtain the total dispersion of MWCNTs is liquid media.[15]

Furthermore, we analyzed by SEM observations, a composite material with equal proportions of CNTs and CoFeB nanoparticles and results are illustrated in Figure 11.8. We can easily identify both multiwalled carbon nanotubes (indicated by arrows) and CoFeB nanoparticles (indicated by circles).

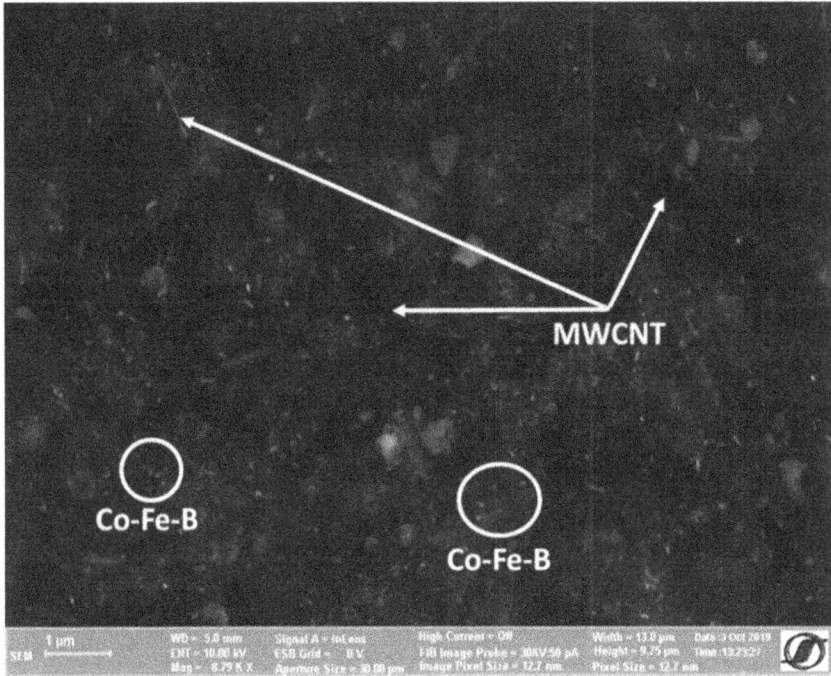

FIGURE 11.8   Cross-sectional SEM image of a composite material obtained using MWCNTs/CoFeB nanoparticles (1/1).\

11.3.3   ELECTROMAGNETIC RADIATION SHIELDING BEHAVIOR OF MWCNTs-BASED COMPOSITE MATERIALS

In order to evaluate the microwave absorption characteristics of our composite materials, we have determined the SE at different microwave frequencies as the ratio of the output energy to the input energy across the shielding materials which are defined as follows:

$$SE = 10 \log P_{inc}/P_{out}, \qquad (11.1)$$

where P_{inc} and P_{out} represent the input and the output energy of the microwave field in the presence of the shielding material to be evaluated.[16-18]

For the MWCNTs composite materials obtained using MWCNTs paper, we measured the S21 transmission parameter S21 (dB) in the frequency range from 8.2 to 12 GHz and results are given in Figure 11.9. We found that the transmissions of the microwave radiation energy through the sample decreases from −15 dB for the composite with s single MWCNTs layer, to −35 dB for that with three MWCNTs layers. Anyway, the efficiency of these composites in terms of electromagnetic radiation shielding was rather poor and we tried a combination of MWCNTs and magnetic nanoparticles as detailed in the next paragraph.

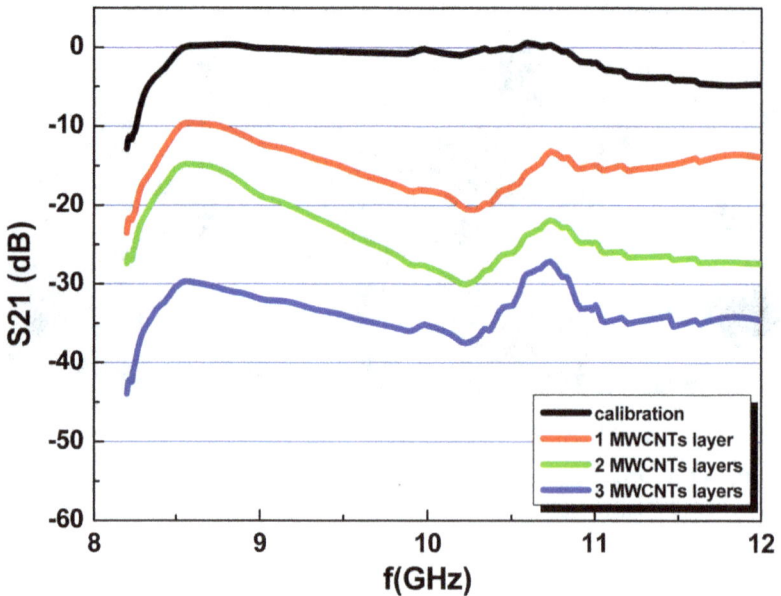

FIGURE 11.9   Absorption spectra of MWCNTs paper-based composite material.

For high frequency measurements of hybrid MWCNTs/CoFeB composites, we prepared toroidal test samples using mixtures of MWCNTs and CoFeB nanoparticles in the ratios of 1:5, 2:5, and 3:5 corresponding to a MWCNTs loading into epoxy resin of 0.2, 0.4 and 0.6 wt.%, respectively. The microwave spectra presented in Figure 11.10 indicated that the prepared composite materials attenuate the microwave radiation and absorb its energy in a frequency range from 1 to 8 GHz.

Using the combination of MWCNTs with CoFeB nanoparticles, we obtained a composite material with two absorption bands: one around

4 GHz given by the presence of CoFeB nanoparticles and another one above 5 GHz due to MWCNTs.[19] Increasing the MWCNTs loading from 0.2 to 0.6 wt.%, the absorption of the composite materials rose by almost 8 dB. This improvement of the SE can be attributed to the higher electrical conductivity of the composite material with higher MWCNTs loading.[20]

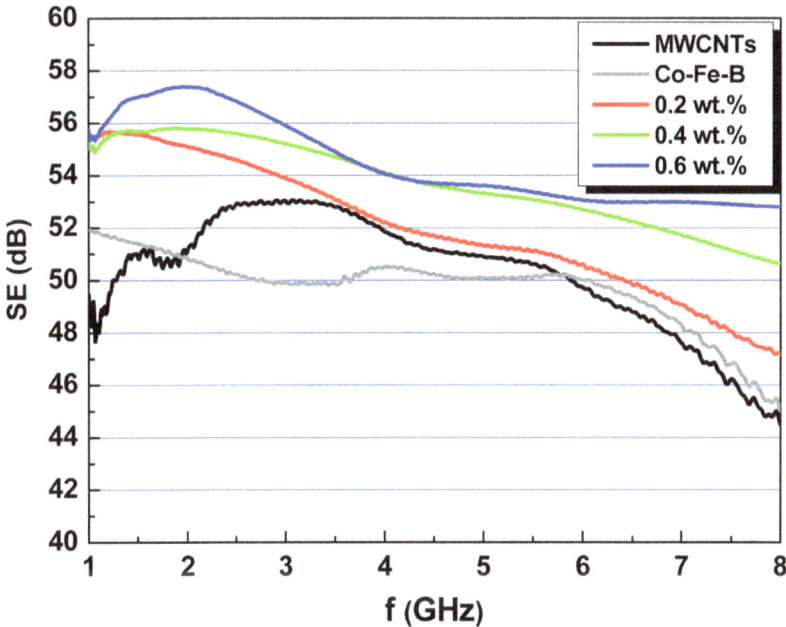

FIGURE 11.10   Shielding effectiveness versus frequency for different MWCNTs loadings.

In order to optimize the electromagnetic shielding characteristics of the composite material, we have fixed the amount of CoFeB nanoparticles and we increased the amount of MWCNTs. Thus, we obtained composite materials with MWCNTs loadings of 0.5%, 1%, and 1.5%. As shown in Figure 11.11, the SE improved slightly up to a value of 58 dB. More importantly, we obtained a broadening and uniform absorption band. At 8 GHz, we measured a SE of 56 dB as compared with only 52 dB at lower MWCNTs loadings. It worth to note here that all these results were obtained for relatively small MWCNTs loadings (around 1%) as compared with other reports.[21] This behavior may be attributed to the high aspect ratio of the nanotubes used in this study and offers a great advantage when it comes to reducing the weight and cost of the composite material.

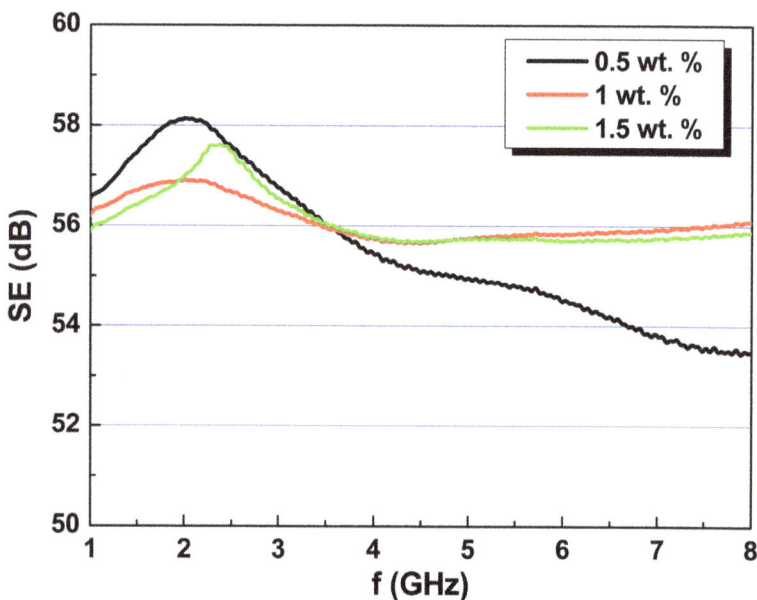

FIGURE 11.11   Shielding effectiveness versus frequency for MWCNTs loadings of 0.5%, 1%, and 1.5%.

11.4   CONCLUSIONS

Vertically aligned MWCNTs with diameters of 30 nm and height of 300 μm were synthesized by a simple chloride-mediated chemical vapor deposition method. Two kinds of MWCNTs-based composite materials were prepared and their electromagnetic SE has been evaluated in the microwave frequency range. The composite material based on MWCNTs paper sandwiched between resin layers shows poor shielding properties. On the other hand, the new hybrid composite material based on MWCNTs and soft magnetic CoFeB nanoparticles presented good electromagnetic radiation shielding characteristics. We found that composite materials with MWCNTs loadings from 0.2 to 0.6 wt.% absorb microwave radiation in a broad frequency range. SE of 57 dB has been achieved at 3 GHz for the composite with MWCNTs loading of 0.6 wt.%. By increasing the MWCNTs loading, we obtained a broadening of the absorption band with a SE of 56 dB at a frequency of 8 GHz for MWCNTs loading higher than 1%.

These results indicate the potential use of the hybrid MWCNT/CoFeB nanoparticles composites for EMI shielding applications.

ACKNOWLEDGMENTS

This work was supported by the Romanian Ministry of Research and Innovation (MCI) under the Nucleu program, Project PN 19 28 01 01.

KEYWORDS

- **carbon nanotubes**
- **composite materials**
- **electromagnetic interference**
- **shielding effectiveness**

REFERENCES

1. Kainz, W.; Alesch, F.; Chan, D. D.; Electromagnetic Interference of GSM Mobile Phones with the Implantable Deep Brain Stimulator, ITREL-III. *Biomed. Eng. Online* **2003,** *2*, 1–9.
2. Amoabeng, D.; Velankar, S. S. A review of Conductive Polymer Composites Filled with Low Melting Point Metal Alloys. *Polym. Eng. Sci.* **2017,** *58* (6), 1010–1019.
3. Jiang, D.; Murugadoss, V.; Wang, Y.; Lin, J.; Ding, T.; Wang, Z.; Shao, Q.; Wang, C.; Liu, H.; Lu, N.; Wei, R.; Subramania, A.; Guo, Z. Electromagnetic Interference Shielding Polymers and Nanocomposites—A Review. *Polym. Rev.* **2019,** *59* (2), 280–337.
4. Iijima, S. Helical Microtubules Of Graphitic Carbon. *Nature* **1991,** *354*, 56–58.
5. Liu, Z.; Bai, G.; Huang Y.; Li, F.; Ma, Y.; Guo, T.; He, X.; Lin, X.; Gao, H.; Chen, Y. Microwave Absorption of Single-Walled Carbon Nanotubes/Soluble Cross-Linked Polyurethane Composites. *J. Phys. Chem. C* **2007,** *111*, 13696–13700.
6. Li, N.; Huang, Y.; Du, F.; He, X.; Lin, X.; Gao, H.; Ma, Y.; Li, F.; Chen, Y.; Eklund, P. C. Electromagnetic Interference (EMI) Shielding of Single-Walled Carbon Nanotube Epoxy Composites. *Nano Lett.* **2006,** *6*, 1141–1145.
7. Singh, B. P.; Saini, P. P.; Gupta, T. K.; Garg, P.; Kumar, G.; Pande, I.; Pande, S.; Seth, R. K.; Dhawan, S. K; Mathur, R. B. Designing of Multiwalled Carbon Nanotubes Reinforced Low Density Polyethylene Nanocomposites for Suppression of Electromagnetic Radiation. *J. Nanopart. Res.* **2011,** *13*, 7065–7074.

8. Gupta, A.; Choudhary, V. Electromagnetic Interference Shielding Behavior of Poly (Trimethylene Terephthalate)/Multi-Walled Carbon Nanotube Composites. *Compos. Sci. Technol.* **2011**, *71*, 1563–1568.

9. Yu, L.; Lan, X.; Wei, C.; Li, X.; Qi, X.; Xu, T.; Li, C.; Li, C.; Wang, Z. MWCNT/ NiO-Fe$_3$O$_4$ Hybrid Nanotubes for Efficient Electromagnetic Wave Absorption. *J. Alloys Compounds* **2018**, 748, 111–116.

10. Fang, J.; Wang, Y.; Wei, W.; Chen, Z.; Li, Y.; Liu, Z.; Yue, X.; Jiang, Z. A MWCNT– Nanoparticle Composite as a Highly Efficient Lightweight Electromagnetic Wave Absorber in the Range of 4–18 GHz. *RSC Adv.* **2016**, *6*, 4695–4704.

11. Chen, C.; Bao, S.; Zhang, B.; Chen, Y.; Chen, W.; Wang, C. Coupling Fe@ Fe3O4 Nanoparticles with Multiple-Walled Carbon Nanotubes with Width Band Electromagnetic Absorption Performance. *Appl. Surf. Sci.* **2019**, *467–468*, 836–843.

12. Wang, B.; Zhang, C.; Mu, C.; Yang, R.; Xiang, J.; Song, J.; Wen, F.; Liu, Z. Enhanced Electromagnetic Wave Absorption Properties of NiCo$_2$ Nanoparticles Interspersed with Carbon Nanotubes. *J. Magnet. Magnet. Mater.* **2019**, *471*, 185–191.

13. Inoue, Y.; Kakihata, K.; Hirono, Y.; Horie, T.; Ishida, A.; Mimura, H. One-Step Grown Aligned Bulk Carbon Nanotubes by Chloride Mediated Chemical Vapor Deposition. *Appl. Phys. Lett.* **2008**, *92*, 213113-1–213113-3.

14. Ababei, G.; Gaburici, M.; Budeanu, L.-C.; Grigoras, M.; Porcescu, M.; Lupu, N.; Chiriac, H. Influence of the Chemically Synthesis Conditions on the Microstructure and Magnetic Properties of the Co-Fe-B Nanoparticles. *J. Magnet. Magnet. Mater.* **2018**, *451*, 565–571.

15. Kharissova, O. V.; Kharisov, B. I.; de Casas Ortiz, E. G. Dispersion of Carbon Nanotubes in Water and Non-Aqueous Solvents. *RSC Adv.* **2013**, *3*, 24812–24852.

16. Savi, P.; Miscuglio, M.; Giorcelli, M.; Tagliaferro, A. Analysis of Microwave Absorbing Properties of Epoxy MWCNT Composites. *Progr. Electromagnet. Res. Lett.* **2014**, *44*, 63–69.

17. Lakshmi, K.; Honey John, H.; Mathew, K. T.; Joseph, R.; George, K. E. Microwave Absorption, Reflection and EMI Shielding of PU–PANI Composite. *Acta Mater.* **2009**, *57*, 371–375.

18. Munir, A. Microwave Radar Absorbing Properties of Multiwalled Carbon Nanotubes Polymer Composites: A Review. *Adv. Polym. Technol.* **2017**, *36*, 21617-1–21617-9.

19. Ghemes, A.; Ababei, G.; Stoian, G.; Lupu, N.; Chiriac, H. New Hybrid MWCNT/ Co-Fe-B Nanoparticles Composite Material for Applications in Microwave Domain. *Romanian Rep. Phys.* **2019**, *71*, 508, 1–7.

20. Ejembi, J. I.; Nwigboji, I. H.; Wang, Z.; Bagayoko, D.; Zhao, G.-L. High Microwave Absorption of Multi-Walled Carbon Nanotubes (Outer Diameter 10–20 nm)-Epoxy Composites in R–Band. *Phys. Sci. Int. J.* **2015**, *8* (4), 1–10.

21. Gonzalez, M.; Mokry, G.; De Nicolas, M.; Baselga, J.; Pozuelo, J.; Carbon Nanotube Composites as Electromagnetic Shielding Materials in GHz Range. *Carbon Nanotubes—Curr. Progr. Polym. Compos.*, IntechOpen **2016**. doi: 10.5772/62508.

CHAPTER 12

CARBON NANOTUBE-BASED NANOCOMPOSITES: PROMISING MATERIALS FOR ADVANCED BIOMEDICAL APPLICATIONS

SIMONA LUMINITA NICA[1*] and DELIA MIHAELA RATA[2]

[1]*"Petru Poni" Institute of Macromolecular Chemistry, Iasi, Romania*

[2]*"Apollonia" University of Iasi, Faculty of Medical Dentistry, Iasi, Romania*

Corresponding author. E-mail: nica.simona@icmpp.ro

ABSTRACT

This chapter focuses on the current literature on the use of carbon nanotubes (CNTs)-based nanocomposite in the healthcare industry. Exceptional and unique properties of CNTs made them candidates to fulfill all the requirements for scaffold constructs in nerve tissue engineering, since they are capable of providing a favorable extracellular environment for cell adhesion. However, toxicity and biocompatibility of CNTs still remain controversial. This chapter also describes the possible utilization of nanocomposite–CNTs in drug delivery systems.

12.1 INTRODUCTION

Living organisms are composed especially of self-assembly biomolecules. A variety of diseases caused by viruses or bacteria disrupt the normal rhythm of life worldwide. Reports given by the World Health Organization (WHO)[1] mention that lower respiratory infection, chronic obstructive pulmonary disease, diarrheal diseases, cardiovascular diseases are in top 10 contributors to morbidity and mortality. Drugs either prevent or cure the diseases. It is worth mentioning that people could develop a certain resistance to the drugs administration. In these cases, patients infected with multidrug-resistant microbes do not recover easily. In most situations, the drug treatment requires different types of antibiotics that are more expensive, less efficient, and toxic.[2] To limit these problems, the WHO made a decision to launch a Global Action Plan on antimicrobial resistance in 2015.[3] Therefore, pharmaceutical nanotechnology and nanomedicine experienced a rapid development. These branches of research have in view the ability to structure materials, devices, and systems at the molecular scale. Their benefits are reflected in the research development of drug delivery systems (DDS)[4,5] and diagnostic methods.[6,7] Carbon nanomaterials received world recognition in these fields.[8] Their diverse structure from renowned allotropic phases, such as amorphous carbon, graphite and diamonds to carbon nanotubes (CNTs), graphene oxide (GO), graphene quantum dots (GQDs), and fullerene[9] convinced researchers to integrate them for diverse biomedical purposes. Among these nanomaterials, we focus on CNTs (either single- or multiwalled CNTs). The different structure of CNTs permits the encapsulation of diverse compounds in their interior. Referring to crystallographic configuration of CNTs (armchair, zigzag, and chiral configuration), it is worth to mention that they are dependent on how the graphene sheet is rolled up.[10] The unique properties of CNTs were correlated by their shape and morphology. Metallic nanoparticles (silver,[11,12] gold,[13,14] platinum[15,16]), or drugs[17,18] integrated into CNTs which act like a unique system for various molecular transports[19] constitute a new class of nanomaterials with unique physicochemical properties (mechanical, optical, electrical).[20] For biomedical applications,that is,biosensors in biomolecules detection,[21] dental implants,[22] contrast agents in biological imaging,[23-25] the high-level purity of CNTs with controlled structure[26] represents a real concern. Row CNTs are insoluble in biological solutions. To overcome this issue, surface functionalization of CNTs seems to be a proper method to determine the nature of nanomaterial interactions with biological fluids.

12.2 BASIC ASPECTS OF CARBON NANOTUBES IN NANOCOMPOSITE FOR BIOMEDICAL APPLICATION

Combination of more than two different materials (a polymer and a nanoscale material) in order to constitute a nanocomposite/hybrid system will result in structural and morphologic variations[27] along with unique changes in physicochemical properties of the final system.[28] It is important to obtain heterogenous materials, the aggregation and orientation of the nanofiller (the fundamental two factors) which influence the structural properties (especially CNTs-reinforced composites) could not be completely excluded. Based on the multitude of research conducted in the biomedical field,[29] spherical nanoparticles are materials which mostly aggregate.[30] In the case of nanofillers with different shape factors, structure, and topology such as CNTs[31] interfacial interaction has a relevant role.[32] An extensive study can be found in the reference.[19] Although a good dispersion of CNTs may show reduced cytotoxicity in nanocomposites-based on CNTs, process of their preparation, their length, the size, and thickness influence their toxicity.[33] Considerable number of in vivo and in vitro studies indicated that long length of CNTs could induce lung fibrinogenic and inflammatory responses.[34] Yang et al.[35] demonstrated the highest toxicity of single-walled CNTs due to their longer length. According to the theory elaborated by Ding et al.,[36] the length of CNTs that provide antimicrobial activity supports differences in different media. As a consequence, for solids, a more antimicrobial activity can be found in the case of short length of CNTs, which is contrary to the liquid media. More recently, Saleemi et al.[37] investigated the cytotoxicity of thermoplastic polyurethane (TPU) nanofibers composite containing functionalized CNTs (double-wall, with length of 10–20 μm and multiwalled CNTs, having the long length of 100 μm) on the human adenocarcinomic lung epithelial cell line (A549). The results indicated that the viability of cells was minimum with a 100 μg/mL concentration of both types (TPU/f-MWCNTs and TPU/f-DWCNTs) of electrospun nanofibers, suggesting a correlation between treated time and dose of both functionalized nanofillers. The length of CNTs can greatly alter the toxicity of CNTs. The susceptibility to produce significant inflammation to cells is attributed to MWCNTs with long length.

The shape of CNTs and diameter size parameters will not be discussed in this chapter.

12.3   APPLICATIONS OF CARBON NANOTUBE-BASED NANOCOMPOSITES IN BIOMEDICAL FIELD

12.3.1   CARBON NANOTUBE-BASED NANOCOMPOSITES IN DRUG DELIVERY SYSTEMS

Drug delivery systems have already an enormous impact on medical technology and are designed to improve the pharmacological and therapeutic profile of a biologically active molecule.[38] An ideal drug delivery system should fulfill several important conditions, namely, to maximize the efficacy and the safety of the drug, to optimize pharmacokinetics, reduce the side effects, and decrease dosing frequency of biological therapeutic agents resulting in an enhanced patient compliance.[39] Among various drug delivery systems (nanospheres, nanocapsules, nanotubes, nanogels, dendrimers, micelles, liposomes, and microemulsions), CNTs that present extraordinary physicochemical properties and special architecture can be a promising alternative for the delivery of various pharmaceutical ingredients (drugs, proteins, antibodies, DNA, enzymes, etc.).[40] A great advantage of CNTs is that they allow the encapsulation of both hydrophilic and lipophilic drugs,[41] and pose the ability to successfully cross various biological barriers, passing through the plasma membrane of cells, thus facilitating the transport and release of the biologically active molecule in a controlled manner at the targeted site.[38] CNTs can be used in multidrug therapy, and their surface functionalization with different ligands makes them suitable to be used in targeted therapy.[42] Depending on the preparation method, the drug molecules can be incorporated into the cores of CNTs, or can be adsorbed onto CNTs surfaces.[43] In order to limit the inconveniences related to the poor biocompatibility of CNTs, and their aqueous instability, considerable efforts were made for the development of new biocompatible systems with improved thermal, electrical, and mechanical properties by embedding CNTs into matrices based on different types of polymers.[44] Also, the hydrophobic property of CNTs makes them difficult to be dispersed in solvents and to obtain stabilized suspensions, and the primary key to solve this problem can be through surface functionalization (covalent and noncovalent bonding) of CNTs, as already mentioned above.[45,46] Noncovalent functionalization can be achieved by the coating of CNTs with amphiphilic surfactant molecules, polymers, DNA as well as carbohydrates and derivatives, while covalent functionalization results due to chemical bond formation between polymer chains and CNTs.[42] CNTs-based nanocomposites can be promising candidates for drug delivery applications and some important directions are presented in Figure 12.1.

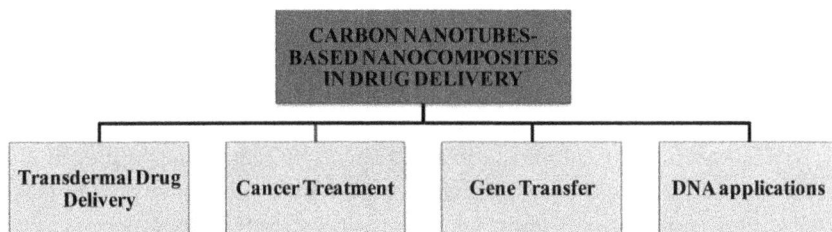

FIGURE 12.1 Schematic representation of carbon nanotube-based nanocomposites with applications in drug delivery systems.

Various studies reported the roles and effect of CNT-based nanocomposites in antitumoral drug delivery. For example, a group of researchers[47] developed a novel system of CNT-based polyvinylalcohol–polyvinylpyrolidone (PVA/PVP) nanocomposite prepared by freeze/thawing method using different amounts of CNTs (from 25 to 100 mg). Incorporation of CNTs into the PVA/PVP hydrogels and chemical structures of polymeric materials has been demonstrated by SEM, FTIR, and DSC analysis. The obtained results revealed that the degree of swelling increases with the increasing of CNTs amount from PVA/PVP hydrogels. Also, the authors investigated the release behavior of 5-fluorouracil (that was chosen as model drug) from CNT-based PVA/PVP nanocomposite hydrogels and found that the amount of drug released from the obtained system was effectively increased with the addition of CNT into the hydrogel matrix. They concluded that the obtained nanocomposite hydrogels can offer a great potential for drug release applications.

Another study reports the development of PCL–CNTs nanocomposites prepared by a facile oil-in water emulsion solvent evaporation method.[48] Mean diameter of obtained PCL–CNTs nanocomposites ranged from 289 to 434 nm and depended on the concentration of the sodium dodecyl sulfate (SDS), CNTs, and PCL. The incorporation of the CNTs in the nanocomposites was demonstrated by atomic force microscopy (AFM) and thermogravimetric (TGA) analysis. The presence of both PCL and CNTs in the nanocomposites was confirmed by Raman measurements. Also, the drug loading and release of docetaxel-DCX (an anticancer drug) from PCL–CNTs nanocomposites was investigated, free PCL and free CNTs were tested as controls. It was found that the entrapment efficiencies for CNTs were 95%, for PCL, the efficiency was 81% and for PCL–CNTs nanocomposites, it was around 89%. PCL–CNTs nanocomposites presented a fast drug release capacity and approximately 90% of DCX was released after 100 h. The authors believe that the results of this study are encouraging, and that PCL–CNTs nanocomposites may have promising applications in drug delivery.

DNA-based nanostructures attracted special attention in drug delivery applications because they can be decorated with a multitude of functionalities, and can be easily fabricated by self-assembly. They present low immunogenicity and allow interesting control over the release.[49,50] An interesting study reports the design of novel programmable drug delivery system based on DNA-functionalized CNTs and silica nanoparticles (SiNP/CNT–DNA nanocomposites) via enzymatic rolling circle amplification.[51] In the present study, two types of DNA nanocomposites, binary S and ternary SC materials were prepared using SiNP with a diameter of 80 nm and CNT with a diameter of 0.83 nm and length 1 μm. SiNP/CNT–DNA nanocomposites were loaded with anthracycline drug doxorubicin (DOX). SEM analysis revealed that the obtained SiNP/CNT–DNA nanocomposite has a hierarchical porous structure (DNA polymers wrap around the nanoparticles incorporated in the nanocomposite materials). The authors concluded that the obtained nanocomposites materials showed targeted drug delivery, high biocompatibility in vitro, that make them suitable for theranostic applications in vivo.

Hollanda et al.[52] designed, evaluated, and applied new graphene and CNT nanocomposite (GCN) in gene transfection of pIRES plasmid conjugated with green fluorescent protein (GFP) in NIH-3T3 and NG97 cell lines. GCN nanocomposite presented a mean diameter that ranged from 50 to 2000 nm, and nanocomposite suspension demonstrated moderate stability behavior with values of zeta potential that ranged from −37.4 to −39.1 mV. The authors concluded that all the transfection results are comparable with the commercial Effectene kit, and the obtained GCN nanocomposite was selectively cytotoxic.

Transdermal delivery represents an alternative route which shows good accessibility improving the therapeutic effects of drug administration by eliminating a series of secondary, undesired effects, determined by the classical administration of pharmaceutical ingredients (like insulin or plasmid DNA). In this context, Guillet et al.[53] developed an interesting nanocomposite device consisting of biocompatible polymer matrix type hydrogel based on agarose, and double-walled CNTs (DWCNTs) that can be used for storing a drug molecule and release it through the skin by electrostimulation. Obtained nanocomposite hydrogel was characterized from several points of view (SEM microscopy, swelling behavior, release properties, electrical characterization, and skin electroporation). SEM microscopy evidenced that the CNTs were well dispersed into polymer matrix, but a few nonuniform agglomerates were observed in the mass of the hydrogel. For the release test, the authors used fluorescein isothiocyanate dextran (FITC-D) molecule, and to mimic the extreme physiological conditions, artificial sweat was used. The obtained results demonstrated that in the first 15 min, the approximately 70% of the FITC-D molecule was slow released, and a maximum release (94% of FITC-D molecule) occurred after 1 h. Also, the skin electroporation tests that used an ex vivo mouse skin model has revealed that the use of DWCNTs improves the transdermal delivery of the biocompatible polymer matrix.

Another study reports the preparation of pH-sensitive polyampholyte nanogels (PANGs) based on poly(vinylimidazole-co-acrylic acid) (PVI-co-AA) and functionalized single-walled CNTs (f-SWCNTs). The obtained PANGs presented good swelling properties and the spectroscopic and microscopic studies revealed the incorporation of nanotubes into nanogels system. Morphological characterization showed that the nanotubes were well dispersed into the polymer matrix and the diameter of nanogels was around 35 nm. The thermal, rheological and MTT assays studies have revealed that the new material is more stable, biocompatible than the polymer gel. Also, the release studies of promethazine (an antidepression drug) from the nanogels evidenced the pH-sensitive behavior of nanogels, and indicated a controlled and delayed drug release profile that make them suitable for use in drug delivery.[54]

12.3.2 CARBON NANOTUBES-BASED NANOCOMPOSITES USED AS DENTAL IMPLANTS

Dental implants defined as "artificial prosthesis" act as a substitute for the tooth root and contribute to maintaining the integrity of facial structure and can represent an alternative for people with missing, damaged, or infected teeth.[55,56] Usually, the materials that are used for obtaining dental implants (whether metallic, ceramic, or polymeric) have some disadvantages, such as problems and concerns regarding osseointegration and differences in mechanical properties between natural bone and the implant, aspects that over time can affect their stability. Also, it is essential that any dental implant must present an excellent performance for resistance in the complex conditions of the oral environment, at least a period of 8–10 years, to prevent inflammatory responses and to promote osseointegration.[57,58] In this context, nanocomposite materials containing CNTs can represent a promising alternative for fabrication of stable and durable dental implants.[59]

Figure 12.2 shows some types of CNTs-based nanocomposites which can be used to obtain high-performance dental implants.

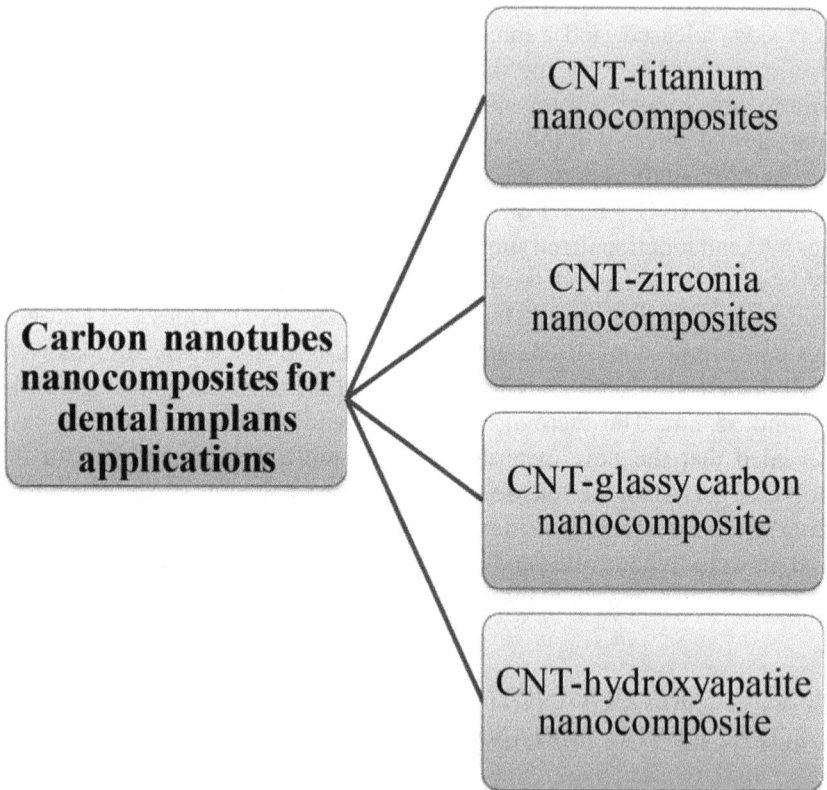

FIGURE 12.2   Types of carbon nanotubes-based nanocomposites used in dental implants.

Vitreous carbon (glassy carbon) is one of the most common implant biomaterials, has the modulus of elasticity close to the dentine and bone, and causes a very minimal response from host tissues.[60] Due to its brittleness, carbon is susceptible to fracture under tensile and stress condition and because of this property other types of materials have been tried (silicon, zirconia, and titanium) or even making different composite materials.[59]

A group of researchers prepared CNT reinforced hydroxyapatite nano-composites (HA-MWCNTs nanocomposite) using the sol–gel technique.[61] In the current study, four different types of nanocomposites have been prepared as follows:

i)   p-MWCNTs-PVA (based on hydroxyapatite-HA, pristine MWCNTs, and polyvinyl alcohol–PVA as surfactant 1).
ii) f-MWCNTs–PVA (based on HA, functionalized MWCNTs and PVA as surfactant 1).
iii) p-MWCNTs–HTAB (based on HA, pristine MWCNTs and hexa-decyl trimethyl ammonium bromide–HTAB as surfactant 2).
iv) f-MWCNTs–HTAB (HA, functionalized MWCNTs, HTAB as surfactant 2).

Pure HA was used as a control. The obtained results showed that HA-MWCNTs nanocomposite was obtained successfully (after TEM analyses), nanocomposites based on f-MWCNTs presented better dispersion and better interaction with the HA particles compared with composites based on p-MWCNTs, and in terms of mechanical properties, a major increase in both tensile and compressive strength of the nanocomposite in comparison with pure HA was observed. Also, it was found that all prepared nanocomposites showed a good biocompatibility with osteoblast but the sample f-MWCNTs–HTAB presented the best biocompatibility with osteoblast cells. The authors concluded that f-MWCNTs–HTAB nanocomposite can potentially be used as dental or small bone graft.

Garmendia et al.[62] designed new zirconia–multiwall CNTs nanocom-posites using hydrothermal synthesis. Yttria-stabilized zirconia (Y-TZP) ceramic is used in dental implants and offers good biocompatibility and mechanical properties, but its major disadvantage is that it lacks stability (it degrade at a low temperature and is prone to aging, especially under humid atmosphere). Therefore, in order to improve the stability of this material, the authors incorporated a small volume fraction of multiwall CNTs (MWCNT) in a polycrystalline nanostructured Y-TZP, and have found that new material exhibited an exceptional balance between aging and crack resistance.

Another research group[63] performed a study on the effects of an assembled CNT composite (CNT-comp) compared with the commercially pure titanium on osteoblasts in vitro and bone tissue in vivo in rats. The CNT-comp was prepared using the layer-by-layer technique. The obtained results showed that CNT-comp allowed better mineral formation compared with the pure titanium substrates, and the surface of CNT-comp was better in promoting matrix mineralization than pure titanium surface. In vivo tests evidenced that CNT-comp was biocompatible and promoted bone formation, which makes these systems ideal for use in dental implants or for repair of bone defects.

12.3.3   NANOCOMPOSITES BASED ON CNTs FOR NERVE TISSUE ENGINEERING APPLICATIONS

The functions of the organs into the body are controlled by the nervous system. From anatomically point of view, nervous system includes both the central nervous system (CNS) and peripheral nervous system (PNS). The CNS includes the brain, spinal cord, optic, olfactory and auditory systems. The PNS includes the cranial nerves, the spinal nerves, and sensory nerve cell bodies.[64] The signals pathway travel from the CNS to the body through PNS. Loss or damage of the nerve can occur due to external trauma, virus infection, hypoglycemia.[65,66] CNS axons do not regenerate after a mammalian adult had suffered an injury[67] which leads to functional deficits. In contrast, PNS axons regenerate, allowing the recovery of organ functions after the damage of peripheral nerve. To improve the regeneration capacity of the neural tissue became a great challenge over the years.

On the basis of literature findings, it could be approved that the degree of peripheral nerves regeneration for humans is determined by the following factors:

- Survival of axotomized neurons.[68]
- The ability of axonal outgrowth.[69]
- The specificity of organs reinnervated by the neural defect.[70]

Biological scaffold materials, the seed cells (including Schwann cells (SCs),[71] stem cells,[72] progenitor cells etc.) and growth factors represent the main elements of peripheral nervous tissue engineering.[73] A comprehensive review about the role of Schwann cells development in nerve tissue engineering can be found in the cited resource.[74] The last two elements will not be discussed here.

Tissue engineering and regenerative medicine in the nervous system mention the use of nerve autograft, allografts (tissue donor of same species) and xenografts (donor of a different species)[75] as current applied treatments. Some difficulties could be recognized such as shortage of donor nerves, donor-site morbidity, improper regeneration, infectious diseases, and immunological issues[76] Biomaterials used in the field of nerve tissue engineering offer the opportunity to develop new innovative therapies for neurodegeneration and traumatic injury.[77] Biomaterials are considered only as cellular growth scaffolds or substrates. An ideal scaffold for nerve tissue engineering must have in view appropriate characteristics, such as porosity, fibers, channels, mechanical properties[78] along with biocompatibility. A comprehensive description of the advantages of scaffold futures can be found in the reference.[79]

Cells interact with the scaffold (which can be either isotropic or anisotropic) making a response to their topography, chemistry, and surface energy.[80] Although some researchers[81] showed the efficiency of a discontinuous anisotropic topography in the directional outgrowth of neural cells, The main difference between the two types of scaffolds relay on the mode of how the cells are arranged and oriented.[82] Good electrical conductivity and mechanical properties of CNTs made them ideal candidates to guide the nerve growth and neurite/axonal elongation.[83] Carbons nanotubes can form nanocomposites with different materials. Several studies have been reported for the formation of hydrogel CNTs-based nanocomposites for nerve tissue engineering. Hydrogels are preferred materials due to their similarity to the extracellular matrix (ECM).

Matrices could be either natural or synthetic polymers. Among the natural polymers which have a good biocompatibility and biodegradability,chitosan, collagen, fibrin, alginate, gelatin, cellulose, starch, etc. are extremely used as scaffold. On the other hand, synthetic biodegradable polymers include poly (hydroxyl butyrate), poly (lactic acid), poly (ethylene glycol), poly (lactide-co-glycolide), etc.

Koppes et al.[84] prepared collagen/SWCNTs composite for the use in nerve tissue engineering. Their results indicated that the SWCNTs loaded in the collagen matrix support neurite outgrowth. According to them, increases in the neurite outgrowth are not influenced by the increase in the electrical conductivity.

Another group of research[85] as materials for nerve stimulation used chitosan–polyurethane/fMWCNTs electrospun nanofiber as scaffolds for neural stimulation. They have studied proliferation of two group cells on aligned oriented and randomly oriented fibers.

In another study, Steel et al.[86] report the development of biocompatible conductive material composite consisting from hyaluronic acid/CNTs nanofibers. They obtained great results on the electrical stimulation of the tissue, improving neuron growth when 200 mV/mm was applied relative to unstimulated neuron.

12.4 CONCLUSIONS

Despite the negative effects associated with CNTs materials, they still have an important role future application in the biomedical field, such as drug delivery systems, dental implants, nerve tissue engineering.

To diminish the toxicity of CNTs in nanocomposite, proper functionalization methods become necessary to overcome toxicity related to these materials for their application in drug delivery system as nanotransporter.

CNTs should be at the proper size, proper length, and functionalized appropriately according to the intended biological end goals.

KEYWORDS

- nanocomposite
- carbon nanotubes (CNTs)
- biomedical applications
- healthcare industry
- tissue engineering
- cell adhesion

REFERENCES

1. World Health Organization. The-top-10-causes-of-death, 2020.
2. Webb, G. F.; D'Agata, E. M.; Magal, P.; Ruan, S. A Model of Antibiotic-Resistant Bacterial Epidemics in Hospitals. *Proc. Natl. Acad. Sci. USA.* **2005,** *102,* 13343–13348.
3. World Health Organization. Global Action Plan on Antimicrobial Resistance, 2015.

4. Quarterman, J. C.; Geary, S. M.; Salem, A. K. Evolution of Drug-Eluting Biomedical Implants for Sustained Drug Delivery. *Eur. J. Pharm. Biopharm.* **2021,** *159,* 21–35.

5. Gea, X.; Fu, M.; Niu, X.; Kong, X. Atomic Layer Deposition of γ-Fe2O3 Nanoparticles on Multi-Wall Carbon Nanotubes for Magnetic Drug Delivery and Liver Cancer Treatment. *Ceram. Int.* **2020,** *46,* 26557–26563.

6. Taroni, T.; Cauteruccio, S.; Vago, R.; Franchi, S.; Barbero, N.; Licandro, E.; Ardizzone, S.; Meroni, D. Thiahelicene-Grafted Halloysite Nanotubes: Characterization, Biological Studies and pH Triggered Release. *Appl. Surf. Sci.* **2020,** *520,* 146351–146359.

7. Xu, S.; Zhang, Y.; Zhu, Y.; Wu, J.; Li, K.; Lin, G.; Li, X.; Liu, R.; Liu, X.; Wong, C.-P. Facile One-Step Fabrication of Glucose Oxidase Loaded Polymeric Nanoparticles Decorating MWCNTs for Constructing Glucose Biosensing Platform: Structure Matters. *Biosens. Bioelectron.* **2019,** *135,* 153–159.

8. Maiti, D.; Tong, X.; Mou, X.; Yang, K. Carbon-Based Nanomaterials for Biomedical Applications: A Recent Study. *Front. Pharmacol.* **2019,** *9,* 1401–1417.

9. Mostofizadeh, A.; Li, Y.; Song, B.; Huang, Y. Synthesis, Properties, and Applications of Low-Dimensional Carbon-Related Nanomaterials. *J. Nanomater.* **2011,** *2011,* 685081.

10. Aqela, A.; Abou El-Nour, K. M. M.; Ammar, R. A. A.; Al-Warthan, A. Carbon Nanotubes, Science and Technology Part (I) Structure, Synthesis and Characterization. *Arab. J. Chem.* **2012,** *5,* 1–23.

11. Lin, J.; Guan, G.; Yang, W.; Fu, H. The Enhanced Raman Scattering of Ag Nanoparticles Decorated on the Carbon Nanotube via a Simple Manipulation. *Opt. Mater.* **2019,** *95,* 109258.

12. Imanzadeh, H.; Bakirhan, N. K.; Habibi, B.; Ozkan, S. A. A Sensitive Nanocomposite Design via Carbon Nanotube and Silver Nanoparticles: Selective Probing of Emedastine Difumarate. *J. Pharm. Biomed. Anal.* **2020,** *181,* 113096.

13. Palomar, Q.; Xu, X. X.; Selegård, R.; Aili, D.; Zhang, Z. Peptide Decorated Gold Nanoparticle/Carbon Nanotube Electrochemical Sensor for Ultrasensitive Detection of Matrix Metalloproteinase-7. *Sens. Act. B Chem.* **2020,** *325,* 128789.

14. Zeng, K.; Wei, D.; Zhang, Z.; Meng, H.; Huang, Z.; Zhang, X. Enhanced Competitive Immunomagnetic Beads Assay with Gold Nanoparticles and Carbon Nanotube-Assisted Multiple Enzyme Probes. *Sens. Act. B Chem.* **2019,** *292,* 196–202.

15. Li, J.; Huang, X.; Shi, W.; Jiang, M.; Tian, L.; Su, M.; Wu, J.; Liu, Q.; Yu, C.; Gu, H. Pt Nanoparticle Decorated Carbon Nanotubes Nanocomposite-Based Sensing Platform for the Monitoring of Cell-Secreted Dopamine. *Sens. Actuat. A Chem.* **2021,** *330,* 129311.

16. Deshmukh, M. A.; Jeon, J.-Y.; Ha, T.-J. Carbon Nanotubes: An Effective Platform for Biomedical Electronics. *Biosens. Bioelectron.* **2020,** *150,* 11919.

17. Cao, M.; Wu, D.; Yoosefian, M.; Sabaei, S.; Jahani, M. Comprehensive Study of the Encapsulation of Lomustine Anticancer Drug into Single Walled Carbon Nanotubes (SWCNTs): Solvent Effects, Molecular Conformations, Electronic Properties and Intramolecular Hydrogen Bond Strength. *J. Mol. Liq.* **2020,** *320,* 114285.

18. Yoosefian, M.; Sabaei, S.; Etminan, N. Encapsulation efficiency of Single-Walled Carbon Nanotube for Ifosfamide Anti-Cancer Drug. *Comput. Biol. Med.* **2019,** *114,* 103433.

19. Garnica-Palafox, I. M.; Estrella-Monroy, H. O.; Vázquez-Torres, N. A.; Álvarez-Camacho, M.; Castell- Rodríguez, A. E.; Sánchez-Arévalo, F. M. Influence of Multi-Walled Carbon Nanotubes on the Physico-Chemical and Biological Responses of Chitosan-Based Hybrid Hydrogels. *Carbohydr. Polym.* **2020,** *236,* 115971.

20. Soni, S. K.; Thomas, B.; Kar, V. R. A Comprehensive Review on CNTs and CNT-Reinforced Composites: Syntheses, Characteristics and Applications. *Mater. Today Commun.* **2020**, *25*, 101546.

21. Patel, B. R.; Imran, S.; Ye, W.; Weng, H.; Noroozifar, M.; Kerman, K.; Simultaneous Voltammetric Detection of Six Biomolecules Using A Nanocomposite of Titanium Dioxide Nanorods with Multi-Walled Carbon Nanotubes. *Electrochem. Acta.* **2020**, *362*, 137094.

22. Sivaraj, D.; Vijayalakshmi, K.; Ganeshkumar, A.; Rajaram, R. Tailoring Cu Substituted Hydroxyapatite/Functionalized Multiwalled Carbon Nanotube Composite Coating on 316L SS Implant for Enhanced Corrosion Resistance, Antibacterial and Bioactive Properties. *Int. J. Pharm.* **2020**, *590*, 119946.

23. Mergen, Ö. B.; Arda, E.; Determination of Optical Band Gap Energies of CS/MWCNT Bio-Nanocomposites by Tauc and ASF Methods. *Synth. Met.* **2020**, *269*, 116539.

24. Sam-Daliri, O.; Faller, L.-M.; Farahani, M.; Roshanghias, A.; Oberlercher, H.; Mitterer, T.; Arae, A.; Zangl, H. MWCNT–Epoxy Nanocomposite Sensors for Structural Health Monitoring. *Electronics* **2018**, *7*, 143–157.

25. Teixeira-Santos, R.; Gomes, M.; Gomes, L. C.; Mergulhão, F. J. Antimicrobial and Anti-Adhesive Properties of Carbon Nanotube-Based Surfaces for Medical Applications: A Systematic Review. *iScience.* **2021**, *24*, 102001.

26. Anzar, N.; Hasan, R.; Tyagi, M.; Yadav, N.; Narang, J. Carbon Nanotube-A Review on Synthesis, Properties and Plethora of Applications in the Field of Biomedical Science. *Sens. Int.* **2020**, *1*, 100003.

27. Camargo, P. H. C.; Satyanarayana, G. K.; Wypych, F. Nanocomposites: Synthesis, Structure, Properties and New Application Opportunities. *Mat. Res.* **2009**, *12*, 1–39.

28. Ates, M.; Aysegul, A. E.; Bulent, E. Carbon Nanotube-Based Nanocomposites and Their Applications. *J. Adhes. Sci. Technol.*, **2017**, *31*, 1977–1997.

29. Dheyab, M. A.; Aziz, A. A.; Jameel, M. S.; Noqta, O. A.; Mehrdel, B. Synthesis and Coating Methods of Biocompatible Iron Oxide/Gold Nanoparticle and Nanocomposite for Biomedical Applications. *Chin. J. Phys.* **2020**, *64*, 305–325.

30. Kefeni, K. K.; Msagati, T. A. M.; Nkambule, T. T.; Mamba, B. B. Spinel Ferrite Nanoparticles and Nanocomposites for Biomedical Applications and Their Toxicity. *Mat. Sci. Eng.* **2020**, *107*, 110314.

31. Saifuddin, N.; Raziah, A. Z.; Junizah, A. R. Carbon Nanotubes: A Review on Structure and Their Interaction with Proteins. *J. Chem.* **2013**, *2013*, 676815 (1–18).

32. Saadat, S.; Pandey, G.; Tharmavaram, M.; Braganza, V.; Rawtani, D. Nano-interfacial decoration of Halloysite Nanotubes for the development of antimicrobial nanocomposites. *Adv. Colloid. Interface Sci.* **2020**, *275*, 102063.

33. Lshehri, R.; Ilyas, A. M.; Hasan, A.; Arnaout, A.; Ahmed, F.; Memic, A.; Carbon Nanotubes in Biomedical Applications: Factors, Mechanisms, and Remedies of Toxicity. *J. Med. Chem.* **2016**, *59*, 8149–8167.

34. Dong, J.; Qiang, M. Type 2 Immune Mechanisms in Carbon Nanotube-Induced Lung Fibrosis. *Front. Immunol.* **2018**, *9*, 1120–1137.

35. Yang, C.; Mamouni, J.; Tang, Y.; Yang, L. Antimicrobial Activity of Single-Walled Carbon Nanotubes: Length Effect. *Langmuir* **2010**, *26*, 16013–16019.

36. Ding, L.; Wang, H.; Liu, D.; Zeng, X.-A.; Mao, Y. Bacteria Capture and Inactivation with Functionalized Multi-Walled Carbon Nanotubes (MWCNTs). *J. Nanosci. Nanotechnol.* **2020**, *20*, 2055–2062.
37. Saleemi, M. A.; Yong, P. V. C.; Wong, E. H. Investigation of Antimicrobial Activity and Cytotoxicity of Synthesized Surfactant-Modified Carbon Nanotubes/Polyure-thane Electrospun Nanofibers. *Nano-Struct. Nano-Objects* **2020**, *24*, 100612.
38. Chen, G. Nanotube-Based Controlled Drug Delivery. *Pharmaceut. Anal. Acta.* **2012**, *3*, 1000e136-1000e138.
39. Cirillo, G.; Iemma, F.; Puoci, F.; Parisi, O. I.; Curcio, M.; Spizzirri, U. G.; Picci, N. Imprinted Hydrophilic Nanospheres as Drug Delivery Systems for 5-Fluorouracil Sustained Release. *J. Drug. Targ.* **2009**, *17*, 72–77.
40. Raval, J. P.; Joshi, P.; Chejara, D. R. Carbon Nanotube for Targeted Drug Delivery. In *Applications of Nanocomposite Materials in Drug Delivery*; Inamuddin, A. M., Asiri, A. M, Eds.; 2018; pp 203–216.
41. Divekar, S. R.; Carbon Nanotubes an Advanced Drug Delivery System—A Review. *Int. J. Pharm. Sci. Res.* **2020**, *11*, 3636–3644.
42. Kaur, J.; Gill, G. S.; Jeet, K.; Kaur, A.; Gill, G. S.; Jeet, K. Chapter 5—Applications of Carbon Nanotubes in Drug Delivery: A Comprehensive Review. In *Micro and Nano Technologies, Characterization and Biology of Nanomaterials for Drug Delivery*; Mohapatra, S. S., Ranjan, S., Dasgupta, N., Mishra, R. K., Thomas, S., Eds.; Elsevier, 2019; 113–135. ISBN 9780128140314.
43. Degim, I. T.; Burgess, D. J.; Papadimitrakopoulos, F. Carbon Nanotubes for Trans-dermal Drug Delivery. *J. Microencapsul.* **2010**, *27*, 669–681.
44. Bouchard, J.; Cayla, A.; Devaux, E. et al. Electrical and Thermal Conductivities of Multiwalled Carbon Nanotubes-Reinforced High-Performance Polymer Nanocom-posites. *Compos. Sci. Technol.* **2013**, *86*, 177–184.
45. Simon, J.; Flahaut, E.; Golzio, M. Overview of Carbon Nanotubes for Biomedical Applications. *Materials* **2019**, *12*, 624.
46. Spitalsky, Z.; Tasis, D.; Papagelis, K.; et al. Carbon Nanotube-Polymer Composites: Chemistry, Processing, Mechanical and Electrical Properties. *Prog. Polym. Sci.* **2010**, *35*, 357–401.
47. Özkahraman, B.; Irmak, E. T. Carbon Nanotube Based Polyvinylalcohol-Polyvinyl-pyrolidone Nanocomposite Hydrogels for Controlled Drug Delivery Applications. *Anadolu Univ. J. Sci. Technol. A—Appl. Sci. Eng.* **2017**, *18*, 543–553.
48. Niezabitowska, E.; Smith, J.; Prestly, M. R.; Akhtar, R.; von Aulock, F. W.; Lavallee, Y.; Ali-Boucettad, H.; McDonald. T. O. Facile Production of Nanocomposites of Carbon Nanotubes and Polycaprolactone with High Aspect Ratios with Potential Applications in Drug Delivery. *RSC Adv.* **2018**, *8*, 16444–16454.
49. Lu, H.; Wang, J.; Wang, T.; Zhong, J.; Bao, Y.; Hao, H.; Recent Progress on Nano-structures for Drug Delivery Applications. *J. Nanomater.* **2016**, 1–12.
50. Vries, J. W. D.; Zhang, F.; Herrmann, A. Drug Delivery Systems Based on Nucleic Acid Nanostructures. *J. Contr. Rel.* **2013**, *172*, 467–483.
51. Hu, Y.; Niemeyer, C. M. Designer DNA–Silica/Carbon Nanotube Nanocomposites for Traceable and Targeted Drug Delivery. *J. Mater. Chem. B.* **2020**, *8*, 2250–2255.

52. Hollanda, L. M.; Lobo, A. O.; Lancellotti, M.; Berni, E.; Corat, E. J.; Zanin, H. Graphene and Carbon Nanotube Nanocomposite for Gene Transfection. *Mater. Sci. Eng. C.* **2014**, *39*, 288–298.

53. Guillet, J. F.; Flahaut, E.; Golzio, M.; A Hydrogel/Carbon-Nanotube Needle-Free Device for Electrostimulated Skin Drug Delivery. *Chem. Phys. Chem.* **2017**, *18*, 2715–2723.

54. Sankar, R. M.; Seeni Meera, K. M.; Samanta, D.; Jithendra, P.; Mandal, A. B.; Jaisankar, S. N. The pH-Sensitive Polyampholyte Nanogels: Inclusion of Carbon Nanotubes for Improved Drug Loading. *Colloids. Surf. B Biointerf.*. **2013**, *112*, 120–127.

55. Zhao, L.; Huo, K.; Chu, P. K. Titania Nanotube Coatings on Dental Implants with Enhanced Osteogenic Activity and Anti-Infection Properties. *Nanobiomat. Clin Dent.* **2013**, 337–357.

56. Das, R.; Bhattacharjee, C. Titanium-Based Nanocomposite Materials for Dental Implant Systems. In *Woodhead Publishing Series in Biomaterials, Applications of Nanocomposite Materials in Dentistry*; Asiri, A. M., Mohammad, I. A., Eds.; Woodhead Publishing, 2019; pp 271–284. ISBN 9780128137420.

57. Choi, A. H.; Ben-Nissan, B.; Matinlinna, J. P.; Conway, R. C. Current perspectives. *J. Dent. Res.* **2013**, *92*, 853–859.

58. Dunne, N.; Mitchell, C. Biomedical/Bioengineering Applications of Carbon Nanotube-Based Nanocomposites. *Polym.–Carbon Nanotube Compos.* **2011**, 676–717.

59. Teh, S. J.; Lai, C. W. Carbon Nanotubes for Dental Implants. *App. Nanocompos. Mater. Dentistry* **2019**, *1*, 93–105.

60. Ananth, H.; Kundapur, V.; Mohammed, H. S.; Anand M.; Amarnath, G. S.; Mankar, S. A Review on Biomaterials in Dental Implantology. *Int. J. Biomed. Sci.* **2015**, *11*, 113–132.

61. Lawton, K.; Le, H.; Tredwin, C.; Handy, R. D. Carbon Nanotube Reinforced Hydroxyapatite Nanocomposites as Bone Implants: Nanostructure, Mechanical Strength and Biocompatibility. *Int. J. Nanomed.* **2019**, *14*, 7947–7962.

62. Garmendia, N.; Grandjean, S.; Chevalier, J.; Diaz, L. A.; Torrecillas, R.; Obieta, I.; Zirconia–Multiwall Carbon Nanotubes Dense Nano-Composites with an Unusual Balance Between Crack and Ageing Resistance. *J. Eur. Ceram. Soc.* **2011**, *31*, 1009–1014.

63. Bhattacharya, M.; Wutticharoenmongkol-Thitiwongsawet, P.; Hamamoto, D. T.; Lee, D.; Cui, T.; Prasad, H. S.; Ahmad, M. Bone Formation on Carbon Nanotube Composite. *J. Biomed. Mater. Res. Part A.* **2010**, *96A*, 75–82.

64. Appler, J. M.; Goodrich, L. V.; Connecting the Ear to the Brain: Molecular Mechanisms of Auditory Circuit Assembly. *Prog. Neurobiol.* **2011**, *93*, 488–508.

65. Masi, E. B.; Levy, T.; Tsaava, T.; Bouton, C. E.; Tracey, K. J.; Chavan, S. S.; Zanos, T. P.; Identification of Hypoglycemia-Specific Neural Signals by Decoding Murine Vagus Nerve Activity. *Bioelectron. Med.* **2019**, *9*, 2019–2029.

66. de Vasconcelos, A. C. P.; Morais, R. P.; Novais, G. B.; da S. Barroso, S.; Menezes, L. R. O.; dos Santos, S.; da Costa, L. P.; Correa, C. B.; Severino, P.; Gomes, M. Z.; Albuquerque Júnior, R. L. C.; Cardoso, J. C. In Situ Photocrosslinkable Formulation of Nanocomposites Based on Multi-Walled Carbon Nanotubes and Formononetin for Potential Application in Spinal Cord Injury Treatment. *Nanomedicine* **2020**, *29*, 102272.

67. Hilton, B. J.; Bradke, F. Can Injured Adult CNS Axons Regenerate by Recapitulating Development? *Development*. **2017,** *144*, 3417–3429.

68. Robinson, G. A.; Madison, R. D. Motor Neuron Target Selectivity and Survival after Prolonged Axotomy. *Restor. Neurol. Neurosci*. **2013,** *31*, 451–460.

69. Guo, W.; Stoklund Dittlau, K.; Van Den Bosch, L. Axonal Transport Defects and Neurodegeneration: Molecular Mechanisms and Therapeutic Implications. *Semin. Cell Dev. Biol.* **2020,** *99*, 133–150.

70. Tuturov, A. O. The Role of Peripheral Nerve Surgery in a Tissue Reinnervation. *Chin. Neurosurg. J.* **2019,** *5*, 5.

71. Zhang, P.-X.; Han, N.; Kou, Y.-H.; Zhu, Q.-T.; Liu, X.-L.; Quan, D. P.; Chen, J.-G.; Jiang, B.-G. Tissue Engineering for the Repair of Peripheral Nerve Injury. *Neural. Regen. Res.* **2019,** *14*, 51–58.

72. Li, H.; Ye, A. Q.; Su, M. Application of Stem Cells and Advanced Materials in Nerve Tissue Regeneration. *Stem. Cells. Int.* **2018,** ID. 4243102.

73. Wongtrakul, S.; Bishop, A. T.; Friedrich, P. F. Vascular Endothelial Growth Factor Promotion of Neoangiogenesis in Conventional Nerve Grafts. *J. Hand. Surg. Am*. **2002,** *27*, 277–285.

74. Lofi, L.; Khakbiz, M.; Moghaddam, M. M.; Bonkdar, S. A Biomaterials Approach to Schwann Cell Development in Neural Tissue Engineering. *J. Biomed. Mat. Res.* **2019,** *107*, 2425–2446.

75. Patel, N. P.; Lyon, K. A.; Huang, J. H. An Update–Tissue Engineered Nerve Grafts for the Repair of Peripheral Nerve Injuries. *Neural. Regen. Res.* **2018,** *13*, 764–774.

76. Hallgren, A.; Björkman, A.; Chemnitz, A.; Dahlin, L. B. Subjective Outcome Related to Donor Site Morbidity after Sural Nerve Graft Harvesting: A Survey in 41 Patients. *BCM. Surg.* **2013,** *13*, 39–46.

77. Khan, F.; Tanaka, M.; Ahmad, S. R. Fabrication of Polymeric Biomaterials: A Strategy for Tissue Engineering and Medical Devices. *J. Mater. Chem. B.* **2015,** *3*, 8224–8249.

78. Rey, F.; Barzaghini, B.; Nardini, A.; Bordoni, M.; Zuccotti G., V.; Cereda, C.; Raimondi, M. T.; Carelli, S. Advances in Tissue Engineering and Innovative Fabrication Techniques for 3-D-Structures: Translational Applications in Neurodegenerative Diseases. *Cells*. **2020,** *9*, 1636–1671.

79. Vila-Parrondo, C.; García-Astrain, C.; Liz-Marzán, L. M. Colloidal Systems toward 3D Cell Culture Scaffolds. *Adv. Colloid. Int. Sci.*, **2020,** *283*, 102237.

80. Hajiali, H.; Contestabile, A.; Mele, E.; Athanassiou, A. Influence of Topography of Nanofibrous Scaffolds on Functionality of Engineered Neural Tissue. *J. Mat. Chem. B.* **2018,** *6*, 930–939.

81. Simitzi, C.; Efstathopoulos, P.; Kourgiantaki, A.; Ranella, A.; Charalampopoulos, I.; Fotakis, C.; Athanassakis, I.; Stratakis, E.; Gravanis, A. Laser Fabricated Discontinuous Anisotropic Microconical Substrates as a New Model Scaffold to Control the Directionality of Neuronal Network Outgrowth. *Biomaterials* **2015,** *67*, 115–128.

82. Dodla, M. C.; Bellamkonda, R. V. Differences between the Effect of Anisotropic and Isotropic Laminin and Nerve Growth Factor Presenting Scaffolds on Nerve Regeneration across Long Peripheral Nerve Gaps. *Biomaterials* **2008,** *29*, 33–46.

83. Lovat, V.; Pantarotto, D.; Lagostena, L.; Cacciari, B.; Grandolfo, M.; Righi, M.; Spalluto, G.; Prato, M.; Ballerini, L. *Nano. Lett.* **2005,** *5*, 107-10.

84. Koppes, A. N.; Keating, K. W.; McGregor, A. L.; Koppes, R. A.; Kearns, K. R.; Ziemba, A. M.; McKay, C. A.; Zuidema, J. M.; Rivet, C. J.; Gilbert, R. J.; Thompson, D. M. Robust Neurite Extension Following Exogenous Electrical Stimulation within Single Walled Carbon Nanotube-Composite Hydrogels. *Acta. Biomater.* **2016,** *15,* 34–43.

85. Shrestha, S.; Shrestha, B. K.; Kim, J.-I.; Koa, S. W.; Park, C. H.; Kim, C. H.; Electrodeless Coating Polypyrrole on Chitosan Grafted Polyurethane with Functionalized Multiwall Carbon Nanotubes Electrospun Scaffold for Nerve Tissue Engineering. *Carbon* **2018,** *136,* 430–443.

86. Steel, E. M.; Azar, J.-Y.; Sundararaghavan, H. G. Electrospun Hyaluronic Acid-Carbon Nanotube Nanofibers for Neural Engineering. *Materialia* **2020,** *9,* 100581.

INDEX

A

Aerogel for wastewater treatment, 30–31
Amidation and esterification, 126
Anticancer
  applications, 130–131
  behavior, 141
Antioxidant behavior, 133
Arsenic
  and heavy metal groundwater, 176–177

B

Biomedical applications, CNTs
  anticancer applications, 130–131
  antioxidant behavior of, 133
  applications, 126, 135
  biocompatibility and toxicity, 148–149
  biological applications, 126–127
  carbon nanodiamonds, 144–145
    as biomarker for cellular imaging,
      146–148
    as delivery vehicle, 145–146
  cell penetration, 127
  characteristics and derivatization, 124–126
  CNHs for drug delivery
    anticancer behavior, 141
    cumulative activity, 143
    enhanced permeability and retention
      (EPR) effect, 140
    MRI applications, 142
    noncovalent/covalent functionalization
      and adsorption, 141
    polyethyleneglycol–doxorubicin
      (PEG–DXR), 140
    SWCNH-binding block (NHBP-1), 142
  gene delivery, 129–130
  in imaging, 133–135
  laser-induced therapeutic agent, 139–140
  neuron interactions with, 131–132
  structures, 124–126

SWCNTS
  amidation and esterification of, 126
  characteristics, and functionalization,
    137–139
  sidewall functionalization, 126
  structure, 137–139
  targeted smart drug delivery, 128–129
  toxic potential of, 135–137
  toxicity of CNHs, 143

C

Carbon nanodiamonds, 144–145
  as biomarker for cellular imaging,
    146–148
  as delivery vehicle, 145–146
Carbon nanomaterial embedded
  membranes
  conventional methods, 103–104
  heavy metal
    natural calamities and human
      activities, 100
    remove, approaches, 103
    separation, objective of, 101–103
    sources, 100
  importance, 110
    carbon nanofibers (CNF), 113
    classification of, 112
    CNTs, 112
    fullerene, 113
    graphene oxide, 113
  interfacial polymerization technique,
    114–115
  membrane methods, significance of,
    106–107
  membrane separation process, 104, 106
  nanomaterials used in, 109–110
  NF membranes
    HPEI (hyperbranched
      polyethyleneimine), 116–117
    Torlon-GO membrane, 116

phase inversion technique, 114
RO membranes
  polyamide-CNT membrane, 115
  TMC (trimesoyl chloride) solution, 116
synthetic/polymer membranes, 107–109
UF membranes
  dimethylformamide (DMF), 117
  MWCNT, 117–118
unique methods to, 113
Carbon nanotubes (CNTs)
bubble retention in fluids, 88
current lacunas, 91–92
enhanced oil recovery and carbon
  sequestration, 88–89
interfacial tension, 83–84
nanocomposite
  biomedical application, 275
  dental implants, 279–282
  drug delivery systems, 276–279
  nerve tissue engineering applications,
    282–284
nanofluids, synthesis & dispersion, 79–82
surface carbon capture, 84–88
Chemical vapor deposition (CVD), 26
Clean water
fundamental aspects
  AOP, 8
  catalytical degradation, 5
  CNT-based membrane, 10
  CNTs, preparation, 4
  morphological structure, 5
  organic and inorganic pollutants, 4
  removal of pollutants, 6–7
  single-walled carbon nanotubes
    (SWCNTs), 3
membranes used, 11

D

Dental implants, 279–282
Dimethylformamide (DMF), 117
Doped carbon nanotubes
characterization, 57–58
elements doped, 55–57
nitrogen doped, 54–55
synthesis, methods, 57
Drug delivery

CNHs for
  anticancer behavior, 141
  cumulative activity, 143
  enhanced permeability and retention
    (EPR) effect, 140
  MRI applications, 142
  noncovalent/covalent functionalization
    and adsorption, 141
  polyethyleneglycol–doxorubicin
    (PEG–DXR), 140
  SWCNH-binding block (NHBP-1), 142
systems, 276–279
Dye remediation, 62
CNTs, use of doped, 63–65

E

Electromagnetic interference (EMI)
carbon nanotubes (CNTs)
  characterization of, 264–267
  CoFeB nanoparticles, 260–261
  composite materials, preparation,
    260–263
  multiwalled carbon nanotubes
    (MWCNTs), synthesis, 260–261
  radiation shielding behavior of,
    267–269
Enhanced permeability and retention
(EPR), 140

F

Fourier-transform infrared spectroscopy
(FTIR spectroscopy)
and carbon nanotubes (CNTs)
  applications, 245–249
  collagen composite, 241–242
  gelatin composite, 238–241
  infrared (IR) spectroscopy, 235
  mid-infrared region (MIR), 236
  MWCNTs, 237–238
  PCL composite, 242–245
  SWCNTs-COOH, 237
  vibration band, 238

G

Graphene oxide-carbon nanotube
composites
aerogel for wastewater treatment, 30–31

computational chemistry, 38–39
GO functionalization, 24
    chemical vapor deposition (CVD), 26
    layer-by-layer (LbL) deposition,
        25–26
    solution casting, 25
    vacuum filtration, 26
GO-CNTs composites, antibacterial
    activity of, 37–38
nanoadsorbents, 21
    carbon-based materials, 22
    graphene oxide, 22–23
    nanocomposites, synthesis and
        functionalization, 23–24
wastewater treatment
    GO-CNTs aerogels for adsorption,
        27–28
    GO-CNTs membranes, 28–29, 32–34,
        35–36
Green engineering, 160
    arsenic and heavy metal groundwater,
        176–177
    carbon nanotubes (CNTs)
        scientific doctrine of, 161–162
    environmental protection science, 161
    environmental sustainability, 163–164
    futuristic recommendations, 179–180
    and green sustainability, 162
    scientific advances in, 166–173
    technological challenges in, 164–166
    water and wastewater treatment, 174–176
    water purification science and
        nanotechnology, 177–179

H

Heavy metal
    natural calamities and human activities,
        100
    remove, approaches, 103
    separation, objective of, 101–103
    sources, 100

I

Infrared (IR) spectroscopy, 235

L

Layer-by-layer (LbL) deposition, 25–26

M

Microbial fuel cells (MFCs), 67–68
Mid-infrared region (MIR), 236

N

NF membranes
    HPEI (hyperbranched
        polyethyleneimine), 116–117
    Torlon-GO membrane, 116

P

Polyethyleneglycol–doxorubicin (PEG–
    DXR), 140

R

Rheological behavior
    carbon nanotubes (CNTs)
        product processing, 195–202
        reinforced fluids, 187–195
    devices, 189
    shear rheological investigations, 191
RO membranes
    polyamide-CNT membrane, 115
    TMC (trimesoyl chloride) solution, 116

S

Single-walled carbon nanotubes
    (SWCNTs), 3
    amidation and esterification of, 126
    characteristics, and functionalization,
        137–139
    sidewall functionalization, 126
    structure, 137–139

T

Targeted smart drug delivery, 128–129
Thermal and electrical transport
    carbon nanotubes (CNTs)
        composites, 210–211, 214–220
        electrical conduction, mechanisms of,
            220–222
        electrical conductivity, composites
            with, 222–225
        thermal conduction, mechanisms,
            212–214

TMC (trimesoyl chloride) solution, 116
Torlon-GO membrane, 116

U

UF membranes
    dimethylformamide (DMF), 117
    MWCNT, 117–118

W

Waste water treatment
    CNTs, 53–54
    doped carbon nanotubes
        characterization, 57–58
        elements doped, 55–57
        nitrogen doped, 54–55
        synthesis, methods, 57

dye remediation, 62
    CNTs, use of doped, 63–65
heavy metal removal
    carbon materials, 58
    chromium, 60–61
    copper and cadmium, 61
    lead, 59
    mercury, 61
    MFCs, 62
    multi-walled CNTs, 60
    wastewater by CNTs, 63
materials pollutants removal, 71
microbial fuel cells (MFCs), 67–68
organic compound removal, 65–67
organic pollutants removal, 69
pathogens, removal of, 69–70
pollutants removal, 70–71

For Product Safety Concerns and Information please contact our EU
representative GPSR@taylorandfrancis.com
Taylor & Francis Verlag GmbH, Kaufingerstraße 24, 80331 München, Germany

* 9 7 8 1 7 7 4 6 3 8 6 3 7 *